Numerical Methods for Differential Equations

A Computational Approach

Library of Engineering Mathematics

Series Editor
Alan Jeffrey, University of Newcastle upon Tyne and University of Delaware

Numerical Methods for Differential Equations

A Computational Approach

John R. Dormand
School of Computing and Mathematics
University of Teesside, UK

CRC Press
Taylor & Francis Group
Boca Raton London New York

CRC Press is an imprint of the
Taylor & Francis Group, an **informa** business

First published 1996 by CRC Press
Taylor & Francis Group
6000 Broken Sound Parkway NW, Suite 300
Boca Raton, FL 33487-2742

Reissued 2018 by CRC Press

Library of Congress Cataloging-in-Publication Data

Catalog record is available from the Library of Congress.

A Library of Congress record exists under LC control number: 95054095

Publisher's Note
The publisher has gone to great lengths to ensure the quality of this reprint but points out that some imperfections in the original copies may be apparent.

Disclaimer
The publisher has made every effort to trace copyright holders and welcomes correspondence from those they have been unable to contact.

ISBN 13: 978-1-315-89600-7 (hbk)
ISBN 13: 978-1-351-07510-7 (ebk)

Visit the Taylor & Francis Web site at http://www.taylorandfrancis.com and the
CRC Press Web site at http://www.crcpress.com

The enclosed disk

The attached disk contains the Fortran 90 code for the programs listed in the book. Some additional data files, not presented in the text, are included to provide coefficients for some high order formulae which are deemed too large for textual display. These formulae are important in accurate work but they are very sensitive to errors of transliteration and thus are best disseminated electronically.

Since all the files are text files in IBM PC compatible format, they are accessible to any text editor and are easy to transfer to a wide range of platforms. Of course, the programs need compilation by a suitable Fortran 90 compiler. During the development phase they have been tested on a PC with the NAG/Salford FTN90 compiler. Other tests with the NAG f90 compiler on a Sun Sparc have been successful.

Only a few differential equations are included as test examples but the computer programs are designed to be easily modified for other applications. Consequently, they are less robust than many codes found in software libraries but a great deal more appropriate for the educational environment.

Preface

Many undergraduate courses in Mathematics contain an introduction to the numerical methods for differential equations, usually as part of a wider study of numerical analysis. At senior undergraduate and at the Master's degree levels, a more specialist study of the subject is appropriate and, with the modern availability of powerful computers, students can be introduced to practical aspects of solving differential equations.

Already there are quite a number of books on differential equations and, for partial differential equations in particular, some good student texts exist. This is not the case for the subject of numerical methods for ordinary differential equations which is well catered for only at the postgraduate and research levels. Such texts are too detailed and expensive to be suitable for the many undergraduates and MSc students who take the subject. Similarly they do not appeal to many engineers and scientists who need to apply numerical methods to differential equations used in the modelling of various processes.

All standard texts on numerical analysis offer an introduction to the subject of differential equations. In many instances the material presented is dated and of very limited use to a practitioner. Almost invariably the reader is given no useful advice regarding which method to employ in a particular type of application.

In this book it is my intention to present the subject of numerical differential equations in a modern fashion, without being mathematically too rigorous, in order to make the material accessible to a wide audience. This will include undergraduates of mathematics or other numerate disciplines and research scientists and engineers who need to find effective methods of solving practical problems. To this end the book will cover the development of methods for various types of ODEs and some PDEs, and many aspects of their application. Some computer programs written in the recently introduced Fortran 90 are presented. This powerful new revision of Fortran, with its new array structures and procedures, is ideal for the coding of numerical processes and will surely become the standard in scientific programming. The programs presented are not intended to be very general. Nevertheless some of them are based on very powerful state-of-the-art algorithms and they will be applicable, with minor modification, over a wide range of problems. The programs are contained on a floppy disk which accompanies this book. The disk also contains some data files not listed in the text.

An important subject which rarely receives detailed attention is that of second order differential equations, which do arise frequently in practical problems in dynamics and in other applications. Special methods for the treatment of these equations will be an important feature of this text.

Consistent with the aim of producing a textbook, each chapter contains a number of exercises of both theoretical and practical type. Some of these have been tested in first degree and in Master's courses over many years, and a few have appeared in the form of examination questions.

A number of people have assisted me in producing this book. I am particularly grateful to my colleague Michael Cummings who has read the text and has made many useful suggestions which have helped to improve the presentation and content. Also I acknowledge the valuable discussions with John Tunstall, Richard Finn and Paul White on the use of the LaTeXtypesetting system with which I am now very familiar!

The Author

John R. Dormand is a Senior Lecturer in the School of Computing and Mathematics at the University of Teesside in the North-East of England. He is author and co-author of many research papers on the numerical solution of differential equations using Runge-Kutta methods. In particular, his work in collaboration with Peter Prince, in constructing embedded Runge-Kutta pairs, is very influential. Much of the present text is motivated by many years of teaching undergraduate and graduate courses in numerical analysis.

The Author also maintains a keen research interest in Astronomy, the subject of his doctorate at the University of York. For many years he has worked on the Capture theory proposed by Michael Woolfson, with whom he is co-author of *The origin of the solar system: The Capture theory*.

Contents

CONTENTS

Chapter 1

Differential equations

1.1 Introduction

Many problems of science and engineering are reduced to quantifiable form through the process of mathematical modelling. The equations arising often are expressed in terms of the unknown quantities and their derivatives. Such equations are called *differential equations*. The solution of these equations has exercised the ingenuity of great mathematicians since the time of Newton, and many powerful analytical techniques are available to the modern scientist. However, prior to the development of sophisticated computing machinery, only a small fraction of the differential equations of applied mathematics were accurately solved. Although a model equation based on established physical laws may be constructed, analytical tools frequently are inadequate for its solution. Such a restriction makes impossible any long-term predictions which might be sought. In order to achieve any solution it was necessary to simplify the differential equation, thus compromising the validity of the mathematical modelling which had been applied.

A good example concerns the gravitational n-body system ($n \geq 3$), whose differential equations are easily constructed but cannot be solved completely. Consequently the problem of predicting the motions of the bodies of the solar system over a long period of time was extremely difficult, despite the attentions of mathematicians as formidable as Gauss. Today, this problem can be solved very easily with the aid of a good numerical method and a cheap desktop computer.

In this chapter, as a prelude to the development of numerical methods for computer application, some simple differential equations and their analytical solutions will be presented. Also, a brief introduction to the application of computers, from the software point of view, is given. This survey is not intended to be comprehensive. The subject is a very large

1

one, and the reader should consult a more appropriate source for a detailed view of any aspect of the analytical approach to differential equations. The intention here is to motivate a study of numerical techniques.

1.2 Classification of differential equations

If the derivatives in any differential system are taken with respect to a single variable (the independent variable), then the system is one of *ordinary* differential equations (ODE). When several independent variables permit the definition of partial derivatives, these would form part of a system of *partial* differential equations (PDE). In this work we shall be concerned chiefly with ODEs, but the numerical solution of PDEs will be considered in Chapter 15.

An equation with a single dependent variable, $y = y(x)$, where x is the independent variable, may be written in the form

$$G(x, y, y', y'', \ldots, y^{(n)}) = 0, \tag{1.1}$$

where

$$y^{(r)} = \frac{d^r y}{dx^r}, \quad r = 1, 2, \ldots, n.$$

If the nth order derivative is the highest present, then (1.1) is said to be an ordinary differential equation of order n. Since higher order equations can be expressed as systems of first order equations, most solution schemes (analytical and numerical) are applied to these. Let us consider the scalar explicit first order equation

$$y'(x) = f(x, y(x)). \tag{1.2}$$

In principle, this equation will have a general solution involving an arbitrary constant arising from a single integration. A particular solution will satisfy an additional relation

$$y(x_0) = y_0,$$

which will determine the arbitrary constant. Such a relation is usually called an *initial value* or *initial condition*, and so the point (x_0, y_0) is then the initial point on the particular solution curve of the differential equation.

Let us consider an example of a first order equation which may be solved easily by integration. The *logistic equation*

$$y' = y(\alpha - \beta y), \tag{1.3}$$

is used to model the growth of certain populations of animal species, where x is the elapsed time, y is the number of individuals, and α, β are constants. Although y is an integer, the large values associated with a population makes the assumption that $y \in \mathbb{R}$ a reasonable one. Since the independent variable x does not appear explicitly in the equation, it is said to be *autonomous*. The model arises from the assumption that the growth rate per individual (y'/y) is the sum of the birth rate (α) and the death rate $(-\beta y)$, where $\alpha, \beta > 0$. For convenience take $\alpha = 2, \beta = 1$, and the initial value $y(0) = 1$. The equation is now

$$\frac{dy}{dx} = y(2 - y),$$

and rearranging and integrating, we have

$$\int \frac{1}{y(2 - y)} \frac{dy}{dx} dx = \int dx.$$

This technique is known as *separating the variables*, and it leads to

$$\frac{1}{2} \int \left(\frac{1}{y} + \frac{1}{2 - y} \right) dy = x + C,$$

where C is an arbitrary constant. Performing the integration on the left-hand side yields, after some manipulation, the general solution

$$y = \frac{2}{1 + Be^{-2x}},$$

where B is arbitrary. Finally, use of the initial point $(0, 1)$ gives $B = 1$, and the particular solution is

$$y(x) = \frac{2}{1 + e^{-2x}}. \tag{1.4}$$

Not all analytical solutions are explicit. For example, the following second order equation models the one-dimensional inverse square law of attraction:

$$y'' = -1/y^2, \quad y(0) = 1, \quad y'(0) = 0. \tag{1.5}$$

Note that there are two initial values in this case. The independent variable x represents time, and the variable y gives the separation of the two bodies whose mutual attraction obeys the inverse square relationship. To solve this equation it is transformed into two first order equations. Writing $v = y'$ (velocity), it is easily seen that $y'' = v\dfrac{dv}{dy}$, and so (1.5) may be written as a first order equation, with the corresponding initial value:

$$v \frac{dv}{dy} = -\frac{1}{y^2}, \quad v = 0 \text{ when } y = 1.$$

Separating the variables yields

$$\int v\,dv = -\int \frac{1}{y^2}\,dy.$$

Performing the integration and applying the initial condition gives another first order differential equation

$$v = y' = -\sqrt{\frac{2(1-y)}{y}}, \quad y(0) = 1, \tag{1.6}$$

which may be solved by a similar technique. The selection of a negative square root is due to the initial condition which implies that the separation y will decrease with time. Making the substitution $y = \cos^2\theta$ and integrating yields the solution

$$\theta + \sin\theta\cos\theta = \sqrt{2}x \tag{1.7}$$

after the initial value has been applied. Although the equation has been solved, y is not determined explicitly in terms of x in this example. Given x, the non-linear equation (1.7) must be solved by an iterative method, such as Newton's method, to yield θ before y, the separation, is determined. In a two-dimensional version of this problem, the solution of the differential equation appears as Kepler's equation, similar to (1.7).

An alternative type of problem in differential equations is known as the *boundary value* problem. In this case a solution is sought to satisfy conditions at more than a single point. A simple example is the second order problem

$$y'' = -\omega^2 y, \quad y(0) = 1, \quad y(1) = 0. \tag{1.8}$$

This has the analytical solution

$$y(x) = \cos\omega x - \frac{\cos\omega\sin\omega x}{\sin\omega}.$$

Numerical methods for the solution of boundary value problems can be based on those developed for the initial value differential equations, although this is not always feasible.

1.3   Linear equations

As with algebraic equations, the property known as linearity is an important factor in the solution of differential equations. An equation of the type (1.1) is said to be *linear* when y and its derivatives occur only to the first degree. Since the logistic equation (1.3) contains a y^2 term it is non-linear. Similarly the inverse square problem of order 2 is non-linear, but the boundary value equation (1.8) is linear.

Consider a scalar linear equation of the form

$$y' + a(x)y = b(x), \quad y(x_0) = y_0. \tag{1.9}$$

This equation may be separable and hence solvable by the method used in the previous section. If this is not possible, the integrating factor method is applicable and the solution may be written as

$$y = e^{-\int a\,dx} \left(C + \int b e^{\int a\,dx}\,dx \right), \tag{1.10}$$

where C is an arbitrary constant.

Linear equations of the type (1.9) occur frequently in the modelling of physical and other processes. One application deals with the absorption of drugs by the human body (Figure 1.1). In this case x and y would represent the drug concentration, b the rate of increase due to intake and metabolic breakdown, and a (constant) would control the rate of absorption of the drug. The problem can be expressed in terms of equation (1.9) with

$$a(x) = a > 0, \text{ a constant, } b(x) = \begin{cases} r + sx, & x \in [0,T], \ r, s > 0 \\ -m, & x > T, \ m > 0 \end{cases},$$

with initial value $y(0) = 0$. Using the formula (1.10), a solution is easily determined:

$$y(x) = \begin{cases} sx/a - (s - ar)(1 - e^{-ax})/a^2, & x \in [0,T] \\ e^{a(T-x)}(y(T) + m/a) - m/a, & x > T \end{cases}.$$

In a similar way, systems of linear equations often can be solved analytically. Consider the system

$$y' = A(x)y + b(x), \quad y(x_0) = y_0, \tag{1.11}$$

where $y, b \in \mathbb{R}^k$ and A is a $k \times k$ matrix. If the elements of A are constant and the system is homogeneous ($b = 0$), the solution can be expressed in terms of the k eigenvalues (λ_i), assumed to be distinct, and k linearly independent eigenvectors (v_i) of A. Thus one obtains

$$y(x) = \sum_{i=1}^{k} C_i e^{\lambda_i x} v_i,$$

where C_i are arbitrary constants to be determined from the initial value. For non-homogeneous cases, a particular integral, to add to the above complementary function, must be found.

There are many physical and biological systems which are modelled in terms of linear differential systems. A common example is that of

Figure 1.1: Absorption of alcohol

the electrical network with a number of coupled, closed loops, for which the laws of Kirchhoff yield differential equations satisfied by the currents flowing in each loop. Consider a transformer network in which each of two loops contains an inductance L and a resistance R, with a mutual inductance M (see Figure 1.2). When a voltage E is applied to one of the loops, the currents I_1, I_2 satisfy the linear differential equations

$$L\frac{dI_1}{dt} + M\frac{dI_2}{dt} + RI_1 = E$$

$$M\frac{dI_1}{dt} + L\frac{dI_2}{dt} + RI_2 = 0$$

where t represents time. Putting $L = 2$, $M = 1$, $R = 1$, $E = 20\cos t$, this system is easily rearranged to give one of the form (1.11):

$$\dot{I} = -\frac{1}{3}\begin{pmatrix} 2 & 1 \\ 1 & 2 \end{pmatrix} I + \frac{20}{3}\cos t \begin{pmatrix} 2 \\ 1 \end{pmatrix}, \quad I = \begin{pmatrix} I_1 \\ I_2 \end{pmatrix}. \qquad (1.12)$$

In this case the eigenvalues and eigenvectors of the coefficient matrix A are found to be

$$\lambda_1 = -1, \ \lambda_2 = -\frac{1}{3}, \ v_1 = \begin{pmatrix} 1 \\ 1 \end{pmatrix}, \ v_2 = \begin{pmatrix} 1 \\ -1 \end{pmatrix},$$

and so the complementary function is

$$I_C(t) = C_1 e^{-t} \begin{pmatrix} 1 \\ 1 \end{pmatrix} + C_2 e^{-\frac{1}{3}t} \begin{pmatrix} 1 \\ -1 \end{pmatrix}.$$

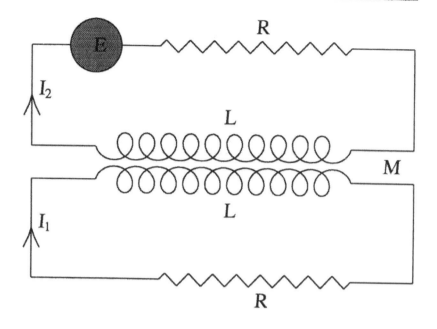

Figure 1.2: A two-loop electrical network

Assuming a particular solution of the form

$$I_P(t) = p\cos t + q\sin t,$$

substitution in the equation (1.12) yields

$$p = \begin{pmatrix} 6 \\ 4 \end{pmatrix}, \ q = \begin{pmatrix} 8 \\ 2 \end{pmatrix}.$$

If the initial value is specified, the constants of integration C_1, C_2 may be determined, and $I(0) = \begin{pmatrix} 0 \\ 0 \end{pmatrix}$ yields the complete solution

$$I(t) = \cos t \begin{pmatrix} 6 \\ 4 \end{pmatrix} + \sin t \begin{pmatrix} 8 \\ 2 \end{pmatrix} - 5e^{-t} \begin{pmatrix} 1 \\ 1 \end{pmatrix} - e^{-\frac{1}{3}t} \begin{pmatrix} 1 \\ -1 \end{pmatrix}.$$

The linear systems in which $A = A(x)$, rather than A constant, are much more difficult to deal with.

1.4 Non-linear equations

Most differential equation problems lead to non-linear systems which, typically, are not amenable to simple analytical procedures such as those illustrated above. It is not uncommon to seek approximate solutions to such problems by adopting approximate physical modelling. A well-known example of this occurs for the equation of motion of the simple pendulum, neglecting the effects of friction and air resistance. If the

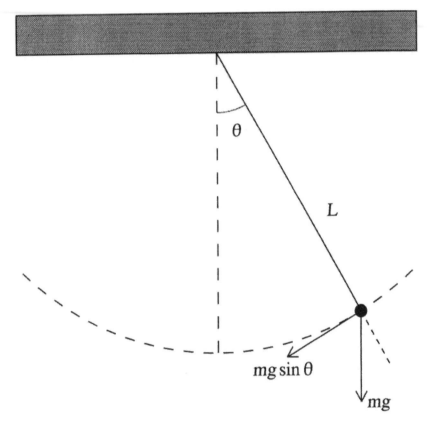

Figure 1.3: A simple pendulum

pendulum has length L and the acceleration due to gravity is g (see Figure 1.3), the equation of motion can be written

$$L\frac{d^2\theta}{dt^2} = -g\sin\theta, \tag{1.13}$$

where θ is the angular displacement of the pendulum. Writing $\omega^2 = g/L$, and using the same technique as that employed for the inverse square law problem (1.5), it is easy to obtain an integral for the velocity $V = d\theta/dt$. For the inital value

$$\theta(0) = 0, \ \ V(0) = V_0,$$

one obtains

$$V^2 = V_0^2 - 2\omega^2(1 - \cos\theta).$$

Unfortunately a second integral, to yield a relationship between displacement and time, cannot be obtained without using special functions. However, the pendulum problem can be further simplified, by assuming oscillations of small amplitude. This implies the replacement of $\sin\theta$ by θ in (1.13), giving a linear equation (1.8). This has a periodic solution (of period $2\pi/\omega$)

$$\theta(t) = C_0 \cos\omega t + C_1 \sin\omega t.$$

For some practical problems this approximate modelling is adequate, but in many others it is not appropriate and numerical methods must be employed.

1.5 Existence and uniqueness of solutions

We have seen various means by which one can seek the solution of differential systems, dependent on the actual classification of the equations. However, it may be that a particular equation does not possess a solution. For example, the first order equation

$$\left(\frac{dy}{dx}\right)^2 = -(y^2 + 1), \ y(0) = 1,$$

has no real solution for $x > 0$. Where an initial value problem does have a solution, the question arises as to whether or not that solution is the only one possible. These considerations are important for analytical and for numerical methods applied to differential equations. This text is not the place for a detailed analysis of existence and uniqueness but the reader should be aware of some relevant theorems.

First, consider the existence of a unique solution to the first order scalar equation (1.2). It can be stated that the differential equation has a unique solution passing through the point (x_0, y_0), provided that $f(x, y)$ is sufficiently *well-behaved* near that point. Using more precise terminology, the solution requires the continuity of f and f_y in a region R containing (x_0, y_0). For the logistic equation (1.3), it is clear that this 'good behaviour' is assured for any initial point but, for the inverse square problem (1.6), the function f is certainly not well behaved at $y = 0$.

If f_y is continuous in R then we can assume that it is bounded:

$$|f_y(x, y)| \le L, \ (x, y) \in R.$$

Using the mean value theorem for derivatives, this leads to the condition

$$|f(x, y_1) - f(x, y_2)| \le L|y_1 - y_2|, \ (x, y_1), (x, y_2) \in R. \tag{1.14}$$

This is known as a Lipschitz condition and it is usually quoted as the necessary condition for the existence of a unique solution to the differential equation in R.

For a system of equations in which $y \in \mathbb{R}^k$, the required Lipschitz condition is written

$$\|f(x, y_1) - f(x, y_2)\| \leq L\|y_1 - y_2\|, \ (x, y_1), (x, y_2) \in R.$$

The bound, L, is known as the Lipschitz constant of f.

1.6   Numerical methods

Since analytical methods are not adequate for finding accurate solutions to most differential equations, numerical methods are required. The ideal objective, in employing a numerical method, is to compute a solution of specified accuracy to the differential equation. Sometimes this is achieved by computing several solutions using a method which has known error characteristics. Rather than a mathematical formula, the numerical method yields a sequence of points close to the solution curve for the problem. Classical techniques sample the solution at equally spaced (in the independent variable) points but modern processes generally yield solutions at intervals depending on the control of truncation error. Of course, it is expected that these processes will be implemented on computers rather than being dependent on hand calculation.

Numerical methods for the solution of ordinary differential equations of initial value type are usually categorized as *single–step* or *multistep* processes. The first method uses information provided about the solution at a single initial point to yield an approximation to the solution at a new one. In contrast, multistep processes are based on a sequence of previous solution and derivative values. Each of these schemes has its advantages and disadvantages, and many practitioners prefer one or the other technique. Such a preference may arise from the requirements of the problem being solved. The view is generally held that different types of numerical processes should be matched to the user objectives.

There is a common tendency for engineers and scientists employing numerical procedures to select an *easy* looking method on the grounds that it is mathematically consistent, and that raw computing power will deliver the appropriate results. This attitude is somewhat contradictory since the methods usually found in text books were developed many years ago when the most advanced computing machine available was dependent literally on manual power. The assumption that such processes can be efficient in modern circumstances is dangerously flawed and quite often it leads to hopelessly inaccurate solutions. A major aim of the present text is to present powerful, *up–to–date*, numerical methods for differential equations in a form which is accessible to non–specialists.

1.7   Computer programming

Any branch of modern numerical analysis is necessarily concerned with computing. How great a concern this should be is a matter of individual judgement but, in the author's view, an effective method for learning about numerical techniques is the study and construction of a large number of relevant computer programs. For this reason many of the chapters ahead contain programs and fragments of programs. These will be of use to any reader who is a student of differential equations and also to the scientist or engineer who needs to work with his own programs. Of course, there are many packages providing a 'friendly' interface for the practitioner. Also there exist many software libraries which offer differential equation solvers, in the form of subroutines, to be referenced by users who require the extra flexibility gained from writing their own programs. Inevitably, such software has to be a compromise between robustness and efficiency, and so there is ample motivation for the researcher, faced with extremely lengthy processing, to seek the ultimate in computational efficiency. The programs in this book will prove valuable from this point of view, although a major aim of the coding has been to achieve clarity rather than to minimise arithmetic operations. Consequently, the program listings should be regarded as starting points for the development of codes tailored to specific applications.

There is no intention to provide a primer for the use of Fortran 90, the high-level language employed here. For the reader who needs to be introduced to this powerful and elegant language, there are a number of excellent texts. The books by Ellis et al. (1994) and Morgan and Schonfelder (1993) will be useful either to the FORTRAN 77 programmer who wishes to upgrade to Fortran 90, or to anyone with little previous experience. The programs can be translated into other scientific programming languages. Successful conversions are likely to be equally successful in converting the programmer to the use of Fortran 90! A major advantage of Fortran 90 is its ability to handle vectors and matrices directly, thus eliminating much explicit coding of loops. As a consequence, programs for application to systems of differential equations can be coded as easily as those dealing with the scalar case. The following illustration demonstrates the facility.

Consider the evaluation of the vector expression:

$$w = y + h \sum_{i=1}^{n} b_i f_i, \quad w, y, f_i \in \mathbb{R}^k, \quad h, b_i \in \mathbb{R}. \qquad (1.15)$$

The FORTRAN 77 code

```
*-------------------------------------
      DO 1 J = 1, K
         SUM = 0.0
         DO 2 I = 1, N
            SUM = SUM + B(I)*F(I, J)
2        CONTINUE
         W(J) = Y(J) + H*SUM
1     CONTINUE
*-------------------------------------
```

is a valid translation for (1.15) but, in Fortran 90 the same result is
obtained from just one statement:

```
!-------------------------------------
      w = y + h*MATMUL(b, f)
!-------------------------------------
```

This simple form assumes that the arrays w, y, f, b are declared to
fit the variables as defined. If this is not so, the one-line translation can
be retained, but the array slices must then be entered in explicit fashion. Since Fortran 90 allows dynamic arrays, the array bounds normally
are declared after execution of the program has started. This facility
permits greater generalisation, and hence greater storage efficiency, than
with earlier Fortran standards. Obviously these features are particularly
welcomed by mathematicians, and the programs which follow take full
advantage of these and other advanced facilities.

1.8   Problems

1. Use the method of separation of variables to solve the following initial value problems.

   (a) $y' = -2xy^2$, $y(0) = 1$

   (b) $y' = -y$, $y(0) = 1$

   (c) $y' = -\frac{1}{2}y^3$, $y(0) = 1$

   (d) $y' = (1 - y/20)y/4$, $y(0) = 1$

   (e) $y' = y^3 - y$, $y(0) = \frac{1}{2}$.

   (f) $y' = y\cos x$, $y(0) = 1$.

2. By making the substitution $y = vx$, transform the homogeneous equation

$$y' = \frac{y - x}{y + x}, \qquad y(0) = 4$$

   into separable form, and hence find the solution.

3. Assume that a tunnel through the centre of the earth has been excavated and evacuated. An object falling through the tunnel will be accelerated by gravity and its motion will satisfy the second order equation

$$\ddot{r} = -\frac{GM(r)}{r^2},$$

   where r is the distance from the centre of the earth, G is the constant of gravitation, $M(r)$ is the mass of the earth contained within r, and the independent variable is time. Using the assumption that the earth is a sphere of uniform density, solve the differential equation satisfying the initial condition

$$r(0) = R, \qquad \dot{r}(0) = 0.$$

   Hence compute the time for the falling object to reach the antipodes.

4. A projectile is fired vertically with speed V from the surface of an airless planet of mass M and radius R. Its instantaneous position r satisfies the same differential equation as in the previous question but the initial condition is now

$$r(0) = R, \quad \dot{r} = V.$$

   Show that the projectile velocity $v = \dot{r}$ satisfies

$$v^2 = V^2 + 2GM\left(\frac{1}{r} - \frac{1}{R}\right).$$

Hence determine the planetary escape velocity. Find the escape velocity for the moon ($M = 7.35 \times 10^{22} kg$, $R = 3.476 \times 10^6 m$).

5. Solve the second order equation

$$y'' + y(y')^3 = 0, \qquad y(0) = 0, y'(0) = 2.$$

6. A man of mass m falls from a stationary helicopter. His rate of fall v satisfies the differential equation

$$m\frac{dv}{dt} = mg - bv^2,$$

where g is the acceleration due to gravity, and b is a parameter depending on the man's shape. Find the terminal speed of descent in terms of the given parameters. Given $g = 9.8$ ms^{-2} and a terminal speed of 48 ms^{-1}, find the value of b. Solve the equation and use the solution to find the time required for the faller to attain 90% of his terminal speed.

7. Solve the linear equation $y' = x + y$ with initial condition
   (a) $y(0) = 1$. (b) $y(0) = 0$.

8. An electrical circuit contains an inductance of L henrys and a resistance of R ohms connected in series with a constant emf of V volts. The current I in the circuit satisfies

$$L\frac{dI}{dt} + RI = V.$$

Find the solution when $I(0) = 0$. Also solve the same initial value problem when $V = \sin t$.

9. The populations of two symbiotic species are governed by the differential system

$$\dot{y}_1 = -y_1 + 2y_2, \quad \dot{y}_2 = \tfrac{1}{2}y_1 - y_2,$$

where y_1, y_2 are the numbers of individuals of the two species. If the initial condition is $y_1(0) = 500$, $y_2(0) = 400$, show that the two populations have limits of 650 and 325 members respectively.

10. Solve the linear system $y' = Ay$, where

$$A = \begin{pmatrix} -2 & 1 & 0 \\ 1 & -2 & 1 \\ 0 & 1 & -2 \end{pmatrix}, \qquad y(0) = \begin{pmatrix} 1 \\ 0 \\ 0 \end{pmatrix}.$$

Chapter 2

First ideas and single–step methods

2.1   Introduction

The first types of numerical methods for ordinary differential equations to be considered here are classified as *single-step* processes. The best-known single-step methods are those based on Taylor series and the Runge-Kutta formulae. Both of these will be introduced in this chapter, together with some important concepts for wider application. The idea of *step-by-step* computation, and that of order reduction, is fundamental to the numerical analysis of differential equations. The importance of Taylor series expansions goes well beyond their direct application as numerical methods, as will become clear as the subject is developed.

2.2   Analytical and numerical solutions

An *Initial Value Problem* in first order ordinary differential equations (ODE) may be written in the form

$$y'(x) = f(x, y(x)), \quad y(a) = y_0, \quad x \in [a, b], \qquad (2.1)$$

where the dash indicates differentiation with respect to x, and y is a k–dimensional vector ($y \in \mathbb{R}^k$). The true solution, which may in some cases be obtained analytically, is expressed as

$$y = y(x), \quad \text{satisfying } (a, y_0).$$

For example, the one–dimensional linear equation

$$y' = x + y, \quad y(0) = 1,$$

which is easily solved by the integrating factor method (1.10), has the solution

$$y = 2e^x - x - 1.$$

Note that a different initial value yields a different $y(x)$. Thus $y(0) = 0$ with the same differential equation gives $y = e^x - x - 1$.

A *numerical* solution to a system of ordinary differential equations (2.1) consists of a set of values $\{y_n\}$, corresponding to a discrete set of values of x,

$$x_n, \ n = 0, 1, 2, \ldots, N.$$

These y values are usually obtained in a **step–by–step** manner, that is, y_1 is followed by y_2, then y_3 and so on. Obviously, it is intended that these values will lie on or near the true solution curve, and so

$$y_n \simeq y(x_n),$$

with $x_{n+1} = x_n + h_n$, $n = 0, 1, 2, \ldots, N - 1$; $x_0 = a$, $x_N = b$, and h_n is called the **step–size**. The step-size is usually varied in practice but, for simplicity, it will be taken sometimes to be constant.

2.3 A first example

Consider the separable first order equation

$$y' = -2xy^2. \tag{2.2}$$

The general solution is

$$y(x) = \frac{1}{x^2 + C}$$

where C is an arbitrary constant to be determined from an initial condition. Setting $y(0) = 1$ gives $C = 1$. This simple equation will be used to demonstrate a numerical method based on a Taylor Series expansion. Note that the application of numerical methods to problems which can be solved by analytical means is a sensible way of gaining experience of these techniques and also of building confidence in them. In order to evaluate any new computational scheme one must ascertain its behaviour when it is applied to problems with known solutions.

Given that a function $y(x)$ is sufficiently differentiable, $y(x + h)$ may be expanded in a Taylor series form

$$y(x + h) = y(x) + hy'(x) + \frac{h^2}{2!}y''(x) + \cdots + \frac{h^p}{p!}y^{(p)}(x) + \cdots, \tag{2.3}$$

where $y^{(p)}(x)$ represents the pth derivative of y with respect to x. If an initial value $y(x_0)$ is specified, and expressions for the derivatives $y', y'', \ldots$

in terms of x and y can be obtained, then these may be substituted into the Taylor series (2.3) above to compute $y(x_0 + h)$ for a given h. Also, if h is small enough, the series should converge quickly making it unnecessary to compute a large number of terms, and hence derivatives. Let us take $x_0 = 0, y(0) = 1, h = 0.2$, and assume that the 3rd order term, involving h^3, and higher order terms are negligible. The first derivative of y is specified by the differential equation and so, additionally, only y'' must be determined. Noting that $y = y(x)$, differentiation yields

$$y'' = -2y^2 - 4xyy'.$$

Substituting for y' gives

$$y'' = 2y^2(4x^2y - 1),$$

but this is not essential since recursive computation, starting from y', is likely to be easier for higher derivatives in general. Thus at the initial point $(0, 1)$, the values $y' = 0$, $y'' = -2$ are obtained. Substituting $x = 0$ in (2.3) now yields

$$
\begin{aligned}
y(0.2) &\simeq y(0) + 0.2y'(0) + 0.02y''(0) \\
&= 1 + 0 - 0.04 \\
&= 0.96 = y_1 \text{ (say)}.
\end{aligned}
$$

The value of $y(0.4)$ can be estimated by repeating the above procedure with initial point $(0.2, 0.96)$. This requires the re-evaluation of the two derivatives as follows:

$$
\begin{aligned}
y'(0.2) &= -2 \times 0.2 \times 0.96^2 \\
&= -0.36864 \\
y''(0.2) &= -1.56008.
\end{aligned}
$$

Substituting $x = 0.2$ in (2.3) now gives

$$
\begin{aligned}
y(0.2 + 0.2) &\simeq 0.96 + 0.2 \times (-0.36864) + \frac{0.2^2}{2} \times (-1.56008) \\
&= 0.8551 = y_2,
\end{aligned}
$$

expressed to four significant figures. It should be remarked at this stage that the values of y', y'', used for the second step, are approximate since they are based on the approximate solution for y at $x = 0.2$.

One could compute an estimate of $y(0.6)$ with a third step of size 0.2, and so on. This **step–by–step** procedure is a feature of all practical numerical schemes for the solution of ODEs.

Normally it would not be possible to compute the accuracy of these results but, in this simple case, the analytical form of the solution is

Table 2.1: Comparison of true and numerical solutions

n	x_n	True $y(x_n) = \frac{1}{(x_n^2+1)}$	Numerical y_n	Error $\varepsilon_n = y_n - y(x_n)$
0	0.0	1.0000	1.0000	0.0000
1	0.2	0.9615	0.9600	−0.0015
2	0.4	0.8621	0.8551	−0.0070
3	0.6	0.7353	0.7248	−0.0104

available. Using this, the Table 2.1 has been constructed. It can be seen that the error ε_2 after two steps is much larger than that of the first step, ε_1. This is because the second step actually is applied to a different initial value problem:

$$Y' = -2xY^2, \quad Y(0.2) = 0.96.$$

The solution of this equation is slightly different to that of the original one (2.2). Applying the new initial value to the general solution yields

$$Y(x) = 1/(x^2 + 1.0017),$$

which gives $Y(0.4) = 0.8608$. Consequently, the error of the second step, obtained by comparing the numerical solution with this *new* true solution, is a bit smaller at −0.0057. This *actual* error at a particular step is called the **local** error of the numerical process. At every step, a **local** true solution can be defined, and hence a local error is also defined. A comparison of the numerical solution with the true solution of the *original* problem yields an overall or **global** error. If, in the general case, the local true solution $u_n(x)$ satisfies

$$u_n' = f(x, u_n), \quad u_n(x_n) = y_n, \quad n = 1, 2, \dots, N,$$

then the local and global errors at $x = x_n$ are

$$\begin{aligned} e_n &= y_n - u_{n-1}(x_n) \\ \varepsilon_n &= y_n - y(x_n). \end{aligned}$$

These local and global phenomena are illustrated in Figure 2.1, where some local true solution curves are plotted alongside the true solution for a suitable problem.

An alternative way of estimating $y(0.4)$ using (2.3) would be to substitute $x = 0$, $h = 0.4$, that is, to increase the step-size by a factor 2; clearly this will cut out the error relating to the local true solution at the second step, and also the original derivative values at $x = 0$ could

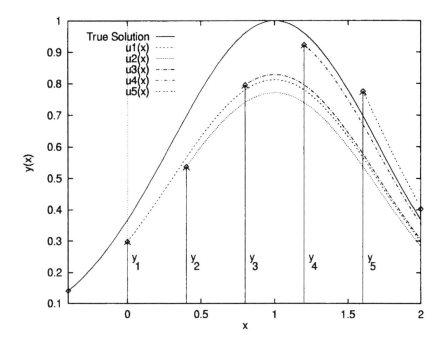

Figure 2.1: True and local true solutions

be reused. With this strategy, the value obtained is 0.84, which has an error -0.0221, over three times as large as in the two-step calculation. It will be obvious that more terms in the Taylor series are required to preserve accuracy for larger values of h. The motivation for a step-by-step approach is clear.

Carrying out two further differentiations for the above example (2.2) yields

$$
\begin{aligned}
y''' &= -4\{2yy' + x((y')^2 + yy'')\} \\
y^{(4)} &= -4\{3((y')^2 + yy'') + x(3y'y'' + yy''')\}.
\end{aligned}
$$

With $h = 0.2$, addition of the third and fourth order terms from (2.3) gives

$$
\begin{aligned}
y(0.2) &\simeq y_1 = 0.9616 \\
y(0.4) &\simeq y_2 = 0.8624.
\end{aligned}
$$

These results are much better than the earlier ones, but the computational cost is much higher. A major disadvantage of this method is the need to find higher order derivatives which can soon become very complicated, even given simple functions such as that in (2.2).

A Fortran 90 program which implements a fourth order Taylor series method for the problem (2.2) is listed below. This can be modified rather easily for application to other scalar problems.

```
!------------------------------------------------------------
PROGRAM Taylorode ! Solves a first order ODE using Taylor
                  ! series of order 4
!-------------------------------------- J.R.Dormand, 6/95
IMPLICIT NONE
INTERFACE
   FUNCTION yd(x, y)
      REAL :: x, y, yd(4)
   END FUNCTION yd
END INTERFACE
INTEGER :: i, n
REAL :: x, y, h
REAL :: der(4), hv(4)
PRINT*, 'Enter initial condition (x,y), h, No of steps '
READ*, x, y, h, n
hv = (/ 1.0, h/2, h**2/6, h**3/24 /) ! powers of h
DO i = 1, n                          ! loop over steps
   der = yd(x, y)                    ! evaluate derivatives
   y = y + h*DOT_PRODUCT(hv, der)    ! compute new step
   x = x + h
   PRINT 100, x, y, 1.0/(x**2 + 1)   ! print results
END DO
100 FORMAT(1X, F10.3, 2F12.6)
END PROGRAM Taylorode
!
FUNCTION yd(x, y)
REAL :: x, y, yd(4)
! Evaluate derivatives to order 4 for ODE: y' = -2xy^2
yd(1) = -2*x*y*y
yd(2) = -2*y*(y + 2*x*yd(1))
yd(3) = -4*(2*yd(1)*y + x*(yd(1)**2 + y*yd(2)))
yd(4) = -4*(3*(yd(1)**2+y*yd(2))+x*(3*yd(1)*yd(2)+y*yd(3)))
END FUNCTION yd
!------------------------------------------------------------
```

2.4   The Taylor series method

An alternative form of equation (2.3) introduces the *increment function*. The Taylor expansion can be written

$$y(x_{n+1}) = y(x_n) + h_n\Delta(x_n, y(x_n), h_n), \tag{2.4}$$

where

$$\Delta(x, y, h) = y' + \frac{h}{2}y'' + \cdots + \frac{h^{p-1}}{p!}y^{(p)} + \cdots$$

and Δ is called the **Taylor Increment Function**. A numerical method is based on the truncation of (2.4) after the h^p term. The result is called a pth order method in which the sequence $\{y_n\}$, corresponding to a discrete set of values of x, x_n, $n = 1, 2, \ldots N$, is computed from

$$y_{n+1} = y_n + h_n y_n' + \frac{h^2}{2}y_n'' + \cdots + \frac{h^p}{p!}y_n^{(p)},$$

or

$$y_{n+1} = y_n + h_n \Phi(x_n, y_n, h_n), \tag{2.5}$$

where

$$\Phi(x, y, h) = y' + \frac{h}{2}y'' + \cdots + \frac{h^{p-1}}{p!}y^{(p)}$$

is the truncated form of the Taylor Increment Function. In principle this method is very straightforward, but the relative difficulty of obtaining and computing derivatives has restricted the popularity of the Taylor series scheme for solving differential equations. However, with the growth in the availability of symbolic computer packages, there may still be a future for this technique.

It is instructive to consider at this stage some further aspects of Taylor differentiation with respect to systems of ODEs (2.1). First let us consider the scalar case of a single differential equation $y'(x) = f(x, y(x))$. The Taylor series requires the higher total derivatives of y with respect to x. Differentiating and using the chain rule,

$$\begin{aligned} y'' &= \frac{d}{dx}f(x, y) \\ &= \frac{\partial f}{\partial x}\frac{dx}{dx} + \frac{\partial f}{\partial y}\frac{dy}{dx} \\ &= \frac{\partial f}{\partial x} + f\frac{\partial f}{\partial y} \end{aligned}$$

since $f = dy/dx$. A more compact expression using suffix notation is

$$y'' = f_x + f f_y.$$

In the same way, the next derivative is

$$\begin{aligned} y''' &= \frac{d}{dx}\{f_x + f f_y\} \\ &= f_{xx} + 2f f_{xy} + f^2 f_{yy} + f_x f_y + f_y^2 f. \end{aligned}$$

The situation for non–scalar cases is much more complicated but some simplification is achieved by considering an autonomous system

$$y' = f(y), \quad y \in \mathbb{R}^M. \tag{2.6}$$

The ith component of this system can be written as

$$^iy' = {}^if({}^1y, {}^2y, \dots, {}^My).$$

The autonomous system is no less general than equation (2.1) since the Mth (or another if required) component can be chosen as $^My = x$, the independent variable, with $^Mf = 1$ as a consequence. Thus a scalar problem can be expressed as a two-component autonomous system of the form

$$\begin{bmatrix} {}^1y' \\ {}^2y' \end{bmatrix} = \begin{bmatrix} f({}^2y, {}^1y) \\ 1 \end{bmatrix}.$$

The higher derivatives of a vector system may be generated as in the scalar case. The ith component has the second derivative

$$
\begin{aligned}
{}^iy'' &= \frac{d}{dx} \, {}^if(y) \\
&= \sum_{j=1}^{M} \frac{\partial \, {}^if}{\partial \, {}^jy} \frac{d \, {}^jy}{dx} \\
&= \sum_{j=1}^{M} \frac{\partial \, {}^if}{\partial \, {}^jy} \, {}^jf.
\end{aligned}
$$

Using the suffix j to denote differentiation with respect to the j component yields

$$^iy'' = \sum_{j=1}^{M} {}^if_j \, {}^jf, \quad i = 1, 2, \dots, M \tag{2.7}$$

and, using tensor notation, the summation over j can be implied. Thus

$$^iy'' = {}^if_j \, {}^jf, \quad i = 1, 2, \dots, M,$$

can replace equation(2.7). An even simpler form, to represent the vector second derivative, is obtained by omitting the component superfix. This is

$$y'' = f_j \, {}^jf. \tag{2.8}$$

In full, the second derivative must be expressed in terms of the Jacobian matrix J of f:

$$y'' = Jf = \begin{pmatrix} {}^1f_1 & {}^1f_2 & \cdots & {}^1f_M \\ {}^2f_1 & {}^2f_2 & \cdots & {}^2f_M \\ \vdots & \vdots & \ddots & \vdots \\ {}^Mf_1 & {}^Mf_2 & \cdots & {}^Mf_M \end{pmatrix} \begin{pmatrix} {}^1f \\ {}^2f \\ \vdots \\ {}^Mf \end{pmatrix}. \tag{2.9}$$

To obtain higher derivatives, the compact form is particularly useful: thus

$$
\begin{aligned}
y''' &= \frac{d}{dx}\{f_i\,{}^i f\} \\
&= f_{ij}\,{}^i f\,{}^j f + f_i\,{}^i f_j\,{}^j f.
\end{aligned}
$$

Generally, the higher total derivatives may be expressed in the form

$$
y^{(p)} = \sum_{i=1}^{n_p} \alpha_i^{(p)} F_i^{(p)}. \tag{2.10}
$$

The $F_i^{(p)}$ are called **elementary differentials**, and for the case $p = 3$, $n_3 = 2$, $\alpha_i^{(3)} = 1$, $i = 1, 2$. The elementary differentials are

$$
\begin{aligned}
F_1^{(3)} &= f_{ij}\,{}^i f\,{}^j f \\
F_2^{(3)} &= f_i\,{}^i f_j\,{}^j f.
\end{aligned}
$$

The corresponding values for $p = 4$ are

$$
n_4 = 4, \quad \alpha_i^{(4)} = 1, \quad i = 1, 3, 4, \quad \alpha_2^{(4)} = 3.
$$

In spite of these simplifications the elementary differential form of higher derivatives becomes very complicated indeed. A detailed discussion of these matters is given by Lambert (1991), who also explains the one–to–one correspondence between elementary differentials and rooted trees from graph theory (Butcher, 1987) which is essential for a study of higher-order processes.

2.5 Runge–Kutta methods

The most practical type of single–step method for the solution of ordinary differential equations is the Runge–Kutta (RK) process. Such a scheme was introduced first about a century ago by Runge, and, a few years later, by Kutta. Because they are robust and easy to implement, Runge–Kutta schemes have enjoyed much popularity with practitioners, but their major development has occurred only since 1960, following pioneering work by Butcher. An example of a **three–stage** RK process is written as follows: given the first order initial value problem

$$
y' = f(x, y), \quad y(x_0) = y_0, \quad y \in \mathbb{R}^M,
$$

$$
y_{n+1} = y_n + \frac{h}{6}(f_1 + 4f_2 + f_3), \quad n = 0, 1, 2, \ldots, \tag{2.11}
$$

where

$$
\begin{aligned}
f_1 &= f(x_n, y_n) \\
f_2 &= f(x_n + \tfrac{1}{2}h, \; y_n + \tfrac{1}{2}hf_1) \\
f_3 &= f(x_n + h, \; y_n + h(2f_2 - f_1)).
\end{aligned}
$$

This yields the sequence of approximations $y_1, y_2, \ldots$ to $y(x_1), y(x_2), \ldots$, where $x_i = x_0 + ih$, assuming equal step–sizes. In practice, the step-size is varied from step to step.

To demonstrate the computational procedure the formula (2.11) is applied to the equation (2.2), for which $f = -2xy^2$, with $y(0) = 1$ and step $h = 0.2$. The calculation is presented in Table 2.2.

Table 2.2: A three–stage Runge–Kutta evaluation

Stage	x	y	$f(x,y)$
	x_0	y_0	f_1
1	0	1	0
	$x_0 + \tfrac{1}{2}h$	$y_0 + \tfrac{1}{2}hf_1$	f_2
2	0.1	1	-0.2
	$x_0 + h$	$y_0 + h(2f_2 - f_1)$	f_3
3	0.2	-0.92	-0.33856

The result of a single step is

$$
\begin{aligned}
y_1 &= 1 + \frac{0.2}{6}(0 + 4 \times (-0.2) - 0.33856) \\
&= 0.9620 \simeq y(0.2).
\end{aligned}
$$

The error of this approximation is $\varepsilon_1 = y_1 - y(x_1) = 0.0005$.

It is natural to compare the three–stage RK method with a Taylor series comprising terms to $h^3 y'''$ since the RK method (2.11) requires three function (f) evaluations rather than the 3 derivatives y', y'', y''' of the Taylor series, making the computational costs of the two methods very similar. However a most important point in favour of the RK method is that no differentiation is needed.

In the example above, the RK method is also more accurate for the first step (N.B. $y'''(0) = 0$ for this problem). Unlike the corresponding Taylor series this 3–stage RK formula is not unique; it will be seen later that it is possible to construct many different three–stage formulae with various properties.

Generally, the explicit RK formula can be expressed as

$$y_{n+1} = y_n + h_n \sum_{i=1}^{s} b_i f_i,$$ (2.12)

where

$$
\begin{aligned}
f_1 &= f(x_n, \ y_n) \\
f_i &= f(x_n + c_i h_n, \ y_n + h_n \sum_{j=1}^{i-1} a_{ij} f_j) \\
i &= 2, 3, \ldots, s.
\end{aligned}
$$

The properties of the formula are determined by the RK parameters c_i, a_{ij}, b_i, and s, the number of stages. The term *explicit* is applied to this kind of formula because each f_i depends only on the previously calculated f_j, $j = 1, 2, \ldots, i-1$ rather than *all* the other f values. This is not an essential property but it does simplify the computational algorithm.

The explicit Runge-Kutta process is very easy to program and an example is given below. Modification for different equations is somewhat easier than in the Taylor case since no differentiation is needed.

```
!---------------------------------------------------
PROGRAM RK3ode      ! Solves a first order ODE using
                    ! Runge-Kutta process of order 3
                    !
!--------------------------------J.R.Dormand, 6/95
IMPLICIT NONE
INTERFACE
   FUNCTION yd(x, y)
      REAL :: x, y, yd
   END FUNCTION yd
END INTERFACE
INTEGER :: i, n, k
REAL :: x, y, w, h, f(3)
REAL, PARAMETER ::  a(2:3, 2) = &
RESHAPE((/ 0.5, -1.0, 0.0, 2.0 /), (/ 2, 2 /)), &
b(3) = (/ 1.0/6.0, 2.0/3.0, 1.0/6.0 /) ,&
c(2:3) = (/ 0.5, 1.0 /)
PRINT*, 'Enter initial condition (x,y), h, No of steps '
READ*, x, y, h, n
DO k = 1, n                      ! loop over steps
   f(1) = yd(x, y)               ! first f evaluation
   DO i = 2, 3
      w = y + h*DOT_PRODUCT(a(i, 1:i-1), f(1:i-1))
```

```
      f(i) = yd(x + c(i)*h, w)      ! new f evaluation
   END DO
   y = y + h*DOT_PRODUCT(b, f)      ! compute new step
   x = x + h
   PRINT 100, x, y, y - 1.0/(x**2 + 1)
END DO
100 FORMAT(1X, F10.3, 2F12.6)
END PROGRAM RK3ode
!
FUNCTION yd(x, y)
REAL :: x, y, yd
! Evaluate derivative: y' = -2xy^2
yd = -2*x*y*y
END FUNCTION yd
!-----------------------------------------------------------
```

The key to the accuracy of the Runge-Kutta process is the number of stages s, often called the number of *function* or *derivative* evaluations. Increasing the value of s allows the order of the RK formula to be raised, just as increasing the number of derivatives improves the accuracy of the Taylor series approximation.

It will be obvious that a one–stage Runge–Kutta formula

$$y_{n+1} = y_n + h_n f_1, \quad f_1 = f(x_n, y_n), \tag{2.13}$$

is identical to a first order Taylor series. This simple formula is usually called Euler's method.

2.6   Second and higher order equations

Thus far, the numerical methods considered apply only to first order ordinary differential equations, but the importance of second order equations has been emphasised in Chapter 1. Fortunately the extension to second order, or higher order, initial value problems is quite straightforward. Consider the system of second order equations with initial values specified:

$$y''(x) = g(x, y(x), y'(x)), \tag{2.14}$$

$$y(a) = y_0, \quad y'(a) = y_0', \quad x \in [a, b], \quad y, y' \in \mathbb{R}^M.$$

The Taylor series formula (2.5) can be applied to this system without modification since y, y' may be substituted directly from the initial condition, and second and higher derivatives may be generated as before prior to numerical substitutions. However, in order to allow second and subsequent steps, a new initial condition for y' must be generated in addition to the y solution. Thus, in addition to using (2.5), the Taylor polynomial for y',

$$y'_{n+1} = y'_n + h_n \Phi(x_n, y'_n, h_n),$$

must be formed. The technique is similar to solving double the number of first order equations. This doubling actually occurs in the application of RK methods to second order equations.

Consider the 3–stage RK applied to the second order equation

$$y'' = -2y(y + 2xy'), \quad y(0) = 1, \quad y'(0) = 0, \quad h = 0.2.$$

It is easily seen that an equivalent pair of first order equations is

$$\begin{aligned}
{}^1y' &= {}^1f = {}^2y \\
{}^2y' &= {}^2f = -2\,{}^1y(\,{}^1y + 2x\,{}^2y)
\end{aligned}$$

with ${}^1y(0) = 1$, ${}^2y(0) = 0$. The calculation of one step is detailed in Table 2.3 and the result is

$$\begin{aligned}
\begin{bmatrix} {}^1y_1 \\ {}^2y_1 \end{bmatrix} &= \begin{bmatrix} 1 \\ 0 \end{bmatrix} + \frac{0.2}{6}\left(\begin{bmatrix} 0 \\ -2 \end{bmatrix} + 4\begin{bmatrix} -0.20 \\ -1.92 \end{bmatrix} + \begin{bmatrix} -0.36800 \\ -1.42195 \end{bmatrix}\right) \\
&= \begin{bmatrix} 0.9611 \\ -0.3701 \end{bmatrix} \simeq \begin{bmatrix} y(0.2) \\ y'(0.2) \end{bmatrix}.
\end{aligned}$$

It will be clear that this second order problem has been constructed by differentiating the equation (2.2) from § 2.3, which was also the subject of the Runge-Kutta example of § 2.5. The result obtained for y is comparable in accuracy to that found earlier. Although the approach illustrated here is perfectly satisfactory a simpler, more direct technique for second order systems of equations is described in Chapter 14.

Table 2.3: A three–stage R–K calculation for a pair of equations

Stage	x	${}^1y$	${}^2y$	${}^1f$	${}^2f$
1	0	1	0	0	-2
2	0.1	1	-0.2	-0.20	-1.92
3	0.2	0.92	-0.368	-0.36800	-1.42195

The splitting technique can be generalised to equations of order greater than 2. The nth order initial value differential equation

$$\begin{aligned}
y^{(n)} &= g(x, y, y', y'', \dots, y^{(n-1)}), \\
y^{(i)}(x_0) = y_0^{(i)}, & \qquad i = 0, 1, \dots, n-1,
\end{aligned}$$

becomes a system of n first order equations. Putting

$$^1y = y, \quad ^2y = y', \ldots, {}^ny = y^{(n-1)},$$

one obtains the system

$$
\begin{bmatrix} ^1y' \\ ^2y' \\ \vdots \\ ^{n-1}y' \\ ^ny' \end{bmatrix}
=
\begin{bmatrix} ^2y \\ ^3y \\ \vdots \\ ^ny \\ g(x, {}^1y, \ldots, {}^ny) \end{bmatrix}
$$

with appropriate initial values.

2.7   Problems

1. Given the differential equation

$$y' = y^3 - y, \quad y(0) = \tfrac{1}{2},$$

use a Taylor series to estimate $y(0.1)$, including terms up to order of h^3. By computing $y^{(iv)}(0)$ estimate the error of your result and compare your estimate with the true error, given the analytical solution

$$y(x) = \frac{1}{\sqrt{1 + 3e^{2x}}}.$$

2. Use the Taylor series method to obtain estimates of y and y' at $x = 0.1, 0.2$ when y satisfies the second order equation

$$y'' = -\frac{1}{y^2}, \quad y(0) = 1, \ y'(0) = 0.$$

Truncate your series after the h^3 terms.

3. Obtain the higher derivatives up to order 5 for the pendulum equation

$$\frac{d^2\theta}{dt^2} = -\omega^2 \sin\theta.$$

Assuming $\omega = 1$, $\theta(0) = 0$, $\dot\theta(0) = 1$, compute a single step of size $h = 0.2$ to estimate the solution and derivative at $t = 0.2$. Modify the computer program in § 2.3 to compute the numerical solution to the pendulum problem for $t \in [0, 10]$ using the step-size $h = 0.2$. Estimate the period of the pendulum.

4. Use the second order two-stage Runge–Kutta formula

$$y_{n+1} = y_n + \frac{h}{2}(f_1 + f_2),$$

$$f_1 = f(x_n, y_n), \quad f_2 = f(x_n + h, y_n + hf_1),$$

to estimate $y(0.1)$ and $y(0.2)$ with $h = 0.1$, for the initial value problem

$$y' = x + y, \quad y(0) = 1.$$

Also estimate $y(0.2)$ with $h = 0.2$.

The analytical solution to this problem is $y(x) = 2e^x - x - 1$; use this to compute the errors of your results.

5. Compute two steps of the solution ($h = 0.1$) of the second order equation of question 2 using the Runge-Kutta formula given in question 3.

6. Modify the computer program in § 2.5 to solve the differential equation

$$y' = (1 - y/20)y/4, \qquad y(0) = 1$$

   for $x \in [0, 20]$ with a step-size $h = 0.2$.

7. If $y' = f(y)$, $y \in \mathbb{R}^3$, with $^3y = x$, the independent variable, obtain formulae for the components of the second derivative y''. Apply your formulae to find the second derivative components for the 2nd order equation of problem 2.

8. Express the third order equation

$$x^3 y''' - 3x^2 y'' + 6xy' - 6y = 0, \ y(1) = 2, \ y'(1) = 1, \ y''(1) = -4,$$

   as a system of three first order equations. Using a Runge-Kutta method, compute a step of size 0.1 to estimate the solution, and two derivatives, at $x = 1.1$.

Chapter 3

Error considerations

3.1 Introduction

For a thorough understanding of any numerical technique it is essential to consider the sources of error and how they are propagated through a calculation. This is particularly important for the solution of differential equations which demands a step-by-step approach. Inevitably, the error at any step will depend on those at earlier steps. In some cases, the properties of the equations and the numerical algorithm will be such that no amplification of error will occur; more often the error will increase as the number of steps increases, although not necessarily in a monotonic fashion.

In the numerical solution of differential equations there are two sources of error:

A: due to the truncation of series, also called mathematical error;

B: due to the finite precision of the computing machinery, also called rounding error.

The development of numerical methods often is concerned more directly with [A] rather than with [B], since the arithmetic precision, or the number of significant digits carried through an arithmetic operation, of a computer is usually beyond the control of the mathematician. Since the rounding error in any lengthy calculation will depend directly on the number of arithmetic operations carried out, it may be reduced by minimising this number. This reduction frequently can be achieved by reducing the mathematical error of a numerical process. This link between the two types of error is not always recognised. Of course, the practitioner must always avoid the loss of significant figures which sometimes occurs as a result of careless program coding.

31

A successful numerical method for differential equations should permit the computation of solutions, from the mathematical point of view, to any desired accuracy. In practice, the accuracy will be subject to the limitations of machine precision. Rounding error will always play its part.

3.2   Definitions

Let us consider again the initial value problem

$$y' = f(x,y), \quad y(x_0) = y_0, \quad y \in \mathbb{R}^k, \tag{3.1}$$

which will be solved numerically using the increment formula

$$y_{n+1} = y_n + h\Phi(x_n, y_n, h), \tag{3.2}$$

yielding a sequence of values

$$\{y_n \mid n = 0, 1, 2, \dots, N\},$$

corresponding to the true solution values $\{y(x_n)\}$. For simplicity of notation and clarity, the step–size h will be assumed to be constant, although extension to the more general variable step application is easily performed. The true solution of the initial value problem is taken to be $y = y(x)$. First, let us introduce some useful terminology.

The **Global error** of the numerical solution is defined as

$$\varepsilon_n = y_n - y(x_n), \quad n = 0, 1, \dots, N \tag{3.3}$$

and a numerical process for solving a differential equation(3.1) is said to be **Convergent** if

$$\lim_{h \to 0} \left(\max_{0 \le n \le N} \|\varepsilon_n\| \right) = 0.$$

Thus, in a convergent process, the global error will tend to zero with the step–size. From a practical point of view, this implies that the global error will be reduced as the step–size is reduced.

Another important concept is that of *consistency*. The numerical method (3.2) is **Consistent** with the differential equation (3.1) if

$$\Phi(x, y, 0) = f(x, y). \tag{3.4}$$

As will be seen later, this is a necessary condition for a numerical process to be convergent. For the Taylor series method of order p, the increment function is

$$\Phi = \sum_{i=1}^{p} \frac{h^{i-1}}{i!} y^{(i)},$$

and so
$$\Phi(x, y, 0) = y' = f(x, y),$$
thus satisfying the consistency property (3.4). Similarly, it is easy to show that any s-stage Runge–Kutta process of the form (2.12) is consistent if

$$\sum_{i=1}^{s} b_i = 1.$$

Although the accumulated or global error is of prime importance in the measurement of solution quality, an easier quantity to assess is the error arising from a single step of the numerical process. The **Local Error** (LE) of step $n + 1$ is defined as

$$e_{n+1} = y_{n+1} - u(x_{n+1}) \tag{3.5}$$

where $u(x)$ is the local true solution satisfying

$$u' = f(x, u), \quad u(x_n) = y_n.$$

This is the error associated with a single step and it is a quantity which plays an important role in the practical computation of solutions.

For analytical purposes it is convenient to introduce a third type of error, called the **Local Truncation Error** (LTE). Suppose the numerical method (Taylor series or Runge–Kutta) has an increment formula (3.2), which may be rearranged as

$$0 = y_n + h\Phi(x_n, y_n, h) - y_{n+1}.$$

This equation is not satisfied by substituting the differential equation's true solution value $y(x_n)$ for y_n, and the discrepancy is defined to be the local truncation error

$$t_{n+1} = y(x_n) + h\Phi(x_n, y(x_n), h) - y(x_{n+1}). \tag{3.6}$$

Thus the local truncation error is the amount by which the true solution fails to satisfy the numerical formula. As will be clear from Figure 2.1, at the first step the three error definitions given here yield identical values. For large n the global error could be much larger than the local errors, but usually the local and local truncation errors are very similar in magnitude. The relationships between these quantities will be examined later.

3.3 Local truncation error for the Taylor series method

Using the equations (2.4), (3.2), and (3.6), the local truncation error for the Taylor series method can written

$$t_{n+1} = y(x_n) + h\Phi(x_n, y(x_n), h) - (y(x_n) + h\Delta(x_n, y(x_n), h)),$$

yielding

$$t_{n+1} = h\{\Phi(x_n, y(x_n), h) - \Delta(x_n, y(x_n), h)\}. \qquad (3.7)$$

In terms of derivatives, and using the elementary differentials (2.10) notation, the pth order method has LTE

$$
\begin{aligned}
t_{n+1} &= -\sum_{i=p+1}^{\infty} \frac{h^i}{i!} y^{(i)}(x_n) \\
&= -\sum_{i=p+1}^{\infty} \frac{h^i}{i!} \sum_{j=1}^{n_i} \alpha_j^{(i)} F_j^{(i)}. \qquad (3.8)
\end{aligned}
$$

Now the Taylor increment can be expressed in finite form as

$$
\begin{aligned}
\Delta(x_n, y(x_n), h) &= y'(x_n) + \frac{h}{2} y''(x_n) + \cdots + \frac{h^{p-1}}{p!} y^{(p)}(x_n) \\
&\quad + \frac{h^p}{(p+1)!} y^{(p+1)}(x_n + \eta h), \quad \eta \in (0, 1),
\end{aligned}
$$

and so the LTE derives from the last term of this polynomial. The value of η is not generally known. Consequently there exists a real positive number A, independent of the step–size h, such that

$$\|t_{n+1}\| \leq A h^{p+1} \qquad (3.9)$$

and, for a pth order method, it is usual to write

$$t_{n+1} = O(h^{p+1}).$$

For any method of type (3.2) the local truncation error may be expressed as

$$t_{n+1} = \sum_{i=p+1}^{\infty} h^i \varphi_{i-1}(x_n, y(x_n)), \qquad (3.10)$$

where the φ_i are called **Error functions**. Using (3.8), the Taylor series scheme has error functions

$$\varphi_i = -\frac{y^{(i+1)}}{(i+1)!}.$$

The *principal* error function is φ_p, and this will be the dominant influence on the local truncation error if h is sufficiently small.

3.4 Local truncation error for the Runge–Kutta method

The analysis of the local truncation error of the Runge–Kutta method is far more complicated than the Taylor series case because the RK formula contains no explicit total derivatives. Instead the RK process is presented in terms of a function of two variables and, as a consequence, all total derivatives must be expanded in terms of the elementary differentials introduced in Chapter 2.

For the sake of simplicity, the local truncation error of a two–stage $(s = 2)$ Runge–Kutta method, applied to the differential equation (3.1), will be considered first. A general 2-stage explicit formula is written

$$y_{n+1} = y_n + h(b_1 f_1 + b_2 f_2), \qquad (3.11)$$

where

$$
\begin{aligned}
f_1 &= f(x_n,\, y_n) \\
f_2 &= f(x_n + c_2 h,\, y_n + a_{21} h f_1).
\end{aligned}
$$

An expression for the local truncation error is obtained by substituting the relevant increment function Φ of formula (3.11) in equation (3.7). To simplify the development of the error expression it is convenient to drop the explicit functional dependence of f and its derivatives where they are to be evaluated at the point (x, y). Thus, in this compact form, the increment function for the RK process may be written

$$\Phi(x, y, h) = b_1 f + b_2 f(x + c_2 h, y + a_{21} h f).$$

As before, the Taylor increment can be expressed as

$$\Delta(x, y, h) = F_1^{(1)} + \frac{h}{2} F_1^{(2)} + \frac{h^2}{6}(F_1^{(3)} + F_2^{(3)}) + \cdots,$$

where the elementary differential notation (see (2.10)) has been employed. Taking the scalar y case and using § 2.4, the elementary differentials are

$$
\begin{aligned}
F_1^{(1)} &= f \\
F_1^{(2)} &= f_x + f f_y \\
F_1^{(3)} &= f_{xx} + 2 f f_{xy} + f^2 f_{yy} \\
F_2^{(3)} &= f_x f_y + f_y^2 f.
\end{aligned}
$$

Expanding the RK increment function Φ in a two variable Taylor series gives

$$
\begin{aligned}
\Phi &= b_1 f + b_2 \{ f + (c_2 h f_x + a_{21} h f f_y) \\
&\quad + \frac{1}{2}(c_2^2 h^2 f_{xx} + 2 a_{21} h c_2 h f f_{xy} + a_{21}^2 h^2 f^2 f_{yy}) + O(h^3) \} \\
&= (b_1 + b_2) f + b_2 \{ h(c_2 f_x + a_{21} f f_y) \\
&\quad + \frac{h^2}{2}(c_2^2 f_{xx} + 2 a_{21} c_2 f f_{xy} + a_{21}^2 f^2 f_{yy}) + O(h^3) \}
\end{aligned}
$$

In order to simplify the formula (3.7) it is essential to identify elementary differentials in the above expression. This process is aided by choosing $a_{21} = c_2$, which gives

$$\Phi = (b_1 + b_2)F_1^{(1)} + b_2 c_2 h F_1^{(2)} + \tfrac{1}{2} b_2 c_2^2 h^2 F_1^{(3)} + O(h^3).$$

Substituting in (3.7) yields a suitable expression for the local truncation error in terms of the elementary differentials and the RK parameters b_1, b_2, and c_2:

$$\begin{aligned}
t_{n+1} &= h(\Phi - \Delta) \\
&= h[(b_1 + b_2) - 1]F_1^{(1)} + h^2[b_2 c_2 - \tfrac{1}{2}]F_1^{(2)} \\
&\quad + h^3\left[\left(\tfrac{1}{2}b_2 c_2^2 - \tfrac{1}{6}\right)F_1^{(3)} - \tfrac{1}{6}F_2^{(3)}\right] + O(h^4)
\end{aligned} \qquad (3.12)$$

Now the elementary differentials are dependent on the problem being solved and so the only parameters which influence the general behaviour of the RK process in (3.12) are b_1, b_2, and c_2. Following the Taylor series experience, the RK parameters should be chosen to maximise the order of the principal local truncation error (3.12). To remove the first two powers of h, the two equations

$$\begin{aligned}
b_1 + b_2 &= 1 \\
b_2 c_2 &= \tfrac{1}{2}
\end{aligned} \qquad (3.13)$$

must be satisfied. Then the local truncation error is reduced to

$$t_{n+1} = h^3\left[\left(\tfrac{1}{2}b_2 c_2^2 - \tfrac{1}{6}\right)F_1^{(3)} - \tfrac{1}{6}F_2^{(3)}\right] + O(h^4). \qquad (3.14)$$

A Taylor polynomial of order 2 has local truncation error

$$t_{n+1} = -\frac{h^3}{6}(F_1^{(3)} + F_2^{(3)}) + O(h^4)$$

and the two–stage RK formula whose parameters satisfy the two equations (3.13) has the same asymptotic behaviour and can be designated order 2 also.

The significance of the above development is that it provides a sensible basis for the choice of Runge–Kutta parameters. In the 2nd order case there are three parameters which must satisfy the two conditions (3.13). There is no unique solution and so one of the parameters must be chosen in an independent manner. An obvious choice, with no justification other than simplicity, is to make $c_2 = 1$, giving

$$y_{n+1} = y_n + \frac{h}{2}(f_1 + f_2), \qquad (3.15)$$

$$f_1 = f(x_n, y_n), \qquad f_2 = f(x_n + h, y_n + hf_1).$$

An alternative choice, $c_2 = \frac{1}{2}$, yields another second order RK formula:

$$y_{n+1} = y_n + hf_2, \tag{3.16}$$
$$f_1 = f(x_n, y_n), \qquad f_2 = f(x_n + \tfrac{1}{2}h, y_n + \tfrac{1}{2}hf_1).$$

A discussion of the optimal choice of a free parameter is left until the next chapter.

In general, the Runge–Kutta increment function can be expressed in the form

$$\Phi = \sum_{i=1}^{\infty} h^{i-1} \left\{ \sum_{j=1}^{n_i} \beta_j^{(i)} F_j^{(i)} \right\}, \tag{3.17}$$

where the $\beta_j^{(i)}$ are functions of the RK parameters a_{ij}, b_i, c_i, and it is assumed that the *row sum* condition

$$c_i = \sum_{j=1}^{i-1} a_{ij}, \quad i = 2, 3, \ldots, s \tag{3.18}$$

is satisfied. Using equations (3.7) and (2.10) with (3.17), the local truncation error for any RK method can be written in the form

$$t_{n+1} = \sum_{i=1}^{\infty} h^i \left\{ \sum_{j=1}^{n_i} \left(\beta_j^{(i)} - \frac{\alpha_j^{(i)}}{i!} \right) F_j^{(i)} \right\}$$

$$= \sum_{i=1}^{\infty} h^i \left\{ \sum_{j=1}^{n_i} \tau_j^{(i)} F_j^{(i)} \right\}, \tag{3.19}$$

where

$$\tau_j^{(i)} = \beta_j^{(i)} - \alpha_j^{(i)}/i!, \quad i = 1, 2, \ldots; \quad j = 1, 2, \ldots, n_i$$

are called error coefficients. Comparing (3.19) with (3.12), the first four error coefficients for the two stage Runge–Kutta process are

$$\begin{aligned}
\tau_1^{(1)} &= b_1 + b_2 - 1 \\
\tau_1^{(2)} &= b_2 c_2 - \tfrac{1}{2} \\
\tau_1^{(3)} &= \tfrac{1}{2} b_2 c_2^2 - \tfrac{1}{6} \\
\tau_2^{(3)} &= -\tfrac{1}{6},
\end{aligned} \tag{3.20}$$

and choosing the RK parameters to eliminate the first two of these error coefficients gives a second order formula. For the 2-stage formula, $\tau_2^{(3)}$ is a constant and so, in general, it is not possible to find parameters for a 2-stage formula to yield third order accuracy. However, increasing the

Table 3.1: Runge–Kutta error coefficients to 3rd order

$$
\begin{array}{ll}
1. & \tau_1^{(1)} = \sum_i b_i - 1 \\[2mm]
2. & \tau_1^{(2)} = \sum_i b_i c_i - \frac{1}{2} \\[2mm]
3. & \tau_1^{(3)} = \frac{1}{2}\sum_i b_i c_i^2 - \frac{1}{6} \\[2mm]
4. & \tau_2^{(3)} = \sum_{ij} b_i a_{ij} c_j - \frac{1}{6}
\end{array}
$$

number of stages to three provides sufficient RK parameters to construct a third order formula.

From (3.19), a Runge-Kutta formula is of order at least p ($t_{n+1} = O(h^{p+1})$) if the *equations of condition*

$$
\tau_j^{(i)} = 0, \quad \left\{ \begin{array}{l} i = 1, 2, \ldots, p \\ j = 1, 2, \ldots, n_i \end{array} \right. \tag{3.21}
$$

are satisfied. The value of n_i increases exponentially with order, so that the number of equations to be satisfied grows rapidly as the order increases. This implies a large number of stages s to provide the degrees of freedom necessary to achieve high order. The general formulae for the error coefficients up to order three are displayed in Table 3.1. Note that $\tau_2^{(3)}$ is not constant in the general case. The error coefficients in this table are valid when the RK formula is applied to general systems of equations and not just the scalar case which was the subject of the earlier analysis. Also it is not necessary for the RK process to be explicit; the more general expression for the ith stage

$$
f_i = f(x_n + c_i h, \; y_n + h \sum_{j=1}^{s} a_{ij} f_j)
$$

can be applied. Coefficients relating to the next two orders are shown in Table 3.2; these orders have 4 and 9 coefficients, respectively.

These tables of coefficients require some further explanation. First, the summations are carried out over the total number of stages s in the RK formula. Also, most of the coefficients involve multiple summations. For example, coefficient 16 from Table 3.2 is actually an abbreviation of

$$
\tau_8^{(5)} = \frac{1}{2} \sum_{i=1}^{s} \sum_{j=1}^{s} \sum_{k=1}^{s} b_i a_{ij} a_{jk} c_k^2 - \frac{1}{120}.
$$

The general row sum condition

$$c_i = \sum_{j=1}^{s} a_{ij} \ , \ i = 1, 2, \ldots, s$$

can replace equation (3.18). Finally, since most RK formulae are of the

Table 3.2: Runge–Kutta error coefficients of orders 4 and 5

5.	$\tau_1^{(4)} = \frac{1}{6}\sum_i b_i c_i^3 - \frac{1}{24}$
6.	$\tau_2^{(4)} = \sum_{ij} b_i c_i a_{ij} c_j - \frac{1}{8}$
7.	$\tau_3^{(4)} = \frac{1}{2}\sum_{ij} b_i a_{ij} c_j^2 - \frac{1}{24}$
8.	$\tau_4^{(4)} = \sum_{ijk} b_i a_{ij} a_{jk} c_k - \frac{1}{24}$
9.	$\tau_1^{(5)} = \frac{1}{24}\sum_i b_i c_i^4 - \frac{1}{120}$
10.	$\tau_2^{(5)} = \frac{1}{2}\sum_{ij} b_i c_i^2 a_{ij} c_j - \frac{1}{20}$
11.	$\tau_3^{(5)} = \frac{1}{2}\sum_{ijk} b_i a_{ij} c_j a_{ik} c_k - \frac{1}{40}$
12.	$\tau_4^{(5)} = \frac{1}{2}\sum_{ij} b_i c_i a_{ij} c_j^2 - \frac{1}{30}$
13.	$\tau_5^{(5)} = \frac{1}{6}\sum_{ij} b_i a_{ij} c_j^3 - \frac{1}{120}$
14.	$\tau_6^{(5)} = \sum_{ijk} b_i c_i a_{ij} a_{jk} c_k - \frac{1}{30}$
15.	$\tau_7^{(5)} = \sum_{ijk} b_i a_{ij} c_j a_{jk} c_k - \frac{1}{40}$
16.	$\tau_8^{(5)} = \frac{1}{2}\sum_{ijk} b_i a_{ij} a_{jk} c_k^2 - \frac{1}{120}$
17.	$\tau_9^{(5)} = \sum_{ijkm} b_i a_{ij} a_{jk} a_{km} c_m - \frac{1}{120}$

explicit type (2.12), in this case we impose $a_{ij} = 0$, $j \geq i$, so that

$$f_i = f(x_n + c_i h, \ y_n + h \sum_{j=1}^{i-1} a_{ij} f_j)$$

and error coefficient 16 from Table 3.2 reduces to

$$\tau_8^{(5)} = \frac{1}{2} \sum_{i=4}^{s} \sum_{j=3}^{i-1} \sum_{k=2}^{j-1} b_i a_{ij} a_{jk} c_k^2 - \frac{1}{120}.$$

Apart from the computational advantage of the explicit method it is obvious that the number of terms contributing to many of the error coefficients is considerably reduced from the most general case. For a 4-stage formula the coefficient 8 in Table 3.2, being a triple summation, contains 4^3 terms. However, imposing the explicit condition $i > j > k > 1$ reduces this to a single term

$$b_4 a_{43} a_{32} c_2 - 1/24.$$

Nevertheless, Tables 3.1 and 3.2 are valid for implicit formulae which will be considered in a later chapter.

At this point it is convenient to introduce the modified Butcher tabular notation, shown in Table 3.3, for displaying the coefficients of the general Runge–Kutta process (2.12). The convention here will be to omit diagonal and upper triangular elements of the matrix whose elements are a_{ij}, when these are zero, as is the case when the RK formula is explicit. This provides a compact and concise method for displaying multistage

Table 3.3: Modified Butcher table for an explicit Runge-Kutta formula

c_i		a_{ij}			b_i
0					b_1
c_2	a_{21}				b_2
c_3	a_{31}	a_{32}			b_3
$\vdots$	$\vdots$	$\vdots$	$\ddots$		$\vdots$
c_s	a_{s1}	a_{s2}	$\cdots$	a_{ss-1}	b_s

RK formulae, and the three stage formula (2.11) introduced in Chapter 2 is given in this form in Table 3.4. This notation will be used extensively in later Chapters.

3.5   Local truncation and global errors

Numerical experiments already have revealed a complicated relationship between the local error, which may be considered as the error per step, and the overall or global error. However, there is a simple and important asymptotic relation connecting the local truncation error with the global error of a numerical process. The analysis of these errors leads to

Table 3.4: A three stage Runge-Kutta formula

c_i	a_{ij}		b_i
0			$\frac{1}{6}$
$\frac{1}{2}$	$\frac{1}{2}$		$\frac{2}{3}$
1	-1	2	$\frac{1}{6}$

an understanding of the consistency requirement of a process and to its relation to convergence.

For greatest clarity the scalar differential equation

$$y' = f(x, y), \quad y(x_0) = y_0, \quad y \in \mathbb{R}, \tag{3.22}$$

is considered here, but an extension of the analysis to the vector case is straightforward. Let us apply a pth order method of the form (3.2) with constant step–size to the equation (3.22). The global error at step $n + 1$ is given by

$$\varepsilon_{n+1} = y_{n+1} - y(x_{n+1})$$

and substituting from equations (3.2) and (3.6) gives

$$\varepsilon_{n+1} = \varepsilon_n + h\{\Phi(x_n, y_n, h) - \Phi(x_n, y(x_n), h)\} + t_{n+1}.$$

Assuming $|t_{n+1}| \leq M$ and $h > 0$, the inequality

$$|\varepsilon_{n+1}| \leq |\varepsilon_n| + h|\Phi(x_n, y_n, h) - \Phi(x_n, y(x_n), h)| + M \tag{3.23}$$

follows immediately. If Φ satisfies the Lipschitz condition with constant L,

$$|\Phi(x_n, y_n, h) - \Phi(x_n, y(x_n), h)| \leq L|y_n - y(x_n)|,$$

over the domain containing the numerical and true solution points, the inequality (3.23) becomes

$$|\varepsilon_{n+1}| \leq |\varepsilon_n|(1 + hL) + M. \tag{3.24}$$

Applying this with $n = 0, 1, 2, \ldots$, it is a simple matter to obtain a bound on the global error $|\varepsilon_n|$ for general n. The inequality (3.24) yields

$$
\begin{aligned}
|\varepsilon_1| &\leq |\varepsilon_0|(1 + hL) + M \\
|\varepsilon_2| &\leq |\varepsilon_1|(1 + hL) + M \\
&\leq |\varepsilon_0|(1 + hL)^2 + M\{1 + (1 + hL)\} \\
\cdots \quad &\cdots \quad \cdots \\
|\varepsilon_n| &\leq |\varepsilon_0|(1 + hL)^n \\
&\quad + M\{1 + (1 + hL) + (1 + hL)^2 + \cdots + (1 + hL)^{n-1}\}.
\end{aligned}
$$

Summing the geometric series on the last line gives

$$|\varepsilon_n| \;\leq\; |\varepsilon_0|(1+hL)^n + M\frac{(1+hL)^n - 1}{hL}$$

$$= (1+hL)^n \left[|\varepsilon_0| + \frac{M}{hL}\right] - \frac{M}{hL}.$$

Using the relation $0 \leq (1+w)^m \leq e^{mw}$, and, putting $\varepsilon_0 = 0$, the global error bound is

$$|\varepsilon_n| \leq \frac{M}{hL}(e^{nhL} - 1). \tag{3.25}$$

To assess the convergence properties of the numerical process it is necessary to identify those parts of (3.25) which depend on the step–size h and also those which are independent of h. If the interval of solution of the differential equation is $[a, b]$, then $x_0 = a$ and $x_n = b$, and since $x_n = x_0 + nh$, the global error bound (3.25) can be expressed as

$$|\varepsilon_n| \leq \frac{M}{hL}(e^{L(b-a)} - 1)$$

which has no h dependence in the term in parentheses. Since a pth order method has $M = Ah^{p+1}$ (from (3.9)), it is clear that there exists a positive number B such that

$$|\varepsilon_n| \leq Bh^p.$$

This result can be summarised in the form

$$t_{n+1} \sim O(h^{p+1}) \implies \varepsilon_n \sim O(h^p).$$

The order of the global error is one less than that of the local truncation error and so a single step numerical method of the form (3.2) which is consistent of order p will be convergent, provided $p \geq 1$.

3.6   Local error and LTE

The local truncation error, or mathematical error, of a numerical method has been used in the analysis of the global error of the process. However, a more practical quantity, from a computational point of view, is the error per step, or local error (3.5), of the method. Fortunately the two local errors are very similar in magnitude. It is easily seen, in the scalar case (3.22), that the local error at step $n+1$ satisfies

$$e_{n+1} = h\{\Phi(x_n, y_n, h) - \Delta(x_n, y_n, h)\},$$

and so, using (3.7), the difference between the two types of local error is

$$e_{n+1} - t_{n+1} = h\{\Phi(x_n, y_n, h) - \Delta(x_n, y_n, h)\}$$
$$- h\{\Phi(x_n, y(x_n), h) - \Delta(x_n, y(x_n), h)\}.$$

Expanding about $(x_n, y(x_n))$ in a two variable Taylor series yields

$$e_{n+1} - t_{n+1} = h\{\varepsilon_n(\Phi_y - \Delta_y) + \frac{\varepsilon_n^2}{2}(\Phi_{yy} - \Delta_{yy}) + \cdots\}.$$

Since $\varepsilon_n \sim O(h^p)$, and also $|\Phi - \Delta_| \leq Ah^p$, it follows from this that

$$|e_{n+1} - t_{n+1}| \leq Dh^{2p+1}, \quad D \geq 0,$$

indicating that the difference between the two errors is p orders higher than the local error itself. As h is reduced, the difference between the two errors will diminish much more rapidly than either local error. Consequently we can regard the local error estimate for a pth order process as being an estimate of the local *truncation* error. Since the local error does not involve the overall true solution, it is much easier to estimate than the LTE for practical problems. Such error estimation is essential for the efficient numerical solution of differential equations.

3.7   Problems

1. Show that the condition required for any s-stage Runge–Kutta method to be consistent with a differential equation is

$$\sum_{i=1}^{s} b_i = 1.$$

2. Compute five steps of Euler's method applied to the equation

$$y' = y - 2y^2, \ y(0) = 1, \ h = 0.1.$$

   Use the true solution and the local true solutions to compute the local and global errors for each step, given that the general solution of the differential equation is $y = 1/(2 + ce^{-x})$.

3. Determine the order of each of the three-stage RK formulae shown in tabular form as

(a)

c_i	a_{ij}		b_i
0			0
$\frac{1}{8}$	$\frac{1}{8}$		0
$\frac{1}{2}$	0	$\frac{1}{2}$	1

(b)

c_i	a_{ij}		b_i
0			$\frac{1}{4}$
$\frac{1}{3}$	$\frac{1}{3}$		0
$\frac{2}{3}$	0	$\frac{2}{3}$	$\frac{3}{4}$

(c)

c_i	a_{ij}		b_i
0			$\frac{5}{12}$
$\frac{1}{2}$	$\frac{1}{2}$		$\frac{1}{2}$
1	$\frac{1}{2}$	$\frac{1}{2}$	$\frac{1}{12}$

4. Write down in expanded form the eight equations of condition for an explicit four–stage fourth order Runge–Kutta formula. Verify that the equations are satisfied by the formula given below

c_i	a_{ij}			b_i
0				$\frac{1}{6}$
$\frac{1}{2}$	$\frac{1}{2}$			$\frac{1}{3}$
$\frac{1}{2}$	0	$\frac{1}{2}$		$\frac{1}{3}$
1	0	0	1	$\frac{1}{6}$

   (Do not solve the equations.)
   Given the initial value problem

$$y' = xy^2, \quad y(0) = 2,$$

   use the above RK4 formula to estimate $y(0.2)$ with $h = 0.2$. Determine the true solution and use it to compute the error of your estimate.

5. Use the RK4 formula of the previous question, with step-size $h = 0.2$, to estimate $y(0.2)$ and $y'(0.2)$ which satisfy the second order equation

$$y'' = -1/y^2, \ y(0) = 1 \ , \ y'(0) = 0.$$

   [NB this must be treated as a pair of 1st order equations.]

6. If $y' = f(x, y) = f_1$, with

$$
\begin{aligned}
f_2 &= f(x + c_2 h,\ y + h a_{21} f_1), \\
f_3 &= f(x + c_3 h,\ y + h(a_{31} f_1 + a_{32} f_2)),
\end{aligned}
$$

obtain f_i, $i = 2, 3$ in terms of the elementary differentials:

$$
\begin{aligned}
F_1^{(2)} &= f_x + f f_y, \\
F_1^{(3)} &= f_{xx} + 2 f f_{xy} + f^2 f_{yy}, \\
F_2^{(3)} &= f_x f_y + f f_y^2.
\end{aligned}
$$

Hence find the equations of condition for a three–stage third order RK formula. (Assume $a_{21} = c_2$ and $a_{31} + a_{32} = c_3$.)

7. Show that, for $w \geq 0$ and $m > 0$,

$$
0 \leq (1 + w)^m \leq e^{mw}.
$$

8. Modify the Runge-Kutta program RK3ode from Chapter 2, which is based on a third order formula, to solve the problem

$$
y' = y(2 - y), \quad y(0) = 1,
$$

with steplength $h = 0.25$ and $x \in [0, 5]$. Using the general solution of the differential equation, compute also the local error e, based on the local true solution, and the global error at each step. By applying the increment formula

$$
w_{n+1} = y(x_n) + h \Phi(x_n, y(x_n), h)
$$

at each step, compute the local truncation error t and compare this with the local error e. Recompute the solutions and errors with a larger step $h = 0.5$. Hence investigate numerically the orders (powers of h) of the various quantities obtained. Do your numerical results confirm the analytical results present in this chapter?

Chapter 4

Runge–Kutta methods

4.1 Introduction

Runge–Kutta methods have been popular with practitioners for many years. Originally developed by Runge towards the end of the nineteenth century and generalised by Kutta in the early twentieth century, these methods are very convenient to implement, when compared with the Taylor polynomial scheme, which requires the formation and evaluation of higher derivatives. This drawback of the Taylor method is particularly serious in the treatment of systems of differential equations but, with the growing popularity of computer software for symbolic manipulation, it seems likely that the series approach will receive further attention.

Modern development of Runge–Kutta processes has occurred since 1960, mainly as a direct result of the advances due to J.C.Butcher in the development and simplification of RK error coefficients. The importance of Butcher's work in this regard cannot be emphasized too greatly. His contributions have paved the way for the development of efficient high order processes on which modern differential equation software depends. However, the availability of powerful computers is itself a motivation for the development of new methods to yield ever greater accuracy.

Some simple Runge-Kutta processes have been introduced in earlier chapters and, in Chapter 3, the basis for the construction of formulae has been presented. An important observation from the last chapter is that, for a specific number of stages, the RK formula is not unique. This means that there exist free parameters which must be chosen. It is essential to adopt a systematic approach in making these choices and, in this chapter, criteria for building new formulae will be introduced and justified.

One of the problems confronting any scientist or engineer with an initial value problem to solve is how to choose the best formula for the required task. With the available alternatives and the range of problems to be solved, this choice is not always straightforward. Although this chapter does not provide the 'best' RK formula, it does give details of the worst numerical process for solving differential equations!

The construction of any numerical methods must be accompanied by practical testing. The tests will often be applied to problems which have known solutions. This enables proper error estimates to be computed, a necessity when different numerical processes are to be compared. A Fortran 90 program applicable to a system of differential equations is presented below. The program is designed to implement *any* explicit Runge-Kutta process, a feature which, although rarely seen in published programs, can be achieved rather easily without significant loss of computational efficiency.

4.2   Error criteria

For a two–stage explicit Runge-Kutta formula

c_i	a_{ij}	b_i
0		b_1
c_2	a_{21}	b_2

an expression for the local truncation error (3.12) was obtained in §3.4. Arising from this were two equations (3.13)

$$b_1 + b_2 = 1$$
$$b_2 c_2 = \tfrac{1}{2}$$

which must be satisfied by the RK parameters (b_1, b_2, c_2) in order to yield a 2nd order formula. Since there are a total of 3 parameters available one of them can be chosen in an arbitrary fashion and still the equations can be satisfied. The *equations of condition* are said to have one degree of freedom in this situation. It is not possible to eliminate the third order term (h^3) in the local truncation error, for which the error function is

$$\varphi_2 = \left(\tfrac{1}{2} b_2 c_2^2 - \tfrac{1}{6}\right) F_1^{(3)} - \tfrac{1}{6} F_2^{(3)},$$

since one error coefficient $(\tau_2^{(3)})$ is constant. As a consequence φ_2 becomes the principal error function and the formula has maximum order 2. However it seems reasonable to use the degree of freedom to remove the variable coefficient since this will reduce the overall magnitude of the principal error term, as evaluated over a wide range of problems. Setting

$$\tfrac{1}{2} b_2 c_2^2 = \tfrac{1}{6},$$

the second condition in (3.13) yields $c_2 = \tfrac{2}{3}$, and so $b_2 = \tfrac{3}{4}$ and $b_1 = \tfrac{1}{4}$. Note, that for a differential equation in which $F_2^{(3)} = 0$, the 2-stage formula here would have order 3. In fact this situation occurs when $f(x, y) = f(x)$; since y does not affect the derivative such a problem is one of quadrature.

This particular 2-stage RK formula, with $c_2 = \frac{2}{3}$, is usually regarded as *optimal* with respect to the local truncation error. Of course it is possible to construct problems for which some other two stage second order RK formula gives greater accuracy. However, practical calculations over a wide range of problems support the above choice of free parameter.

When more than two RK stages are present the order can be increased. Since the number of error coefficients increases also, so does the complexity of the principal error function. For a method of order p, this is written as

$$\varphi_p = \sum_{j=1}^{n_{p+1}} \tau_j^{(p+1)} F_j^{(p+1)}, \tag{4.1}$$

and reference to Table 3.2 shows that $n_4 = 4$ and $n_5 = 9$. For higher orders the numbers of error coefficients increase rapidly with order, while the degrees of freedom depend on the number of stages. Consequently it becomes inappropriate to try to optimise a Runge-Kutta formula by choosing free parameters to eliminate particular error coefficients.

In an influential paper Dormand and Prince (1980) proposed that, having achieved a particular order of accuracy, the best strategy for practical purposes would be to choose the free RK parameters to *minimise* the error norm

$$A^{(p+1)} = \|\tau^{(p+1)}\|_2 = \sqrt{\sum_{j=1}^{n_{p+1}} \left(\tau_j^{(p+1)}\right)^2}. \tag{4.2}$$

It will be clear that this is consistent with the 2–stage optimal formula. The minimisation should not be taken as a perfectly strict condition (hence the italics), because it applies to a single term in an infinite series. This term will dominate the series for small enough h, but may not do so at practical step-sizes when the error coefficients are strictly minimised.

In some cases of practical importance, a strict minimisation according to equation (4.2) could result in the secondary term in the series for the local truncation error becoming dominant for typical step–sizes. Obviously this will invalidate the arguments above. Dormand and Prince have proposed a number of criteria to be used in the construction of RK formulae. Achieving a small principal error function is perhaps the key feature, but others are concerned with aspects of practical computation to be encountered later in this text.

4.3   A third order formula

Let us consider now the derivation of a third order RK formula which will have $t_{n+1} \sim O(h^4)$. The equations of condition to be satisfied will be

$$\tau_j^{(i)} = 0, \quad i = 1, 2, 3; \quad j = 1, 2, \ldots, n_i,$$

and a 3-stage process

c_i	a_{ij}		b_i
0			b_1
c_2	a_{21}		b_2
c_3	a_{31}	a_{32}	b_3

is assumed. Using Table 3.1, and noting that $a_{ij} = 0$, $j \geq i$, the four equations of condition may be expressed as

$$b_1 + b_2 + b_3 = 1 \tag{4.3}$$
$$b_2 c_2 + b_3 c_3 = \tfrac{1}{2} \tag{4.4}$$
$$b_2 c_2^2 + b_3 c_3^2 = \tfrac{1}{3} \tag{4.5}$$
$$b_3 a_{32} c_2 = \tfrac{1}{6}. \tag{4.6}$$

These equations have six unknowns and so there are two degrees of freedom. It is convenient to choose the free parameters to be c_2 and c_3, and to solve the equations in terms of these. Taking $(4.4) \times c_2 - (4.5)$ yields

$$b_3 c_3 (c_2 - c_3) = \tfrac{1}{2} c_2 - \tfrac{1}{3},$$

giving

$$b_3 = \frac{3c_2 - 2}{6c_3(c_2 - c_3)}.$$

Similarly

$$b_2 = \frac{3c_3 - 2}{6c_2(c_3 - c_2)},$$

and, from equation (4.6), this leads to

$$a_{32} = \frac{c_3(c_2 - c_3)}{c_2(3c_2 - 2)}.$$

From equation (4.3), $b_1 = 1 - b_2 - b_3$, and the row sum condition (3.18) yields $a_{21} = c_2$ and $a_{31} = c_3 - a_{32}$. This solution assumes that

$$c_2, c_3 \neq 0, \quad c_2 \neq \tfrac{2}{3}, \quad c_2 \neq c_3.$$

Other forms of solution, including the singular case with $c_2 = c_3$, are possible.

In order to choose the free parameters c_2, c_3, the principal term of the local truncation error must be considered. Using equation (4.1) and Table 3.2, the principal term is

$$\varphi_3 = \sum_{j=1}^{4} \tau_j^{(4)} F_j^{(4)},$$

where

$$\tau_1^{(4)} = \tfrac{1}{6}(b_2 c_2^3 + b_3 c_3^3) - \tfrac{1}{24}$$
$$\tau_2^{(4)} = b_3 c_3 a_{32} c_2 - \tfrac{1}{8}$$
$$\tau_3^{(4)} = \tfrac{1}{2} b_3 a_{32} c_2^2 - \tfrac{1}{24}$$
$$\tau_4^{(4)} = -\tfrac{1}{24}.$$

Substituting for b_2, b_3, a_{32} from above gives the truncation error coefficients in terms of the free parameters c_2 and c_3. These simplify to

$$\tau_1^{(4)} = -\frac{6c_2 c_3 - 4(c_2 + c_3) + 3}{72}$$

$$\tau_2^{(4)} = \frac{4c_3 - 3}{24}$$

$$\tau_3^{(4)} = \frac{2c_2 - 1}{24}$$

$$\tau_4^{(4)} = -\tfrac{1}{24}.$$

Up to two of these may be eliminated by choosing the free parameters, but the last coefficient will remain a constant, similar to the two–stage case. Such elimination is not strictly in accord with (4.2) but, in this case, it is easy to do. Substituting $c_2 = \tfrac{1}{2}$ in $\tau_3^{(4)}$ and $c_3 = \tfrac{3}{4}$ in $\tau_2^{(4)}$ eliminates these two coefficients. Again with $c_2 = \tfrac{1}{2}$ it is easily shown that $\tau_1^{(4)} = 0$ if $c_3 = 1$.

These two alternative choices for the free parameters give the two third–order three–stage Runge–Kutta formulae whose parameters are most conveniently displayed in the tabular forms:

c_i	a_{ij}		b_i
0			$\frac{2}{9}$
$\frac{1}{2}$	$\frac{1}{2}$		$\frac{1}{3}$
$\frac{3}{4}$	0	$\frac{3}{4}$	$\frac{4}{9}$

(4.7)

c_i	a_{ij}		b_i
0			$\frac{1}{6}$
$\frac{1}{2}$	$\frac{1}{2}$		$\frac{2}{3}$
1	-1	2	$\frac{1}{6}$

(4.8)

The formulae represented by (4.7) and (4.8) have principal error norms $A^{(4)} = 0.042$ and 0.059, respectively, and so it would seem that (4.7) is

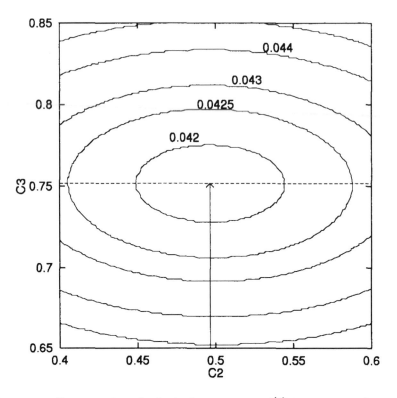

Figure 4.1: Contour plot of principal error norm $A^{(4)}$ for a 3-stage RK3

likely to be the better of the two. Actually this one is very close to the absolute optimum for this model since the minimum of φ_3 occurs at $c_2 = 0.49650$, $c_3 = 0.75175$, being about 1% smaller than the value for (4.7), which, in the future, will be designated RK3M. A contour plot, showing the variation of $A^{(4)}$ with the two free parameters, is shown in Figure 4.1. Note that the formulae (4.7, 4.8) have coefficients which are single digit fractions. Thus it would be impossible to find a formula with a simpler appearance. To attain the absolute minimum the fractions $\frac{1}{2}$, $\frac{3}{4}$ would have to be replaced by $\frac{993}{2000}$, $\frac{3007}{4000}$, with corresponding changes in the b values, giving a more ugly appearance than (4.7). Such aesthetic considerations would not affect a computer program, but any improvement in efficiency gained by use of the strict optimum would be negligible since the error behaviour is not sensitively dependent on the values of c_2 and c_3.

4.4   Fourth order formulae

The derivation of a fourth order Runge–Kutta process is more lengthy than the lower order cases above. Consideration of Table 3.2 shows that constructing a 4th order RK process requires the elimination of 8 error coefficients and so, in general, more than three stages are necessary. From Table 3.3, it can be seen that the number of independent parameters in an explicit RK formula is given by

$$\eta_s = \frac{s(s+1)}{2}, \tag{4.9}$$

where s is the number of stages. With $s = 4$, this gives $\eta_4 = 10$, implying that the 8 equations of condition will have two free parameters, as was the case with the 3rd order RK model of the previous section.

Setting the coefficients $(1)\ldots(8)$ to zero, from Tables 3.1 and 3.2, yields the 8 equations of condition:

$$
\begin{aligned}
b_1 + b_2 + b_3 + b_4 &= 1 & (E1)\\
b_2 c_2 + b_3 c_3 + b_4 c_4 &= \tfrac{1}{2} & (E2)\\
b_2 c_2^2 + b_3 c_3^2 + b_4 c_4^2 &= \tfrac{1}{3} & (E3)\\
b_3 a_{32} c_2 + b_4 (a_{42} c_2 + a_{43} c_3) &= \tfrac{1}{6} & (E4)\\
b_2 c_2^3 + b_3 c_3^3 + b_4 c_4^3 &= \tfrac{1}{4} & (E5)\\
b_3 c_3 a_{32} c_2 + b_4 c_4 (a_{42} c_2 + a_{43} c_3) &= \tfrac{1}{8} & (E6)\\
b_3 a_{32} c_2^2 + b_4 (a_{42} c_2^2 + a_{43} c_3^2) &= \tfrac{1}{12} & (E7)\\
b_4 a_{43} a_{32} c_2 &= \tfrac{1}{24} & (E8)
\end{aligned}
$$

Direct solution of these equations is quite straightforward, but a more powerful approach is based on the simplifying relations introduced by Butcher in 1963,

$$\sum_{i=1}^{s} b_i a_{ij} = b_j (1 - c_j), \quad j = 2, 3, 4, \tag{4.10}$$

which affect the expressions for $\tau_2^{(3)}, \tau_3^{(4)}, \tau_4^{(4)}$. Substituting (4.10) in the expression for $\tau_2^{(3)}$ yields

$$
\begin{aligned}
\tau_2^{(3)} &= \sum_j b_j (1 - c_j) c_j - \tfrac{1}{6}\\
&= \tau_1^{(2)} + \tfrac{1}{2} - 2(\tau_1^{(3)} + \tfrac{1}{6}) - \tfrac{1}{6}\\
&= \tau_1^{(2)} - 2\tau_1^{(3)}.
\end{aligned}
$$

Similarly, $\tau_3^{(4)}$ and $\tau_4^{(4)}$ can be expressed in terms of earlier error coefficients. It is easily verified that

$$\tau_3^{(4)} = \tau_1^{(3)} - 3\tau_1^{(4)}$$

$$\tau_4^{(4)} \;=\; \tau_1^{(2)} - 2\tau_1^{(3)} - \tau_2^{(4)}.$$

Thus the simplifying conditions (4.10) will ensure that the error coefficients (4), (7), and (8) will be made zero when the other five are eliminated. In other words, the three equations of condition $E4, E7$, and $E8$ reduce to three simpler conditions. In this particular instance the total number of equations to be solved remains the same, whether or not the Butcher conditions are employed, but, for higher orders involving more stages, the substitution of (4.10) will reduce the number of independent equations to be satisfied. Such simplification is essential for the construction of high order Runge-Kutta processes.

The solution now proceeds as follows: taking $j = 4$ in (4.10), and setting $a_{ij} = 0$, $j \geq i$, one obtains

$$b_4(1 - c_4) = 0 \implies c_4 = 1,$$

since a 4-stage formulae requires $b_4 \neq 0$. As in the 3rd order case, c_2, c_3 are chosen to be the two free parameters. Then the three equations $E2, E3$, and $E5$, which are linear in the b_i, yield

$$b_2 \;=\; \frac{1 - 2c_3}{12c_2(1 - c_2)(c_2 - c_3)},$$

$$b_3 \;=\; \frac{1 - 2c_2}{12c_3(1 - c_3)(c_3 - c_2)},$$

$$b_4 \;=\; \frac{6c_2c_3 - 4(c_3 + c_2) + 3}{12(1 - c_2)(1 - c_3)},$$

and then, using $j = 3$ in (4.10), one obtains

$$a_{43} = \frac{(1 - c_2)\,(2\,c_2 - 1)\,(1 - c_3)}{c_3\,(c_2 - c_3)\,(6\,c_2\,c_3 - 4\,(c_3 + c_2) + 3)}.$$

Finally, the equations $E8$ and (4.10) with $j = 2$ give

$$a_{32} \;=\; \frac{c_3\,(c_2 - c_3)}{2\,c_2\,(2\,c_2 - 1)},$$

$$a_{42} \;=\; \frac{(1 - c_2)\{2(1 - c_3)(1 - 2c_3) - (c_2 - c_3)\}}{2c_2(c_2 - c_3)\{6\,c_2\,c_3 - 4(c_2 + c_3) + 3)\}}.$$

The other parameters follow from $E1$ and the row sum condition (3.18), giving a two–parameter family of formulae subject to the conditions on c_2, c_3 listed above. This solution assumes that

$$c_2 \neq 0, 1, \quad c_3 \neq 0, 1, \quad c_2 \neq c_3, \quad c_2 \neq \tfrac{1}{2}.$$

Other possibilities exist, and a well-known formula is actually based on the singular case $c_2 = c_3$, which will be the subject of an exercise problem below. Using the modified Butcher table, one member ($c_2 = \frac{1}{3}$, $c_3 = \frac{2}{3}$) of this RK4 family takes the form

$$
\begin{array}{c|ccc|c}
c_i & & a_{ij} & & b_i \\
\hline
0 & & & & \frac{1}{8} \\
\frac{1}{3} & \frac{1}{3} & & & \frac{3}{8} \\
\frac{2}{3} & -\frac{1}{3} & 1 & & \frac{3}{8} \\
1 & 1 & -1 & 1 & \frac{1}{8}
\end{array}
\tag{4.11}
$$

As with the third order Runge-Kutta process, it is natural to seek values for the free parameters which will define the optimal fourth formula. In this case, the principal error function φ_4 has 9 error coefficients but, since the RK parameters satisfy the Butcher conditions (4.10), these can be used to simplify the coefficients. For example, substituting (4.10) in the expression for $\tau_5^{(5)}$ gives

$$
\begin{aligned}
\tau_5^{(5)} &= \tfrac{1}{6} \sum_j b_j (1 - c_j) c_j^3 - \tfrac{1}{120} \\
&= \tau_1^{(4)} - 4\tau_1^{(5)} \\
&= -4\tau_1^{(5)},
\end{aligned}
$$

since $\tau_1^{(4)} = 0$ for any fourth order process. In addition, it is easy to show that

$$
\tau_6^{(5)} = -\tau_9^{(5)} = 1/120, \quad \tau_7^{(5)} = -2\tau_2^{(5)}, \quad \tau_8^{(5)} = -\tau_4^{(5)}.
$$

Since there are only 5 independent coefficients, the minimisation process can be applied to

$$
A^{(5)} = \sqrt{17\{\tau_1^{(5)}\}^2 + 5\{\tau_2^{(5)}\}^2 + \{\tau_3^{(5)}\}^2 + 2\{\tau_4^{(5)}\}^2 + \tfrac{1}{7200}}.
$$

In terms of the free parameters c_2 and c_3, the relevant error coefficients can be written in the form

$$
\begin{aligned}
\tau_1^{(5)} &= \frac{(10c_2 - 5)c_3 - 5c_2 + 3}{1440} \\
\tau_2^{(5)} &= \frac{3 - 5c_3}{240} \\
\tau_3^{(5)} &= \frac{10c_3^2(3c_2 - 2) + 2c_2^2(3 - 7c_3) + 3c_3(9 - 14c_2) + 3(5c_2 - 3)}{480(2c_2 - 1)\{6\,c_2\,c_3 - 4(c_2 + c_3) + 3\}} \\
\tau_4^{(5)} &= \frac{2 - 5c_2}{240}.
\end{aligned}
$$

Figure 4.2 shows some contours of the function $A^{(5)}$ near the minimum,

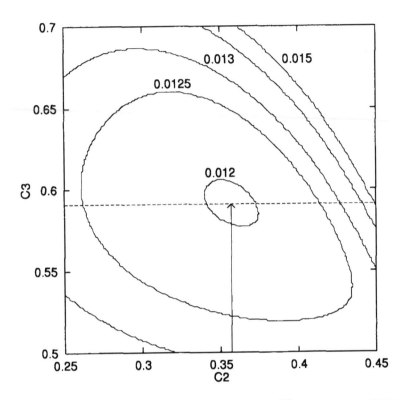

Figure 4.2: Contour plot of principal error norm $A^{(5)}$ for a 4-stage RK4

which occurs close to $c_2 = \frac{5}{14}$, $c_3 = \frac{13}{22}$. However the function has very shallow gradients near the minimum and a formula close to optimal (RK44M) is presented in Table 4.1 below. Since two of the error coefficients are constant it is not possible to gain substantial improvements over the formula (4.11). The error norms for the two given formulae are

$$A^{(5)}(\text{RK44M}) = 0.0123, \quad A^{(5)}(4.11) = 0.0127,$$

and so the performances of these two RK processes are not likely to differ substantially.

Adding a fifth stage ($\eta_5 = 15$) provides an extra 5 RK parameters, and therefore more opportunity to reduce the magnitude of $A^{(5)}$. The drawback is increased complexity and larger computational cost per step. Let us consider such a formula. As in the 4-stage process, the Butcher simplifying equations (4.10) will be applied, this time with an extra member for $j = 5$. This eliminates the same error coefficients as in the 4-stage process, at the cost of an extra constraint. Even greater simplification is

Table 4.1: A near-optimal 4-stage RK4 (RK44M)

c_i	a_{ij}			b_i
0				$\frac{11}{72}$
$\frac{2}{5}$	$\frac{2}{5}$			$\frac{25}{72}$
$\frac{3}{5}$	$-\frac{3}{20}$	$\frac{3}{4}$		$\frac{25}{72}$
1	$\frac{19}{44}$	$-\frac{15}{44}$	$\frac{10}{11}$	$\frac{11}{72}$

provided by imposing a second Butcher condition

$$\sum_{j=1}^{i-1} a_{ij}c_j = \tfrac{1}{2}c_i^2, \quad i = 3, 4, 5. \tag{4.12}$$

Substituting (4.12) in the coefficient $\tau_2^{(4)}$ yields

$$\tau_2^{(4)} = \sum_{i=3}^{5} b_i c_i(\tfrac{1}{2}c_i^2) - \tfrac{1}{8}$$

$$= 3\tau_1^{(4)} \iff b_2 = 0.$$

The equations remaining for the 5-stage formula of order 4 are

$$\tau_1^{(i)} = 0, \quad i = 1, 2, 3, 4,$$

together with the 8 Butcher conditions. In fact the number of equations has increased by 2, but they have a much simpler form than the 'raw' conditions derived from the RK error coefficients. Three degrees of freedom remain, and these can be chosen to minimize the principal error function $A^{(5)}$, which may be written as

$$A^{(5)} = \sqrt{206\{\tau_1^{(5)}\}^2 + 2\{\tau_4^{(5)}\}^2 + 2\{\tau_6^{(5)}\}^2}.$$

Although it may appear that the three independent error coefficients may be eliminated by a suitable combination of the three free parameters, it is not possible, in this case, to achieve a fifth-order formula. It may be proved (Hairer, Nørsett, and Wanner, 1993) that no 5-stage fifth order RK formula exists. An example of a 5–stage RK4 developed by Prince (1979), designated here as RK45M, is shown in Table 4.2. With a principal error norm, $A^{(5)} = 0.00199$, much smaller than that for the 4–stage process RK44M, this 5-stage formula usually will give better accuracy for a given step-size. This improvement is more than adequate to compensate for the extra computational cost per step and results presented below will show that the 5–stage formula is more efficient than the 4–stage process.

Table 4.5   A 5–stage RK4 (RK45M)

c_i	a_{ij}				b_i
0					$\frac{13}{96}$
$\frac{1}{5}$	$\frac{1}{5}$				0
$\frac{2}{5}$	0	$\frac{2}{5}$			$\frac{25}{48}$
$\frac{4}{5}$	$\frac{6}{5}$	$-\frac{12}{5}$	2		$\frac{25}{96}$
1	$-\frac{17}{8}$	5	$-\frac{5}{2}$	$\frac{5}{8}$	$\frac{1}{12}$

4.5   Fifth and higher order formulae

Moving to order 5, it is seen from Table 3.2 that 17 equations of condition need to be satisfied. If s is the number of stages in the RK process, then $\eta_s > 17$, and s must satisfy

$$\frac{1}{2}s(s+1) > 17.$$

The smallest value of s to satisfy this inequality is 6. The use of simplifying conditions, such as (4.10), is essential in the construction of 5th and higher order formulae. No example of a fifth order formula is presented here although an example will be given later.

For higher orders (> 5) the numbers of error coefficients increase rapidly, as shown in the Table 4.3. Fortunately many of the higher order

Table 4.3: Numbers of RK error coefficients for orders to 10

Order p	2	3	4	5	6	7	8	9	10
$\displaystyle\sum_{i=1}^{p} n_i$	2	4	8	17	37	85	200	486	1205

coefficients may be made dependent by imposing the Butcher conditions (Butcher, 1963)

$$\sum_{i=1}^{s} b_i a_{ij} = b_j(1 - c_j), \quad j = 2, 3, \ldots, s \tag{4.13}$$

and, additionally, $b_2 = 0$ with

$$\sum_{j=1}^{i-1} a_{ij} c_j = \frac{1}{2} c_i^2, \quad i = 3, 4, \ldots, s. \tag{4.14}$$

These valuable relations actually reduce the total number of conditions arising in higher orders. Consider the derivation of a 6th order process for which 37 equations arise. For $\eta_s > 37$, equation (4.9) yields $s \geq 9$. With the reduction of conditions due to the Butcher relationships, the minimum number of stages required for a 6th order formula is 7. The reduction for order 8 is even more drastic. An 8th order RK can be achieved with only 11 stages. It must be admitted that, for very high order processes, the conditions (4.13) and (4.14) are supplemented by extra simplifying relations. This is not to suggest that construction of such formula is simple!

4.6 Rationale for high order formulae

Why do we seek high–order formulae? A formula of order 5 requires twice as many stages as one of order 3 and so the computational cost per step is doubled by increasing the algebraic order by 2. This can only be justified by a gain in computational efficiency. Following §3.5, assume that the global error of a pth order numerical solution estimating $y(b)$, $b = x_N$, is $\varepsilon(x_N) = Ch^p$, where h is the step–size. Taking logarithms gives

$$\log \varepsilon = \log C + p \log h.$$

Assuming an integration with constant steplength on $[a, b]$, the step-size is $h = (b - a)/N$, and so

$$\begin{aligned} \log \varepsilon &= B - p \log N, \\ B &= \log C + p \log(b - a). \end{aligned}$$

This relationship is illustrated for two formulae of different orders p and q, $(q > p)$, in Figure 4.3.

If the linear relationship is valid, then there exists a value of N above which the higher order formula yields smaller error, and below which the lower order process is more accurate. In the diagram (Figure 4.3) the critical value of N occurs at $\log \varepsilon = -4$. This implies that the qth order formula is preferable when high accuracy is sought.

With Runge-Kutta methods the amount of computational effort per step is proportional to the number of stages s (function evaluations) per step. Since the total number of stages for the integration is $F = Ns$ there

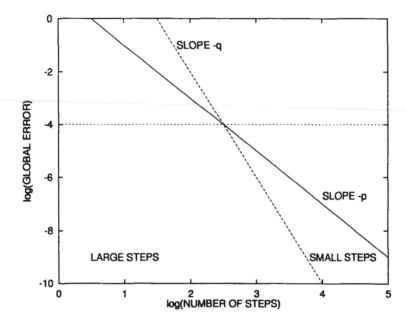

Figure 4.3: Global error for different orders

arises the *Error–Cost* relation

$$\log \varepsilon \;=\; D - p \log F,$$
$$D \;=\; B + p \log s,$$

which is similar to the last equation. From this, one concludes that, in comparing RK formulae of different orders, the highest order method must be the most efficient when the step–size is small enough. Of course, problems with a moderate error requirement may be best approached with a formula of intermediate order. These arguments have neglected the effect of rounding error which will depend on the arithmetic precision of the computer being employed. The propagation of rounding error is not as easy to quantify as is the mathematical error. However, one can assume that it will be proportional to the number of steps and, for large enough N, eventually it will dominate the solution. The arithmetic precision will define a lower limit for the attainable global error.

4.7   Computational examples

The preceding sections in this chapter have dealt with the derivation of Runge–Kutta formulae of increasing orders. Now it is necessary to check, by direct computation, the properties of the methods constructed.

In particular, the assertions concerning optimal RK processes, and also the importance of asymptotic behaviour, should be investigated. The computer program RKc, listed below, was designed to compute the solution of a system of differential equations, using a Runge-Kutta method with a constant step-size. The version shown here has been applied to the gravitational 2–body orbit problem, written here as four first order equations

$$
\begin{aligned}
{}^1y' &= {}^3y \\
{}^2y' &= {}^4y \\
{}^3y' &= -{}^1y/r^3 \\
{}^4y' &= -{}^2y/r^3, \\
r &= \sqrt{{}^1y^2 + {}^2y^2},
\end{aligned}
$$

with initial conditions

$$y(0) = [1, 0, 0, 1]^T.$$

The true solution represents a circular orbit of period 2π, which can be expressed as

$$y(x) = [\cos x, \sin x, -\sin x, \cos x]^T.$$

By applying a numerical method to a problem with a known true solution, the error at any step may be determined. Clearly this allows a comparison of different RK processes, and the program RKc is capable of reading as data the coefficients of any explicit Runge–Kutta formula.

Actually this particular orbit problem is not a severe test for any numerical differential equation solver, but it can be made more difficult by changing the initial values. In this case, a constant step-size is selected to carry out the integration over a suitable range of x. Both quantities are entered as data.

For solving differential equations the Fortran 90 language provides very convenient programming structures. In particular, the vector operations permit simple coding. This is largely free of the unsightly loops which disfigure programs coded in most other languages. The program RKc should be compared with the earlier RK3ode from Chapter 2. In RKc many of the operations involve vectors but the coding has almost the same simple structure seen in the earlier program. The arrays containing the vector quantities are dynamic, or ALLOCATABLE, using the Fortran 90 syntax. This means that it is unnecessary to fix the array dimensions until the program is executed and, in the example here, the array sizes increase automatically with the number of RK stages, and with the size of the system of differential equations being solved.

To modify the program RKc for application to any other system of equations, it is necessary only to change the subprograms. To output the solution at each step, some extra WRITE statements could be added, but

the main objective in this case is to measure the quality and the efficiency
of a particular RK process. The program prints the number of function
evaluations rather than the number of steps, in order to present a measure
of cost which can be used to compare methods with different numbers
of stages. A typical data file, for the application of the fourth order
RK44M to the test problem, is given in Table 4.4. The results obtained

Table 4.4: Data file for RKc

$$
\begin{array}{llll}
4 & & & \\
2,5 & & & \\
-3,20 & 3,4 & & \\
19,44 & -15,44 & 10,11 & \\
11,72 & 25,72 & 25,72 & 11,72
\end{array}
$$

from three of the formulae developed in this chapter (RK3M, RK44M,
RK45M) are presented graphically in Figure 4.4. The orbit equations
were integrated in [0,10] (about 1.5 orbital periods) and the error plotted
is that at the end point. The number of function evaluations is used
for abscissa, rather than the number of steps computed. This enables
the relative cost of different formulae to be assessed. The scales are
logarithmic and each data point represents a full integration over the
interval [0,10]. The curves are piecewise linear.

As predicted in the last section the two formulae of order 4 produce
graphs with slope -4, while the third order formula (RK3M) yields the
less steep gradient. A feature seen in Figure 4.4, but not predicted in
Figure 4.3, is the non-linear behaviour at large step values. This should
not be too surprising since proper asymptotic behaviour depends on a
small quantity, the step-size, being sufficiently small. The top left-hand
corner of the diagram is based on relatively large step-sizes.

The truncation error minimisation is well justified by the performance
of the 5-stage formula matched against the 4-stage RK44M. In spite of
the extra cost (25%) in computing a single step, the 5-stage RK45M is
actually the more economical of the two for any specified global error
achieved. In other words, the smaller error norm permits a larger step-
size for a given error. These results are derived from a single problem but
more tests over a wide range of systems of differential equations would
support the same conclusion.

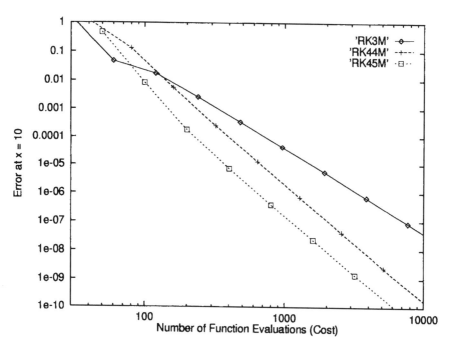

Figure 4.4: Error of integration of orbit problem with three formulae

```
!----------------------------------------------------------
PROGRAM RKc      ! Solves a system of ODEs with constant step
                 ! using a Runge-Kutta process whose
                 ! coefficients are entered from a data file.
                 !
!-----------------------------------------------J.R.Dormand, 6/95
   IMPLICIT NONE
   INTERFACE             ! Explicit interface required here
     FUNCTION Fcn(x, y)
       REAL (KIND = 2) :: x, y(:), Fcn(SIZE(y))
     END FUNCTION Fcn
     FUNCTION True(neq, x)
       INTEGER :: neq
       REAL (KIND = 2) :: x, True(neq)
     END FUNCTION True
   END INTERFACE
   INTEGER :: s, i, j, mcase, nstep, neq, n, k, Noeqns
   REAL (KIND = 2), PARAMETER :: z = 0.0D0, one = 1.0D0
   REAL (KIND = 2) :: x, h, xend
!----------------------------- Dynamic Arrays
   REAL (KIND = 2), ALLOCATABLE :: y(:), w(:), err(:), &
```

```
                                 f(:, :)
     REAL (KIND = 2), ALLOCATABLE :: a(:, :), b(:), c(:), &
                                 to(:), bo(:)
     CHARACTER(LEN = 30) :: rkname
!
! Read Runge-Kutta coefficients from a data file
!
     PRINT*, 'Enter file containing RK coefficients'
     READ '(a)', rkname
     OPEN(13, FILE = rkname)
     READ(13, *) s                    ! No of stages
     ALLOCATE(a(2:s, 1:s-1), b(s), c(2:s), to(s), bo(s))
     c = z
     DO i = 2, s
        READ(13, *) (to(j), bo(j), j = 1, i - 1)
        a(i, 1: i-1) = to/bo     ! Compute a(i,j)
        c(i) = SUM(a(i, 1: i-1)) ! Apply row sum condition
     END DO
     READ(13, *) (to(i), bo(i), i = 1, s)
     b = to/bo                        ! Compute b(i)
     neq = Noeqns()                   ! Determine dimensions
     ALLOCATE( y(neq), w(neq), err(neq), f(s, neq))
     PRINT*, 'Enter xend, initial No of steps, No of cases '
     READ*, xend, nstep, mcase
     PRINT '(a)', 'No of evals  Error at end'
     DO k = 1, mcase                  ! Loop over step-sizes
        CALL Initial(neq, x, y)
        h = (xend - x)/nstep
        DO n = 1, nstep               ! Loop over steps
          f(1, :) = Fcn(x, y)
          DO i = 2, s                 ! Loop over stages
             w = y + h*MATMUL(a(i, 1: i-1), f(1: i-1, :))
             f(i, :) = Fcn(x + c(i)*h, w)
          END DO
          y = y + h*MATMUL(b, f)         ! New solution for y
          x = x + h
        END DO
        err = y - True(neq, x)         ! Error at end point
        PRINT '(1x, i10, e14.4)', nstep*s, MAXVAL(ABS(err))
        nstep = 2*nstep                ! Double for next case
     END DO
     END PROGRAM RKc
!
     FUNCTION Noeqns()    ! Assign number of equations
```

```
    INTEGER :: Noeqns
      Noeqns = 4
  END FUNCTION Noeqns
!
  SUBROUTINE Initial(neq, x, y)  ! Specify initial values
    INTEGER :: neq
    REAL (KIND = 2) :: x, y(neq)
      x = 0.0d0
      y(1) = 1.0d0; y(2) = 0.0d0
      y(3) = 0.0d0; y(4) = 1.0d0
  END SUBROUTINE Initial
!
  FUNCTION Fcn(x, y)             ! Compute derivatives
    REAL (KIND = 2) :: x, y(:), Fcn(SIZE(y)), r3
      r3 = SQRT(y(1)**2 + y(2)**2)**3
      Fcn = (/ y(3), y(4), -y(1)/r3, -y(2)/r3 /)
  END FUNCTION Fcn
!
  FUNCTION True(neq, x)          ! Compute true solution
    INTEGER :: neq
    REAL (KIND = 2) :: x, True(neq)
      True = (/ COS(x), SIN(x), -SIN(x), COS(x) /)
  END FUNCTION True
!------------------------------------------------------------
```

Although the 3rd order and 4th order error curves do intersect, as predicted in the Figure 4.3, it will be noted that the accuracy for which the lower order formula is most economical is rather low. Again this is a typical result. The use of low order formulae is not likely to be efficient in accurate work, and the only justification for employing a third order formula would be for obtaining a rough solution over a small interval of the independent variable. In practice, the RK4 formulae would be considered also to be of low order.

The above comments may be surprising to those readers who favour 'simple' ODE solvers. In these days of powerful and convenient micro-computers, why should we not save ourselves the trouble of implementing a complicated algorithm in favour of computing a large number of steps with, say, Euler's method? After all, it is well known that any consistent RK process is convergent, and to compute a million steps may take only a few seconds of time. To answer this question it is best to solve a simple test problem. Consider the solution of a single equation

$$y' = y\cos x, \quad y(0) = 1, \quad x \in [0, 10] \tag{4.15}$$

using a modified version of the program RKc in which all the REAL (KIND = 2) declarations, indicating double precision of about 15 significant fig-

Table 4.5: Euler's method applied to equation $y' = y \cos x$

No of steps	Error at $x = 10$
100	$0.9176E{-}01$
200	$0.4746E{-}01$
400	$0.2416E{-}01$
800	$0.1218E{-}01$
1600	$0.6065E{-}02$
3200	$0.3089E{-}02$
6400	$0.1753E{-}02$
12800	$0.6772E{-}03$
25600	$0.4783E{-}03$
51200	$0.5543E{-}03$
102400	$0.3570E{-}02$

ures on a typical machine, have been changed to plain REAL. Thus the problem is being solved in single precision, which carries about 7 significant figures with many Fortran compilers. Note that the problem–dependent routines, Noeqns, Initial, Fcn, True, have to be changed, but the Euler method is easily specified from a data file and does not require program changes. Table 4.5 shows the error results obtained with Euler's method compared with the true solution $y(x) = \exp(\sin x)$. The symbol E is used in this and future tables to denote an exponent; thus $0.3570E - 02 \equiv 0.3570 \times 10^{-2}$. Halving the step-size has the predictable

Table 4.6: RK45M applied to equation $y' = y \cos x$ with single precision and with double precision

Steps	Cost (F.E.)	Error at $x = 10$	
		Single Precision	Double Precision
10	50	$0.2993E{-}02$	$0.2993E{-}02$
20	100	$0.1094E{-}03$	$0.1093E{-}03$
40	200	$0.3040E{-}05$	$0.3276E{-}05$
80	400	$0.1192E{-}06$	$0.9653E{-}07$
160	800	$0.1013E{-}05$	$0.2723E{-}08$
320	1600	$0.4768E{-}06$	$0.6780E{-}10$
640	3200	$0.1550E{-}05$	$0.1053E{-}11$
1280	6400	$0.1729E{-}05$	$0.3419E{-}13$

effect of halving the global error, at least for the first seven occurrences. Thereafter, the error increases and the minimum error achieved is somewhat smaller than 10^{-3} using about 25,000 steps for $x \in [0, 10]$. The reason for this upturn in the error magnitude is the use of finite precision arithmetic, so that after a very large number of operations, rounding error overwhelms the truncation error. Thus, if a method requires a very large number of steps, it will be incapable of producing accurate results. In this case the problem would have been alleviated to some extent by the use of higher precision but, eventually, rounding error would dominate the solution even with 15 significant figures.

Since higher order formulae take far fewer arithmetic operations, they are less badly affected by rounding errors. The results of applying the 4th order formula RK45M to equation (4.15) are shown in Table 4.6. First, it should be noted that the fourth order formula yields a more accurate solution in only 20 steps than is achievable at all with the Euler process (taking 25,000 steps). With single precision a smallest possible error of less than 10^{-6} is attained at about a sixtieth of the computational cost of the most accurate Euler integration. To demonstrate the effect of higher precision, the results of a double precision calculation are displayed in the same table. The implication of the above results is quite clear: very low order methods ($p < 4$) are unlikely to yield accurate results, regardless of computational effort, and they should be avoided if at all possible.

4.8   Problems

1. Find numerical solutions at $x = 0.4$, using step sizes $h = 0.2$ and $h = 0.4$, to the differential equation

$$y' = 1 - y^2 , \quad y(0) = 0,$$

with (a) RK2, (b) RK3, and (c) RK4 integrators. In each case compute the global error ϵ using the true solution

$$y(x) = \frac{e^{2x} - 1}{e^{2x} + 1},$$

and confirm that this is consistent with the prediction that $\epsilon = O(h^p)$, for a formula of order p. Also consider the local errors at the first step in each case.

2. Show that the number of *independent* RK parameters in a formula with s stages is $\eta_s = s(s + 1)/2$.

3. Construct a three–stage RK3 with $c_2 = c_3$.

4. Obtain the RK4 parameters from the general formulae derived in §4.4 for the case $c_2 = 5/14$, $c_3 = 13/22$.

5. Apply the Butcher relations (4.13) to the 5th order error coefficients to show that, under these conditions, $\tau_i^{(5)}$, $i = 5, 7, 8, 9$, are dependent on other coefficients. Also apply the second set of simplifying conditions (4.14) to other suitable 5th order coefficients. In particular, show that

$$\tau_6^{(5)} = \tau_4^{(5)} - \tfrac{1}{2} c_2^2 \sum_{i=3}^{s} b_i c_i a_{i,2}.$$

6. Consider the equations of condition for a 4th order RK in 4 stages in the singular case $c_2 = c_3$; show that $c_2 = \tfrac{1}{2}$ and then obtain the other parameters in terms of b_3. You should find that the formula

c_i	a_{ij}			b_i
0				$\frac{1}{6}$
$\frac{1}{2}$	$\frac{1}{2}$			$\frac{1}{3}$
$\frac{1}{2}$	0	$\frac{1}{2}$		$\frac{1}{3}$
1	0	0	1	$\frac{1}{6}$

$$(4.16)$$

is a member of this class. Find the principal truncation error coefficients of this formula and hence compare its error norm with other 4th order schemes in the text.

7. The three tables below show the number of function evaluations $F = Ns$ and the corresponding global error ϵ at the end of three Runge–Kutta solutions of the same differential equation. Estimate the order of the Runge–Kutta formula used for each table.

F	$\log_{10}\epsilon$	F	$\log_{10}\epsilon$	F	$\log_{10}\epsilon$
2221	-2.31	3801	-4.93	4533	-7.22
2833	-3.44	4689	-6.04	5632	-7.72
4039	-4.43	6801	-7.17	6018	-8.45
6385	-5.44	9953	-8.27	7648	-9.56
10105	-6.48	14593	-9.36	10108	-10.65

8. Obtain formulae for a family of 5-stage RK4 processes according to the scheme described in §4.4. To simplify the solution, set
$$\sum_{i=3}^{5} b_i c_i a_{i,2} = 0,$$
and show that this makes the error coefficients independent of c_2. Hence, or otherwise, verify that your 5-stage formula cannot be fifth order. Find a formula with a smaller error norm than RK45M.

9. Write program units to substitute in the Fortran 90 program RKc in order to solve the equations

   (a) $y' = \frac{1}{4}y(1 - \frac{1}{20}y)$, $x \in [0, 20]$, $y(0) = 1$;

   True solution: $y(x) = 20/(1 + 19e^{-x/4})$

   (b) $y'' = 0.032 - 0.4(y')^2$, $x \in [0, 20]$, $y(0) = 30$, $y'(0) = 0$;

   True solution: $\begin{cases} y(x) &= \frac{5}{2}\log(\cosh(2\sqrt{2}x/25)) + 30, \\ y'(x) &= \frac{\sqrt{2}}{5}\tanh(2\sqrt{2}x/25) \end{cases}$

   Apply third and fourth order RK formulae to each problem, in the specified interval, over a range of diminishing step-sizes starting with $h = 1$. Include the process from problem (4.16) in your testing. Tabulate the maximum global error, over all steps and variables, and the computational cost in terms of CPU time or function evaluations.

10. Find the smallest possible global error at $x = 10$ when the Euler method is applied to the equation
$$y' = y\cos x, \quad y(0) = 1,$$
using arithmetic precision given by REAL (KIND = 2).

11. Apply RK45M to the system of equations modelling the 2-body gravitational problem with the initial condition

$$y(0) = (0.5, 0.0, 0.0, \sqrt{3})^T,$$

using step-size $h = 0.1$, and $x \in [0, 2\pi]$. Plot graphs of ^2y against ^1y, and ^3y, ^4y against x. *(The orbit should be an ellipse with eccentricity 0.5.)*

Chapter 5

Step–size control

5.1 Introduction

In the previous chapter we have examined some of the properties of Runge–Kutta methods applied in constant step–size fashion. Also, the value of the step–size was chosen in an arbitrary fashion, although its variation did illustrate the order properties of the various schemes.

The use of high precision arithmetic with small step-sizes led to accurate solutions for the simple problems attempted. For many practical problems, the constant steplength approach is not recommended because the derivatives can vary significantly in magnitude. If the step-size were kept constant, the RK, or other process, is likely to deliver local errors which vary greatly from step to step according to the variability of the error functions. Intuition suggests the undesirability of such a feature. A more attractive proposition would be a solution whose local errors vary only slightly or, perhaps, a solution whose local errors vary in proportion to the step-size. From the mathematical form of the local truncation error, one can see that each of these properties implies variations in the step-size.

For any method, the relation between the local error and the step-size is well-known and most schemes for the prediction of step-size make use of this. Instead of fixing the steplength, an attempt is made to bound the local (truncation) error, and this bound is used to determine the step-size. Ideally, one would prefer to bound the global error but, in practice, the local error is much easier to estimate. Practical results confirm that local error control acually leads to control of the global error. This chapter will be concerned with schemes for error estimation and step–size control.

Another constraint which can affect step–size is a specified output frequency. In some applications the solution of the system of differential equations may be required at intervals whose size is unconnected with any truncation error criterion. This requirement is easily satisfied at the cost of computing extra steps, but a more sophisticated approach to this problem will be considered in the next chapter.

5.2   Steplength prediction

In controlling the step–size for the integration of a system of equations $y' = f(x, y)$, it is necessary to predict a value for the length of the step following that which has already been computed. Let us suppose that a step of length h_n has just been completed, and also suppose that it is possible to estimate the local error e_{n+1} of the computed solution, y_{n+1}. Generally, the local error (see Chapter 3) for a method of order p can be expressed as

$$e_{n+1} = h_n^{p+1} \varphi_p(x_n, y_n) + h_n^{p+2} \varphi_{p+1}(x_n, y_n) + \cdots \qquad (5.1)$$

For example, in the Taylor polynomial method the principal term is

$$\varphi_p = \frac{y^{(p+1)}(x_n)}{(p+1)!},$$

making the error estimation process very straightforward. Based on the computation of one extra term of the series, the cost is not excessive in relation to that of the main process.

The local error of the next step, of length h_{n+1}, will be specified in a similar manner. Assuming positive steplengths, one can write

$$\|e_{n+1}\| = h_n^{p+1} \|\varphi_p(x_n, y_n) + h_n \varphi_{p+1}(x_n, y_n) + \cdots \|,$$
$$\|e_{n+2}\| = h_{n+1}^{p+1} \|\varphi_p(x_{n+1}, y_{n+1}) + h_{n+1} \varphi_{p+1}(x_{n+1}, y_{n+1}) + \cdots \|.$$

These relations yield

$$h_{n+1}^{p+1} = \frac{h_n^{p+1} \|\varphi_p(x_n, y_n) + h_n \varphi_{p+1}(x_n, y_n) + \cdots \|}{\|\varphi_p(x_{n+1}, y_{n+1}) + h_{n+1} \varphi_{p+1}(x_{n+1}, y_{n+1}) + \cdots \|} \times \frac{\|e_{n+2}\|}{\|e_{n+1}\|}$$

$$= h_n^{p+1} \theta \frac{\|e_{n+2}\|}{\|e_{n+1}\|},$$

where $\theta = 1 + O(h_n)$. Now suppose that the tolerated error bound for step $n + 1$, called the *Tolerance*, is $\|e_{n+2}\| = T$. Then the appropriate step-size is predicted by

$$h_{n+1} = h_n \left\{ \frac{\theta T}{\|e_{n+1}\|} \right\}^{\frac{1}{p+1}}. \qquad (5.2)$$

For small h_n, the factor θ is close to unity but, to be conservative, usually a smaller value is chosen. Thus, a widely accepted formula for predicting a step–size h_{n+1}, following a successful step of size h_n, is

$$h_{n+1} = 0.9 h_n \left\{ \frac{T}{\|\delta_{n+1}\|} \right\}^{\frac{1}{p+1}} , \quad \|\delta_{n+1}\| \leq T, \qquad (5.3)$$

where δ_{n+1} is an estimate of the local error e_{n+1}. Any norm could be chosen but usually the max norm is preferred. In this case the ratio of consecutive principal error functions has been chosen so that

$$\theta^{\frac{1}{p+1}} = 0.9.$$

The value '0.9' is a typical safety factor and it is effective in many applications. Not all authors would make the same choice but the precise value of θ serves only to modify the effective tolerance. The terms *strategy parameter* and *optimal reduction factor* are applied in the same context. Of course, the need for such a parameter stems from the non-asymptotic nature of the numerical process for large values of the step-size h_n arising from lax tolerances. This is particularly true when high order (large p) processes are selected.

The integration proceeds with the step $n + 1$ of size given by the formula (5.3) only when $\|\delta_{n+1}\| \leq T$. If $\|\delta_{n+1}\| > T$, the local error is too large to be tolerated, and the step must be rejected. Fortunately the same formula (5.3) provides a useful estimate of the reduced step-size for a repeated step, which then must be smaller than h_n. For large values of $\|\delta_{n+1}\|/T$, one would assume that the step is too big to justify asymptotic behaviour, and then reduce the step-size in an arbitrary manner, perhaps by a factor of 2 or even by an order of magnitude.

More sophisticated schemes, based on error estimates at two or more successive steps, have been devised for step–size control. These are employed in some computer packages for the solution of differential equations, but the simple method described above is very robust and reasonably efficient, particularly if the step amplification factor is restricted. An example of this type of restriction will be given later.

5.3   Error estimation

With the Taylor series method the estimation of local error is very easy since the principal term in the local error can be determined in the same way as any of the terms contributing to the Taylor polynomial.

To demonstrate the error estimation–step control procedure, consider two steps of the numerical solution of

$$y' = y^3 - y, \quad y(0) = \tfrac{1}{2},$$

using a Taylor series of order 2 with trial steplength $h_0 = 0.2$, subject to a local error tolerance $T = 10^{-3}$. The true solution of this problem is $y(x) = 1/\sqrt{1 + 3e^{2x}}$. The second order Taylor polynomial formula is

$$y_{n+1} = y_n + h_n y_n' + \frac{h_n^2}{2} y_n''$$

and the local error has principal term

$$\delta_{n+1} = \frac{h_n^3}{6} y_n''',$$

which will serve as the error estimator. To implement the scheme, two higher derivatives are required; these are

$$y'' = (3y^2 - 1)y', \quad y''' = 6y(y')^2 + (3y^2 - 1)y''.$$

At $(0, \frac{1}{2})$, the derivative values are

$$y_0' = -\frac{3}{8}, \quad y_0'' = \frac{3}{32}, \quad y_0''' = \frac{51}{128},$$

and so the second order solution is

$$y_1 = 0.5 - \tfrac{3}{8} \times (0.2) + \tfrac{3}{32} \times \tfrac{1}{2}(0.2)^2 = 0.426875,$$

with principal local error

$$\delta_1 = \tfrac{51}{128} \times \tfrac{1}{6}(0.2)^3 = 0.531 \times 10^{-3}.$$

Since $|\delta_1| < T$, the step is acceptable and so formula (5.3) is employed to predict a value h_1 for the size of the next step. This gives

$$h_1 = 0.9 h_0 (10^{-3}/0.531 \times 10^{-3})^{1/3} = 0.222284.$$

The second step requires a new evaluation of the derivatives at the new initial condition $(0.2, 0.426875)$. Thus

$$y_1' = -0.349089, \quad y_1'' = 0.158254, \quad y_1''' = 0.240380,$$

yielding the second step solution

$$y_2 = 0.353188 \simeq y(0.422284) = 0.353975.$$

The local error estimate and third step-size prediction are

$$\delta_2 = 0.440 \times 10^{-3}, \quad h_2 = 0.263027.$$

All quantities in the above calculation have been rounded to six decimal places. The step–size control procedure has operated successfully for the first two steps at least. Further steps can be computed in a similar manner, but it will be clear that the solution points $x_4, x_5, \ldots$ are not predictable at this stage.

In the simple example above the step-size control was successful but in some cases the estimated error exceeds the tolerance, indicating a solution of unacceptable accuracy. When this happens the step must be rejected, and the formula (5.3) is used to compute a reduced steplength, with which the integration is restarted. This procedure may be illustrated by considering the same problem as before but with a reduced (more stringent) tolerance $T = 10^{-4}$. With the trial step $h_0 = 0.2$, the local error estimate $\delta_1 = 0.531 \times 10^{-3} > T$ is now too large, indicating that the step should be rejected. A reduced step-size

$$h_0^* = 0.9 h_0 (10^{-4}/0.531 \times 10^{-3})^{1/3} = 0.103175$$

is now predicted to give sufficient accuracy. The revised first step is

$$
\begin{aligned}
y_1 &= 0.5 - \tfrac{3}{8} \times (0.103175) + \tfrac{3}{32} \times \tfrac{1}{2}(0.103175)^2 \\
&= 0.461808 \simeq y(0.103175) = 0.461878,
\end{aligned}
$$

and a new local error estimate, $\delta_1 = 0.729 \times 10^{-4} < T$, which confirms that it is acceptable.

The type of step-size control practised here leads to numerical solutions being obtained for values of the independent variable not predictable for more than a single step ahead. This is sometimes considered to be a disadvantage. However, the local error control will generate smaller step-sizes where they are required, hence providing more information where this would be needed to characterise the solution. In those regions where the error coefficients are very small, less solution points are necessary. If a solution at a specific point is needed an extra step could be inserted. Suppose $y(X)$ is a specifed solution value. Then one could integrate the problem with variable steps yielding the sequence of output points $x_1, x_2, \ldots$. Eventually a value of n will arise such that

$$x_n < X < x_{n+1}, \quad h_n = x_{n+1} - x_n,$$

suggesting that h_n is replaced by $h_n^* = X - x_n$. Subsequent steps can be computed according to the variable step-size scheme, until the next specified point is reached.

5.4 Local extrapolation

Since error estimation in the case above implies the computation of second and third order solutions, it is natural to consider which of the *two* estimates of $y(x_{n+1})$ should be retained to act as the initial value for

the next step. In the example of the last section, the second order value
(0.2, 0.426875) was selected as the initial value for the 2nd step but, with
this choice, a third order value, which can be obtained by adding δ_1, was
deliberately neglected. On the basis of the numerical investigations in
Chapter 4, with formulae of different orders, the higher order formula
should be more accurate, and so it would serve as the better choice for
an initial condition. If this were not so, the error estimate itself would
be inaccurate and a poor basis for step-size prediction. The third order
Taylor polynomial can be expressed as

$$\hat{y}_1 = y_1 + \delta_1 = 0.427406$$

and use of this value to initialize the next step usually will give better
results than those obtained earlier. The procedure based on the higher
order of a formula pair is called *Local Extrapolation*.

Returning to the numerical example from the previous section, evaluating the derivatives at (0.2,0.427406) gives

$$y_1' = -0.349329, \quad y_1'' = 0.157887, \quad \text{and} \quad y_1''' = 0.241580,$$

and substituting in the third order or extrapolated Taylor series yields
$\hat{y}_2 = 0.354099$. The estimated error for the second order result is
$\delta_2 = 0.0004422$, which is smaller than the tolerance and hence acceptable. Substitution of the error estimate in formula (5.3) gives the next
steplength, $(h_2 = 0.262590)$, and so on. It is important to compare the
errors of the two estimates of $y(0.422284)$. These are

$$\hat{\varepsilon}_2 \quad = \quad \hat{y}_2 - y(x_2) = 0.000123,$$
$$\varepsilon_2 \quad = \quad y_2 - y(x_2) = -0.000788,$$

indicating a marked improvement for the extrapolated version. Direct
comparisons at other points are not possible since the step sequences
are not identical for the two schemes. However, the local extrapolation
process is generally found to be superior to that which assigns the lower
order result as the preferred solution.

By plotting the two types of solution, the superiority of the local
extrapolation technique becomes clear. Consider the solution of the differential equation

$$y' = y^3 - y, \quad y(0) = 1/\sqrt{1 + 3e^{-10}}, \quad x \in [0, 10]. \tag{5.4}$$

This problem differs from the earlier test problem only in respect of the
initial condition. The true solution is

$$y(x) = 1/\sqrt{1 + 3e^{2(x-5)}}.$$

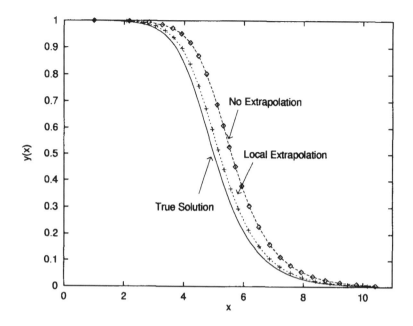

Figure 5.1: Comparison of the true solution and the two modes of numerical solutions based on two formulae

As before, the 2nd/3rd order Taylor series is applied in each of the two modes. Figure 5.1 depicts the two numerical solutions and the true solution, when $T = 10^{-3}$ and $h_0 = 1$. The superiority of the local extrapolation mode is clearly visible.

A feature of the test problem is that it demands increasing and decreasing the step–size in the given interval of solution. Also a number of step rejections occur. In the extrapolated case, 25 successful steps are computed in addition to 8 rejections; the corresponding figures for the unextrapolated calculation are 25 and 7. The two sequences of step–sizes, for integration over $[0, 10]$, are very similar, and these are shown in Figure 5.2. The program used to generate the data for the Figures 5.1 and 5.2 is listed in Appendix A. The computational costs in the two cases are almost the same.

With the local extrapolation strategy, it should be emphasised that one does not estimate the local error of the formula used for updating the solution. However, it is likely that the true local error is less than the working tolerance. If this were not true, it would spell trouble for non-extrapolated operations applied to the same equation.

Each of the two modes of operation can be expressed in terms of two formulae of different orders. For local extrapolation the formula pair can

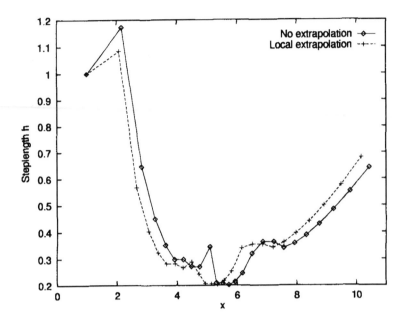

Figure 5.2: Variation of steplengths for problem (5.4)

be written in increment form as

$$\widehat{y}_{n+1} = \widehat{y}_n + h_n \widehat{\Phi}(x_n, \widehat{y}_n, h_n),$$
$$y_{n+1} = \widehat{y}_n + h_n \Phi(x_n, \widehat{y}_n, h_n), \qquad (5.5)$$

where the 'caps' indicate extrapolated quantities. With no extrapolation the corresponding formulae are

$$\widehat{y}_{n+1} = y_n + h_n \widehat{\Phi}(x_n, y_n, h_n),$$
$$y_{n+1} = y_n + h_n \Phi(x_n, y_n, h_n). \qquad (5.6)$$

In both cases, the error estimate is

$$\delta_{n+1} = y_{n+1} - \widehat{y}_{n+1} = h_n \{\Phi - \widehat{\Phi}\} = e_{n+1} - \widehat{e}_{n+1},$$

and for asymptotic viability the requirement is $e_{n+1} \gg \widehat{e}_{n+1}$. In other words, the principal error term must be dominant in order to justify the step prediction formula (5.3). This condition will hold if the step–size is small enough, but, when a lax (large) tolerance is specified, inaccurate predictions may cause many step rejections.

5.5 Error estimation with RK methods

The procedure outlined above is natural in the Taylor series context, but the different nature of an RK formula might seem to preclude such a technique. However, the same strategy can be applied if two RK formulae of different orders are used to represent y_{n+1} and $\widehat{y}_{n+1}$. An obvious drawback is the implied extra cost of this procedure when compared with the Taylor series case, in which a higher order result is available following the evaluation of one extra term. Fortunately the remedy turns out to be rather simple, although its discovery was made only fairly recently. To economize on applying a second RK formula it is *embedded* in the first one by forcing it to use the same function evaluations ('f's). This technique was introduced by Merson (1957), and another pioneer was England (1969), but Fehlberg (1968, 1969) played a very important role in popularising Runge-Kutta embedding. Let us consider an example in

Table 5.1: An embedded RK3(2) pair of formulae

c_i	a_{ij}		$\widehat{b}_i$	b_i
0			$\frac{1}{6}$	$\frac{1}{2}$
$\frac{1}{2}$	$\frac{1}{2}$		$\frac{2}{3}$	0
1	-1	2	$\frac{1}{6}$	$\frac{1}{2}$

which a second order RK formula is embedded in the 3-stage third order process (4.8) developed earlier. To satisfy the embedding criterion in this case, the RK2 must be based on the same three function values as the third order formula. Thus the equations of condition must satisfy

$$a_{21} = \tfrac{1}{2}, \quad a_{31} = -1, \quad a_{32} = 2.$$

Using Table 3.1, the relevant equations of condition for a second order formula with $s = 3$ are

$$
\begin{aligned}
b_1 + b_2 + b_3 &= 1, \\
b_2 c_2 + b_3 c_3 &= \tfrac{1}{2},
\end{aligned}
$$

which have one free parameter when the values of c_2 and c_3 are determined. For a simple formula, take $b_2 = 0$, then the equations yield $b_1 = b_3 = \tfrac{1}{2}$. The resulting RK3(2) embedded pair in modified Butcher notation is defined in Table 5.1. To depict the embedded pair, an extra column containing the b_i for the lower order formula is added on the right-hand side of the table for the RK3.

With the embedded approach, the error estimation process to control step-size is available at very small computational cost. The RK3(2) is applied to the same differential equation as used to illustrate the Taylor series error computation in §5.3. The equation is

$$y' = y^3 - y, \quad y(0) = \tfrac{1}{2}$$

and the trial step-size is $h_0 = 0.2$. The three function values are found to be

$$(f_1, f_2, f_3) = (-0.375, -0.363568, -0.350302),$$

and then the third and second order solutions are

$$\widehat{y}_1 = 0.5 + \frac{0.2}{6}(f_1 + 4f_2 + f_3) = 0.427347,$$

$$y_1 = 0.5 + \frac{0.2}{2}(f_1 + f_3) = 0.427470.$$

The error estimate is $\delta_1 = y_1 - \widehat{y}_1 = 0.000123$. With the same tolerance as before ($T = 10^{-3}$), the step–length formula (5.3) yields $h_1 = 0.361941$.

The reader may suspect that the above choice of free parameter has no particular justification other than that of convenience. Also it will be clear that any other choice for c_2 would have produced a different error estimate and hence a different step–size prediction. In this respect the error estimation process differs from the corresponding Taylor series scheme, whose polynomial for a given order is unique. So, how should any free parameter affecting the lower order formula of the pair be chosen? The criterion will depend on the intended mode of operation of the pair. If local extrapolation is not employed, it will be appropriate to reduce the size of the error coefficients in the lower order member. These must not be made too small, or the error estimate could be correspondingly small, causing a large step–size to be predicted. In these circumstances, the accuracy of the solution would be poor in relation to the tolerance. Viewed another way, one would have the acceptance of an inaccurate result.

When local extrapolation is intended, there is no particular merit in improving the lower order solution since it is a purely local phenomenon and will not be used to initiate the next step. Of course, an optimal high order formula will always give the best error estimate for the embedded lower order process, as well as providing the best global solution. To some extent, the size of the error coefficients can be balanced by the tolerance specified, although one would not wish a tolerance of, say, 10^{-12} to be the requirement for an actual local error as large as 10^{-2}. In developing embedded formulae it may be necessary to adopt an experimental approach in order to determine a reliable pair. One important requirement is that the error coefficients for the lower order process must be complete. If any of these coefficients are zero it is possible that, for certain problems, the difference in orders for the pair may not be maintained. This would invalidate the step-size control formula (5.3). Generally, the embedded RK pair employing local extrapolation can be expressed as

$$\widehat{y}_{n+1} \;\; = \;\; \widehat{y}_n + h_n \sum_{i=1}^{s} \widehat{b}_i f_i,$$

$$y_{n+1} \;\; = \;\; \widehat{y}_n + h_n \sum_{i=1}^{s} b_i f_i, \qquad\qquad (5.7)$$

$$f_i \;\; = \;\; f(x_n + c_i h_n, \widehat{y}_n + h_n \sum_{j=1}^{i-1} a_{ij} f_j), \;\; i = 1, 2, 3, \ldots, s.$$

The 'cap' on the y_n indicates a value computed with the higher (qth) order formula, and the uncapped y_n is based on the pth order RK where, usually, $q = p + 1$. Both formulae use the same 'f' evaluations, and the pair is designated RK$q(p)$.

5.6 More Runge–Kutta pairs

The ease with which a second order embedding is achieved encourages an attempt to form an 4(3) pair based on an existing fourth order process, examples of which can be found in Chapter 4. First, consider a four stage RK4 such as the one defined in the Butcher table (4.11). Since any embedded formula must utilize the same a_{ij} as the four stage process, there will be only four parameters (b_1, b_2, b_3, b_4) to satisfy the new equations of condition. For a third order RK, four equations derived from Table 3.1 are applicable. These are linear in the b_i and so, assuming linear independence, the unique solution is the one already obtained to yield a fourth order formula. Consequently, a distinct third order embedded formula does not exist.

To obtain the necessary non-unique solution to the third order equations of condition, a fifth stage is essential. This adds five parameters to the four already available from the four existing stages. Of course, the extra stage implies a 25% increase in computational cost in order to provide the step-size control statistic. A simple device which offers a discount in this respect is called the FSAL (*first same as last*) procedure. In this case, FSAL specifies the fifth stage to be the same as the first stage at the next step $(f_1)_{n+1}$, giving

$$f_5 = (f_1)_{n+1} \;\; = \;\; f(x_{n+1}, \widehat{y}_{n+1})$$

$$= \;\; f(x_n + h_n, \widehat{y}_n + h_n \sum_{i=1}^{4} \widehat{b}_i f_i)$$

$$= \; f(x_n + c_5 h_n, \widehat{y}_n + h_n \sum_{j=1}^{4} a_{5j} f_j),$$

where the 'caps' indicate values from the fourth order formula. The fifth stage RK parameters are evidently

$$c_5 = 1, \quad a_{5j} = \widehat{b}_j, \quad j = 1, 2, 3, 4.$$

Note that the row sum condition (3.18) is consistent with the FSAL[1] device.

The FSAL procedure was proposed for a four stage RK4 by Dormand and Prince (1978) and its application to the formula (4.11) is now considered. Suppose that the third order embedded formula has weights b_i, $i = 1, \ldots, 5$, and that FSAL is specified. Then the four equations of condition become

$$\begin{pmatrix} 1 & 1 & 1 & 1 & 1 \\ 0 & \frac{1}{3} & \frac{2}{3} & 1 & 1 \\ 0 & \frac{1}{9} & \frac{4}{9} & 1 & 1 \\ 0 & 0 & \frac{1}{3} & \frac{1}{3} & \frac{1}{2} \end{pmatrix} \begin{pmatrix} b_1 \\ b_2 \\ b_3 \\ b_4 \\ b_5 \end{pmatrix} = \begin{pmatrix} 1 \\ \frac{1}{2} \\ \frac{1}{3} \\ \frac{1}{6} \end{pmatrix}$$

and the embedded RK4(3), with b_5 as a free parameter, is defined in Table 5.2. The value of b_5 needs to be chosen to provide error estimates

Table 5.2: The embedded RK4(3)4F formula pair

c_i	a_{ij}				$\widehat{b}_i$	b_i
0					$\frac{1}{8}$	$\frac{1}{8} + \frac{5}{4}b_5$
$\frac{1}{3}$	$\frac{1}{3}$				$\frac{3}{8}$	$\frac{3}{8} - \frac{3}{4}b_5$
$\frac{2}{3}$	$-\frac{1}{3}$	1			$\frac{3}{8}$	$\frac{3}{8} - \frac{3}{4}b_5$
1	1	-1	1		$\frac{1}{8}$	$\frac{1}{8} - \frac{3}{4}b_5$
1	$\frac{1}{8}$	$\frac{3}{8}$	$\frac{3}{8}$	$\frac{1}{8}$	0	b_5

of an acceptable order of magnitude. Using such a process the first step will occupy five stages, but subsequent steps will require only four, which is the minimum for a fourth order method. Following a rejected step, four computed stages are wasted, as compared with only three for a non-FSAL four stage formula. The example given here is not recommended

---

[1] Pronounce it *effsall* rather than *fuzzle*. Perhaps LAFS would have been a better acronym!

for general use since the third order formula has an incomplete set of principal error coefficients. It is easily verified that $\tau_1^{(4)} = 0$ for this formula, meaning that for some problems it is fourth order. Since the step-size control depends on the embedded pair having different orders, the reliability of the pair in Table 5.2 is not to be trusted.

For a five stage RK4 such as that in Table 4.2 the construction of a lower order embedded process is straightforward. An example of such a formula, developed by Prince (1979), is given in Table 5.3. In the third order process, none of the error coefficients are zero, and so the difference in orders will be maintained for any system of differential equations.

Table 5.3: The embedded RK4(3)5M formula pair

c_i	a_{ij}				$\widehat{b}_i$	b_i
0					$\frac{13}{96}$	$\frac{23}{192}$
$\frac{1}{5}$	$\frac{1}{5}$				0	0
$\frac{2}{5}$	0	$\frac{2}{5}$			$\frac{25}{48}$	$\frac{55}{96}$
$\frac{4}{5}$	$\frac{6}{5}$	$\frac{-12}{5}$	2		$\frac{25}{96}$	$\frac{35}{192}$
1	$-\frac{17}{8}$	5	$-\frac{5}{2}$	$\frac{5}{8}$	$\frac{1}{12}$	$\frac{1}{8}$

Many higher order pairs have been constructed. In particular, Fehlberg (1968, 1969) has presented pairs with orders 4(5), 5(6), 6(7), 7(8), 8(9), which were designed to be implemented in lower order mode. Verner (1978) has also developed some popular high order embedded pairs. The earliest RK pairs designed from the outset to operate in local extrapolation mode were presented by Prince and Dormand in 1980 and 1981. Of these, the most widely used pairs, and generally acknowledged as the most efficient in their respective classes, are the 7–stage RK5(4)7FM (Table 5.4) and the 13–stage RK8(7)13M, sometimes referenced as DOPRI5 and DOPRI8, respectively. A feature of DOPRI5 is the use of the FSAL technique, which in this case requires

$$\widehat{b}_7 = 0, \quad a_{7j} = \widehat{b}_j, \quad j = 1, 2, \ldots, 6.$$

The coefficients of higher order processes are not usually obtained as simple fractions, although their derivation is usually carried out symbolically or with rational arithmetic to avoid any rounding errors. Consequently, no process of order greater than five is displayed in this chapter. It is best not to transcribe manually high order coefficients since errors are most likely. These RK methods can be very sensitive to what may

Table 5.4: The RK5(4)7FM embedded pair (DOPRI5)

c_i	a_{ij}						$\hat{b}_i$	b_i
0							$\frac{35}{384}$	$\frac{5179}{57600}$
$\frac{1}{5}$	$\frac{1}{5}$						0	0
$\frac{3}{10}$	$\frac{3}{40}$	$\frac{9}{40}$					$\frac{500}{1113}$	$\frac{7571}{16695}$
$\frac{4}{5}$	$\frac{44}{45}$	$-\frac{56}{15}$	$\frac{32}{9}$				$\frac{125}{192}$	$\frac{393}{640}$
$\frac{8}{9}$	$\frac{19372}{6561}$	$-\frac{25360}{2187}$	$\frac{64448}{6561}$	$-\frac{212}{729}$			$-\frac{2187}{6784}$	$-\frac{92097}{339200}$
1	$\frac{9017}{3168}$	$-\frac{355}{33}$	$\frac{46732}{5247}$	$\frac{49}{176}$	$-\frac{5103}{18656}$		$\frac{11}{84}$	$\frac{187}{2100}$
1	$\frac{35}{384}$	0	$\frac{500}{1113}$	$\frac{125}{192}$	$-\frac{2187}{6784}$	$\frac{11}{84}$	0	$\frac{1}{40}$

appear to be insignificant errors in their coefficients. If a computation is to be performed in double precision arithmetic, the parameters of the RK formula must be accurate to that degree. Otherwise there is no possibility of achieving the higher accuracy. Electronic transmission is the safest mechanism for receiving Runge-Kutta coefficients. For work which would demand a very large number of steps of a 5th order formula, great cost savings and/or improved accuracy will be achieved using an eighth order process.

5.7 Application of RK embedding

Most modern differential equation software aims to control the global error of the solution by bounding the local error, and the estimation of this with RK embedded pairs is very straightforward. Figure 5.3 shows a few lines of additional code which can be added to the program RKc from Chapter 4 in order to control the step–size according to estimates of the local error. The step-size formula (5.3) could be applied in its standard form at every step, but experience shows that it is preferable to incorporate some extra controls, which will alleviate problems encountered when operating with lax tolerances. These can occur when an assumption, that the principal error term for the method is dominant, is not fully justified. Also, the system being solved may possess very large higher derivatives, which necessitates large step-size variations within a short interval. In either of these cases, the standard step–size predictor formula may fail, causing a step rejection with concomitant loss of efficiency. To avoid this, the program fragment in Figure 5.3 sets arbitrary limits on the ratio of

```
done = .FALSE.
DO                        ! Loop over steps
  f(1, :) = Fcn(x, y)
  DO i = 2, s             ! Loop over stages
    w = y + h*MATMUL(a(i, 1:i-1), f(1:i-1, :))
    f(i, :) = Fcn(x + c(i)*h, w)
  END DO
  delta = h*MAXVAL(ABS(MATMUL(b1 - b, f))) ! Error estimate
  alpha = (delta/ttol)**oop               ! Step-size ratio
  IF(delta < tol) THEN       ! Accepts step
    y = y + h*MATMUL(b, f)   ! Update y
    x = x + h
    IF(done) EXIT            ! End-point reached
    h = h/MAX(alpha, 0.1D0)  ! Predict next step
    IF(x + h > xend) THEN
        h = xend - x         ! Reduce step to hit end-point
        done = .TRUE.
    ENDIF
  ELSE                       ! Rejects step
    h = h/MIN(alpha, 10.0D0) ! Reduce step-size
    IF(done) done = .false.
  END IF
END DO
```

Figure 5.3: A program fragment demonstrating variable step Runge–Kutta implementation

successive step–sizes:

$$\tfrac{1}{10} \leq \frac{h_{n+1}}{h_n} \leq 10.$$

More sophisticated algorithms can be devised, particularly involving the prediction of the initial step, and also dealing with consecutive step failures. Some computational aspects of step–size control will be discussed later in this book.

An important change in the new code is the unlimited DO loop over the steps. The number of steps to reach the end–point xend is not predictable when step–sizes are controlled by error estimates. Thus the algorithm keeps incrementing until the end–point is detected. In the example, the tolerance tol is applied to the absolute error and, using formula (5.2), ttol= $\theta \times$ tol. The two embedded formulae are characterised by the

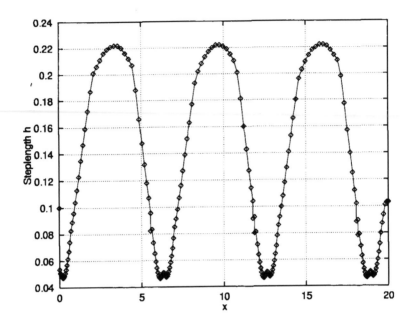

Figure 5.4: Variation of steplengths for orbit problem $e = 0.5$

vectors b and b1, which represent $\widehat{b}$ and b, respectively. The complete program, which also caters efficiently for FSAL formulae, is listed as Rkmbed in the Appendix A.

To illustrate the application of RK embedding, let us consider again the orbit problem

$$
\begin{aligned}
{}^{1}y' &= {}^{3}y \\
{}^{2}y' &= {}^{4}y \\
{}^{3}y' &= -{}^{1}y/r^3 \\
{}^{4}y' &= -{}^{2}y/r^3, \\
r &= \sqrt{{}^{1}y^2 + {}^{2}y^2},
\end{aligned}
$$

this time with initial conditions

$$
y(0) = \left[1 - e, 0, 0, \sqrt{\frac{1+e}{1-e}}\right]^T .
$$

The solution represents motion in an elliptic orbit of eccentricity $e < 1$, with period 2π. The earlier solution of this problem in Chapter 4 concerned a circular orbit ($e = 0$), for which the true solution was easily quoted. In the eccentric case the Runge–Kutta solution can be compared with one based on the solution of Kepler's equation,

$$E - e \sin E = x.$$

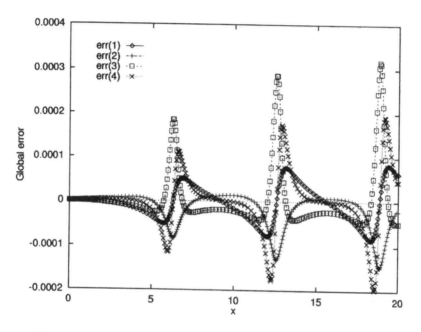

Figure 5.5: Global errors for orbit problem $e = 0.5$, $x \in [0, 20]$

In terms of the eccentric anomaly E, the solution may be expressed as

$$
\begin{aligned}
{}^1y &= \cos E - e \\
{}^2y &= \sqrt{1 - e^2} \sin E \\
{}^3y &= \sin E / (e \cos E - 1) \\
{}^4y &= \sqrt{1 - e^2} \cos E / (1 - e \cos E).
\end{aligned}
$$

For this problem it is not possible to express the solution for y as a function of the independent variable x in an explicit form. Nevertheless, the solution can be determined to any specified accuracy by substituting the appropriate value of x in Kepler's equation, which is solved iteratively for E by Newton's method.

It will be clear that large eccentricities e will be accompanied by large variations in the two 'velocity' components ${}^3y$, ${}^4y$. In turn this implies a greater variation in step–sizes for large values of e than for small eccentricity.

Figure 5.4 shows the variation in step–size for $x \in [0, 20]$, which is just over three times the orbital period, when the embedded pair shown in Table 5.3 is applied with a tolerance $T = 10^{-5}$. The smallest steps are encountered near the perifocal point, which is the point on the orbit with minimum r, at intervals of 2π in x. Although the extreme steplengths vary by a factor of about 5, the control procedure works well in this case. Only a few of the steps are rejected and need to be recomputed. These are detectable in Figure 5.4 by vertical line segments, which indicate the use of two different step–sizes at the same point. On each orbit, one or two rejections occur as the step–size is being reduced just prior $(x \simeq 6, 6 + 2\pi, \ldots)$ to the perifocus. The errors in the four solution components are shown in Figure 5.5. As would be expected, the error grows as the number of orbits increases but the growth is not monotonic. All four components show rapid variation near the perifocal point. The

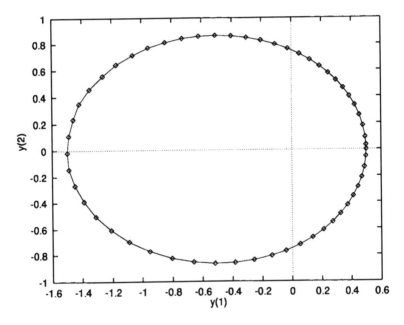

Figure 5.6: Step points on orbit $e = 0.5$

distribution of step points around the orbit is shown in Figure 5.6. The variable step strategy has provided a fairly even distribution of points on the trajectory. With a velocity ratio of 3 at the two extremes of the orbit, a constant steplength would have given a very different picture. To yield similar accuracy, a constant step integration would have taken far more steps. With the orbit problem, the step–size control becomes more difficult as the eccentricity e increases. The value $e = 0.5$ is a modest one, and a simple control mechanism copes with it very easily.

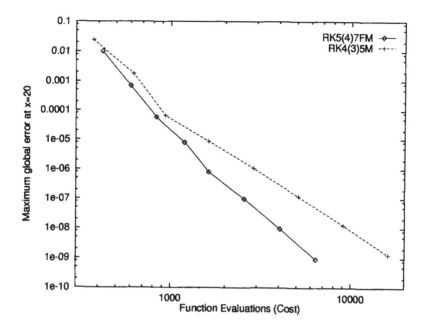

Figure 5.7: Variation of global error with cost in orbit problem $e = 0.5$

Larger eccentricities, as will be encountered in cometary orbits, imply much larger and more rapidly varying step–size changes.

It might be expected that the variation of the global error of a numerical solution will be more complicated in the variable step application than with constant steplengths. Fortunately this is not the case, and it is easy to show that the inequality (3.24)

$$|\varepsilon_{n+1}| \leq |\varepsilon_n|(1 + hL) + M, \quad |t_{n+1}| \leq M = Ah^{p+1},$$

still holds if $h = \max_n(h_n)$ is the largest step–size. Thus, for a method of order p, we have

$$|\varepsilon_n| \leq Bh^p,$$

where B is independent of h, as before. This result can be verified by solving the orbit problem, defined above, over a range of tolerances. Each tolerance will yield a different maximum step–size h, and so the global error can be plotted against the total function evaluations, which will be proportional to the computational cost. This is illustrated in Figure 5.7 for the fourth and fifth order embedded formulae considered earlier, and the curves should be compared with those in Figure 4.4, which were produced from a constant step integration. The new curves, often called *efficiency curves*, are less regular than may be obtained from a constant step case, but they do include the effect of rejected steps. Table 5.5

shows the data used for the efficiency curves; N_f and N_r are the number of function evaluations and step rejections, respectively.

Table 5.5: Results from orbit problem with two embedded pairs

RK4(3)5M				RK5(4)7FM			
$\log T$	N_f	N_r	$\|\varepsilon\|$	$\log T$	N_f	N_r	$\|\varepsilon\|$
-3.0	375	15	$0.2406E{-}01$	-4.0	421	18	$0.1007E{-}01$
-4.0	622	23	$0.1795E{-}02$	-5.0	601	25	$0.6926E{-}03$
-5.0	934	6	$0.6392E{-}04$	-6.0	829	27	$0.5759E{-}04$
-6.0	1619	1	$0.8673E{-}05$	-7.0	1189	26	$0.8074E{-}05$
-7.0	2879	1	$0.1068E{-}05$	-8.0	1615	1	$0.8047E{-}06$
-8.0	5114	1	$0.1194E{-}06$	-9.0	2551	1	$0.9798E{-}07$
-9.0	9093	2	$0.1264E{-}07$	-10.0	4039	1	$0.9663E{-}08$
-10.0	16158	2	$0.1306E{-}08$	-11.0	6403	2	$0.9164E{-}09$

At lax tolerances a fairly high proportion of steps is rejected, but the step prediction process becomes reliable when stringent tolerances are specified. For example, with $T = 10^{-4}$ in the RK5(4)7FM, 18 steps are rejected against 52 accepted. The proportion of rejections diminishes as the tolerance decreases, becoming negligible at stringent tolerances. A more sophisticated control procedure than the one used here can be effective in reducing step rejections.

An important property of the results shown here is known as *tolerance proportionality*. For sufficiently small T, reducing this by an order of magnitude produces the same reduction factor in the global error. This type of behaviour is typical and would be expected from the assumptions implicit in the step-size formula (5.3). Although the error at $x = 20$ has been selected for comparison purposes, a similar conclusion would derive from a consideration of the maximum global error over *all* steps and variables.

5.8   Problems

1. Use the embedded RK3(2) formula given in Table 5.1 to compute two steps of the solution for

$$y' = y^3 - y, \qquad y(0) = 0.5,$$

   using a trial step $h = 0.1$ and absolute error tolerance $T = 10^{-4}$.

2. Write down the principal error coefficients for any three stage 2nd order formula. Assuming that the formula is embedded in the 3rd order formula (4.8) solve the relevant equations of condition for your 2nd order process in terms of b_2 and hence simplify the error coefficients. Consider the effect of the choice of b_2 on the step-size control procedure. Use the differential equation from the previous problem to illustrate the situation.

3. Consider the construction of a third order formula to be embedded in a 4-stage 4th order RK. Show that the expressions for the parameter a_{32} in the three stage third order and in the four stage fourth order processes are incompatible. Reconsider the problem when an FSAL stage is added to the RK44M formula (4.1).

4. Show that all the lower order error coefficients in the RK4(3)5M and the DOPRI5 formulae are non-zero (complete).

5. Show that the inequality for the global error,

$$|\varepsilon_{n+1}| \leq |\varepsilon_n|(1 + hL) + M,$$

   where M is a bound on the local truncation error, L is a Lipschitz constant, and h is the *maximum* step-size, is satisfied for a variable step solution of a scalar differential equation. Hence justify the global error bound

$$|\varepsilon_n| \leq Bh^p,$$

   for a pth order method.

6. The Lorenz attractor equations

$$\frac{dx}{dt} = -ax + ay,$$
$$\frac{dy}{dt} = bx - y - xz,$$
$$\frac{dz}{dt} = -cz + xy$$

have chaotic solutions which are sensitively dependent on initial conditions. Carry out a numerical solution for $a = 5$, $b = 15$, $c = 1$, with initial conditions

$$x(0) = 2, \ y(0) = 6, \ z(0) = 4, \ t \in [0, 20],$$

with tolerance $T = 10^{-4}$. Repeat the integration with

(a) $T = 10^{-5}$;

(b) $x(0) = 2.1$.

In each case compare the results with those obtained previously.

7. The progress of an epidemic of influenza in a population of N individuals is modelled by the system of differential equations

$$\frac{dx}{dt} = -\beta xy + \gamma,$$

$$\frac{dy}{dt} = \beta xy - \alpha y,$$

$$\frac{dz}{dt} = \alpha y - \gamma,$$

where x is the number of people susceptible to the disease, y is the number infected, and z is the number of immunes, which includes those recovered from the disease, at time t. The parameters α, β, γ are the rates of recovery, transmission and replenishment (per day), respectively. It is assumed that the population is fixed so that new births are balanced by deaths.

Modify the program Rkmbed in Appendix A to solve the equations with initial conditions $x(0) = 980$, $y(0) = 20$, $z(0) = 0$, given the parameters $\alpha = 0.05$, $\beta = 0.0002$, $\gamma = 0$. You should terminate the simulation when $y(t) > 0.9N$. Determine approximately the maximum number of people infected and when this occurs.

Investigate the effect of (a) varying the initial number of infected individuals on the progress of the epidemic, and (b) introducing a non–zero replenishment factor.

8. Captain Kirk and his crew, aboard the starship *Enterprise*, are stranded without power in orbit around the earth–like planet Capella III, at an altitude of 127 km. Atmospheric drag is causing the orbit to decay, and if the ship reaches the denser layers of the atmosphere, excessive deceleration and frictional heating will cause irreparable damage to the life–support system. The science officer, Mr Spock, estimates that temporary repairs to the impulse engines will take 29 minutes provided that they can be completed before

the deceleration rises to $5g$ ($1g = 9.81ms^{-1}$). Since Mr Spock is a mathematical genius, he decides to simulate the orbital decay by solving the equations of motion numerically with the DOPRI5 embedded pair. The equations of motion of the starship, subject to atmospheric drag, are given by

$$\frac{dv}{dt} = \frac{GM\sin\gamma}{r^2} - c\rho v^2$$

$$\frac{d\gamma}{dt} = \left(\frac{GM}{rv} - v\right)\frac{\cos\gamma}{r}$$

$$\frac{dz}{dt} = -v\sin\gamma$$

$$\frac{d\theta}{dt} = \frac{v\cos\gamma}{r},$$

where v is the tangential velocity (m/s),
γ is the re–entry angle
     (between velocity vector and horizontal),
z is the altitude (m),
M is the planetary mass (6×10^{24} kg),
G is the constant of gravitation (6.67×10^{-11} SI),
c is a drag constant ($c = 0.004$),
r is the distance to the centre of the planet
     ($z + 6.37 \times 10^6$m),
ρ is the atmospheric density ($1.3\exp(-z/7600)$),
θ is the longitude,
and t is the time (s).

At time $t = 0$, the initial values are $\gamma = 0, \theta = 0$, and $v = \sqrt{GM/r}$. Mr Spock solved the equations numerically to find the deceleration history and the time and the place of the impact of the *Enterprise* should its orbital decay not be prevented. Repeat his simulation using a variable step Runge–Kutta code in Fortran 90, and also estimate approximately the maximum deceleration experienced during descent and the height at which this occurs. Should Captain Kirk give the order to 'Abandon Ship'?

9. Halley's comet last reached perihelion (closest to the sun) on February 9th 1986. Its position and velocity components at this time were

$$(x, y, z) = (0.325514, -0.459460, 0.166229)$$

$$\left(\frac{dx}{dt}, \frac{dy}{dt}, \frac{dz}{dt}\right) = (-9.096111, -6.916686, -1.305721),$$

where the position is measured in astronomical units (the earth's mean distance from the sun), and the time in years. The equations of motion are

$$\frac{d^2x}{dt^2} = -\frac{\mu x}{r^3}, \quad \frac{d^2y}{dt^2} = -\frac{\mu y}{r^3}, \quad \frac{d^2z}{dt^2} = -\frac{\mu z}{r^3},$$

where $r = \sqrt{x^2 + y^2 + z^2}$, $\mu = 4\pi^2$, and the planetary perturbations have been neglected. Solve these equations numerically to determine approximately the time of the next closest approach to the sun.

10. In a study of the dynamics and control of bovine tuberculosis in New Zealand possums, Roberts (1992) formulated the system of differential equations

$$\begin{aligned}
\frac{dN}{dt} &= [r(N) + s(N)]N - \alpha I \\
\frac{dE}{dt} &= pr(N)I + \beta SI - [\sigma + s(N)]E \\
\frac{dI}{dt} &= -\sigma E - [\alpha + s(N)]I
\end{aligned}$$

where the total number of animals per hectare is $N = S + E + I$, and S is the number of susceptible animals, E is the number of infected, but not infectious animals, and I is the number of infectious animals. The birth rate is $r(N) = 0.305$ and the death rate is $s(N) = 0.105 + 0.0002N^3$. The parameter $p \in [0, 1]$ accounts for pseudo-vertical transmission, σ is the rate at which infected hosts become infectious, α is the increase in mortality rate due to the disease and β is the infectious contact rate.

With the disease parameters and initial condition

$$(\alpha, \beta, \sigma, p) = (0.3, 0.11, 2.0, \tfrac{1}{2}), \quad N(0) = 5, \quad E(0) = I(0) = 1,$$

use the RK5(4)7FM embedded pair to compute a numerical solution for the above system for $t \in [0, 50]$. The solution should converge to the equilibrium values

$$(N, E, I) = (7.09, 0.724, 3.04).$$

Also find the equilibrium solution for an acute disease with parameters

$$(\alpha, \beta, \sigma, p) = (3.0, 0.7, 5.0, \tfrac{1}{2}).$$

Try a range of initial conditions.

11. If a proportion q of possums from the previous problem are vaccinated each year, the density X of vaccine–immune animals satisfies the equation

$$\frac{dX}{dt} = -[p + s(N)]X + qS, \quad N = S + E + I + X,$$

where $1/\rho$ is the effective period of vaccination and the other symbols are as defined above. Given $\rho = 2$ and $q = 0.33$, determine a new equilibrium solution for each of the two previous sets of disease parameters.

Chapter 6

Dense output

6.1 Introduction

The use of error estimates to control step–size has one apparent disadvantage; the resulting solution points $\{x_n\}$ cannot be predicted or specified at the start of a calculation. Suppose the solution of the differential equation is required at $x = x^*$, $x^* \in [x_n, x_{n+1}]$, where $x_{n+1} = x_n + h_n$ is the predicted end-of-step. One obvious way of achieving this would be to choose a smaller step-size $h_n^* = x^* - x_n$ rather than the predicted value from the step-size formula (5.3). From the point of view of error tolerance this would be almost certainly acceptable, but there would be a loss of efficiency due to extra steps being necessary. If *dense output* is desired, implying a large number of intermediate points in $[x_n, x_{n+1}]$, the extra computational cost could be considerable. Fortunately, this unpleasant prospect can be avoided by forming an interpolant or dense output formula, sometimes termed a *continuous extension* to the RK formula. Of course, a conventional interpolant based on the values of two or more step points could be constructed, but it is possible, and also preferable, to avoid this multistep approach by deriving directly a continuous Runge–Kutta solution.

As well as providing solutions at specific values of the independent variable x, a continuous RK process also will make it more convenient to solve the inverse problem, in which is sought the value of x for a specified y solution. This is an important advantage for the determination of discontinuity points. These must be 'hit' by the numerical solution when it is necessary to continue the solution beyond such values. More generally the continuous extension can be used to solve algebraic equations connecting the solution variables.

6.2   Construction of continuous extensions

Consider the s stage Runge–Kutta formula

$$y_{n+1} = y_n + h \sum_{i=1}^{s} b_i f_i, \tag{6.1}$$

where the usual notation applies. In this chapter the embedded form is assumed and, although formula (6.1) should be taken to be one of a pair, 'caps' have been left off some parameters to simplify their appearance. However it is assumed that all formulae will be used in local extrapolation mode. Consider a new formula based on the same initial value

$$y_{n+1}^* = y_n + h^* \sum_{i=1}^{s^*} b_i^* f_i^*,$$

with a step–size $h^* = \sigma h$, $\sigma \in [0,1]$. This provides a numerical solution at $x^* = x_n + \sigma h$. As usual, the function evaluations are defined to be

$$f_i^* = f(x_n + c_i^* h^*, y_n + h^* \sum_{j=1}^{i-1} a_{ij}^* f_j^*), \quad i = 1, 2, \ldots, s.$$

Great simplification is achieved by setting

$$c_i^* = c_i/\sigma, \quad a_{ij}^* = a_{ij}/\sigma, \quad i = 2, \ldots, s^*; \quad j = 1, 2, \ldots, i-1,$$

which ensures that $f_i^* \equiv f_i$. Consequently the new solution employs the same function evaluations as formula (6.1) and can be written

$$y_{n+1}^* = y_{n+\sigma} = y_n + \sigma h \sum_{i=1}^{s^*} b_i^* f_i. \tag{6.2}$$

Assuming that the value of y_{n+1} has been computed from formula (6.1), the f_i will be available for use in (6.2). If $s^* \leq s$, no new function evaluations will be necessary although new stages usually are required for all but very low order cases. Solution of the appropriate equations of condition will yield weights $b_i^* = b_i^*(\sigma)$, which depend on the interpolation parameter σ, and hence must be recomputed for each new output point x^*. For many problems this is not a costly computation compared to the function evaluations. Also the b_i^* are valid for all components of the system being solved.

Let us illustrate the construction of a dense output formula by considering an extension to the well-known optimal RK2 process given in the tabular form (6.3).

$$
\begin{array}{c|c|c}
c_i & a_{ij} & b_i \\
\hline
0 & & \frac{1}{4} \\
\frac{2}{3} & \frac{2}{3} & \frac{3}{4}
\end{array}
\tag{6.3}
$$

It seems natural to match the order of the interpolant formula (6.2) with that of the discrete RK2 process above, and this can be achieved with $s^* = 2$. The new parameters must satisfy the two equations of condition

$$
\begin{aligned}
b_1^* + b_2^* &= 1 \\
b_2^* c_2^* &= \tfrac{1}{2},
\end{aligned}
$$

and setting $c_2^* = c_2/\sigma$, to ensure function evaluations in common with the discrete formula (6.3), the second equation becomes

$$
b_2^* c_2 = \tfrac{1}{2}\sigma.
$$

Substituting the given value $c_2 = \frac{2}{3}$ from (6.3) yields the solution

$$
b_2^* = \tfrac{3}{4}\sigma, \quad b_1^* = 1 - \tfrac{3}{4}\sigma.
$$

The RK formula, complete with continuous extension, is defined in Table 6.1.

Table 6.1: RK2 with dense output formula

$$
\begin{array}{c|c|c|c}
c_i & a_{ij} & b_i & b_i^*(\sigma) \\
\hline
0 & & \frac{1}{4} & 1 - \frac{3}{4}\sigma \\
\frac{2}{3} & \frac{2}{3} & \frac{3}{4} & \frac{3}{4}\sigma
\end{array}
$$

To compute an intermediate solution, the appropriate value of σ is substituted in the right-hand column. Thus, for a 'half–way' point $x^* = x_n + \frac{1}{2}h$, the appropriate value is $\sigma = \frac{1}{2}$, giving the second order formula

$$
y_{n+\frac{1}{2}} = y_n + \tfrac{1}{2}h \left(\tfrac{5}{8}f_1 + \tfrac{3}{8}f_2 \right),
$$

where f_1, f_2 will have been computed already for the normal step. A numerical example is illustrated in Table 6.2 where the solutions at four points are given. All are based on the same pair of 'f' values.

For higher order processes the benefit of a continuous extension becomes much greater since step-sizes will be larger. However, the derivation of these formulae is more difficult because the equations of condition

Table 6.2: A second order dense output example

$y' = y(1 - 2y), y(0) = 1, h = 0.15$			
Stage	x	y	$f(x,y)$
	x_0	y_0	f_1
1	0	1	-1
	$x_0 + \frac{2}{3}h$	$y_0 + \frac{2}{3}hf_1$	f_2
2	0.1	0.9	-0.72

$y_{n+\sigma} = y_n + \sigma h((1 - \frac{3}{4}\sigma)f_1 + \frac{3}{4}\sigma f_2)$			
x	σ	$y_{n+\sigma}$	$y(x)$
0.15	1	0.8815	0.8777
0.075	$\frac{1}{2}$	0.9329	0.9326
0.05	$\frac{1}{3}$	0.9535	0.9535
0.10	$\frac{2}{3}$	0.9140	0.9131

increase much more rapidly than the number of stages employed. Since the a_{ij} are already determined, only the b_i^* are available to satisfy the equations of condition. This suggests that at least 4 stages will be necessary for any formula to receive a 3rd order continuous extension. In fact, the construction of dense output formulae usually demands additional stages to provide necessary parameters. A third order case serves to demonstrate this feature. Consider a third order continuous extension to the third order formula RK3M, tabulated in (4.7), and repeated below.

c_i	a_{ij}		b_i
0			$\frac{2}{9}$
$\frac{1}{2}$	$\frac{1}{2}$		$\frac{1}{3}$
$\frac{3}{4}$	0	$\frac{3}{4}$	$\frac{4}{9}$

The four equations to be satisfied by the b_i^* are

$$\sum_{i=2}^{s^*} b_i^* c_i^k = \sigma^k/(k+1), \quad k = 0, 1, 2;$$

$$\sum_{i=3}^{s^*}\sum_{j=2}^{i-1} b_i^* a_{ij} c_j = \sigma^2/6$$

and so the minimum number of stages must be $s^* = 4$. The most convenient choices for the 4th stage parameters are

$$c_4 = 1, \quad a_{4j} = b_j, \quad j = 1, 2, 3,$$

thus creating the FSAL stage, a device also used in DOPRI5 from Table 5.4. An important advantage of this choice is that it implies very little extra computational cost, certainly less than the 33% increase required by a completely new stage. Substituting the parameters from RK3M in the equations of condition gives the linear system

$$b_1^* + b_2^* + b_3^* + b_4^* = 1$$
$$\tfrac{1}{2}b_2^* + \tfrac{3}{4}b_3^* + b_4^* = \tfrac{1}{2}\sigma$$
$$\tfrac{1}{4}b_2^* + \tfrac{9}{16}b_3^* + b_4^* = \tfrac{1}{3}\sigma^2$$
$$\tfrac{3}{8}b_3^* + \tfrac{1}{2}b_4^* = \tfrac{1}{6}\sigma^2$$

which has a unique solution. The complete FSAL process with dense output formula is given in Table 6.3.

Table 6.3: RK3 with 3rd order dense output formula

c_i	a_{ij}			b_i	$b_i^*(\sigma)$
0				$\tfrac{2}{9}$	$1 - \tfrac{4}{3}\sigma + \tfrac{5}{9}\sigma^2$
$\tfrac{1}{2}$	$\tfrac{1}{2}$			$\tfrac{1}{3}$	$\sigma(1 - \tfrac{2}{3}\sigma)$
$\tfrac{3}{4}$	0	$\tfrac{3}{4}$		$\tfrac{4}{9}$	$\tfrac{4}{3}\sigma(1 - \tfrac{2}{3}\sigma)$
1	$\tfrac{2}{9}$	$\tfrac{1}{3}$	$\tfrac{4}{9}$	0	$\sigma(\sigma - 1)$

The inclusion of the FSAL stage has an important advantage over the alternative since it leads in this case to a smooth dense output solution. The continuity of the third order dense formula is clear since

$$b_i^*(1) = b_i, \quad i = 1, \dots, 4.$$

The same property is seen also in the second order case in Table 6.1. Differentiating the third order dense formula with respect to x gives

$$y'_{n+\sigma} = \left(1 - \tfrac{8}{3}\sigma + \tfrac{5}{3}\sigma^2\right) f_1 + 2\sigma(1 - \sigma)f_2 + \tfrac{8}{3}\sigma(1 - \sigma)f_3 + \sigma(3\sigma - 2)f_4,$$

and so

$$y'_{n+0} = f_1 = f(x_n, y_n)$$
$$y'_{n+1} = f_4 = f(x_{n+1}, y_{n+1}).$$

Thus the derivative of the dense output solution is continuous. This property is a necessary one for the FSAL model chosen above but, in other cases where the solution of the equations of condition is non-unique, it need not be satisfied. For a non-FSAL model the C^1 property cannot be obtained generally. The extra computational cost for the FSAL formula in comparison with non-FSAL is usually very small; only one more function evaluation, which occurs at the first step, is necessary. However, step rejections will imply further cost when the extra stage is used also for error estimation as in DOPRI5.

The general conditions to be satisfied by the $b_i^*(\sigma)$ so that the dense interpolant (6.2) is smooth (C^1) are

$$b_i^*(1) = \begin{cases} \widehat{b}_i, & i \le s \\ 0, & s < i \le s^* \end{cases} \tag{6.4}$$

and

$$d_i^*(1) = \begin{cases} 0, & i \ne s_F \\ 1, & i = s_F \end{cases} \tag{6.5}$$

where

$$d_i^*(\sigma) = \frac{d}{d\sigma}\{\sigma b_i^*(\sigma)\},$$

and $i = s_F$ indicates the FSAL stage of the RK pair which employs local extrapolation.

6.3   Choice of free parameters

With higher order RK formula the problem of forming continuous extensions is aggravated by the relative lack of free parameters. Since the a_{ij} for existing stages are already set, only the b_i^* are free to satisfy the RK equations of condition. The number of these is the same as the number of stages s, compared with the total number of parameters in the tableau which is proportional to s^2. The situation is alleviated to some extent by the local nature of the dense output calculation; the dense solution has no global impact. For a formula of order p the global error is $O(h^p)$, but this is the same order as the local error of a formula of order $p - 1$. Consequently a valid continuous extension to the RKp will have order $p^* \ge (p - 1)$.

To set the scene for this type of interpolant let us consider the construction of an interpolant RK2 to use with the RK3M above. Since there are only 2 equations to satisfy, the FSAL stage will not be utilized. With $s^* = 3$, the dense output weights b_i^* must satisfy

$$
\begin{aligned}
b_1^* + b_2^* + b_3^* &= 1 \\
b_2^* c_2 + b_3^* c_3 &= \tfrac{1}{2}\sigma.
\end{aligned}
$$

In terms of b_3^* the solution is

$$
b_1^* = 1 - \sigma + \tfrac{1}{2}b_3^*, \quad b_2^* = \sigma - \tfrac{3}{2}b_3^*. \tag{6.6}
$$

This solution cannot achieve C^1 continuity, but the choice $b_3^* = \tfrac{4}{9}\sigma$ for the free parameter will ensure that the interpolant is C^0. If the FSAL stage had been added it would have been possible to ensure the first derivative continuity but then there would be no reason to avoid satisfying the extra order conditions, thus increasing the order to 3.

The choice $b_3^* = \tfrac{4}{9}\sigma$ certainly yields a globally continuous solution but it may not give a particularly small principal error function for $\sigma \in [0, 1]$. Recalling the construction of discrete Runge–Kutta formulae in Chapter 4 it was found that there was an advantage in choosing any free parameters to minimise the principal error coefficients. For a continuous extension the local truncation error can be expressed as

$$
t_{n+\sigma} = \sum_{i=p^*+1}^{\infty} (\sigma h)^i \varphi_{i-1}^*(x_n, y(x_n), \sigma). \tag{6.7}
$$

Following equation (4.1), and assuming that $p^* = p - 1$, the principal error function for the dense formula will satisfy

$$
\sigma^p \varphi_{p-1}^* = \sum_{j=1}^{n_p} \tau_j^{(p)*} F_j^{(p)}, \tag{6.8}
$$

where $\tau_j^{(p)*}$ is an error coefficient based on the b_i^* weights. An optimal formula will minimise the error norm

$$
A^{(p)}(\sigma) = \sigma^p \| \tau^{(p)*} \|_2. \tag{6.9}
$$

For the second order dense formula (6.6) the relevant error norm is

$$
A^{(3)}(\sigma) = \sigma^3 \sqrt{\left(\tau_1^{(3)*}\right)^2 + \left(\tau_2^{(3)*}\right)^2}.
$$

Only a single parameter b_3^* is involved here and so, following Shampine (1985), the error coefficients can be written

$$
\tau_j^{(3)*} = \zeta_j(\sigma) + \rho_j(\sigma)b_3^*, \quad j = 1, 2,
$$

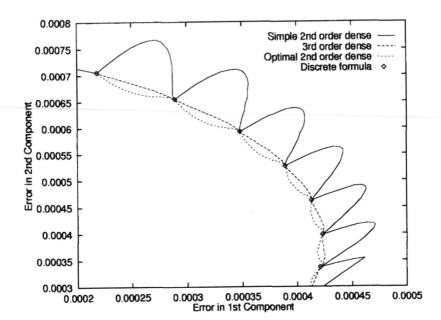

Figure 6.1: Components of global error for $x \in [1.5, 2.5]$ in the gravi-tational two–body problem with $e = 0.2$, using three dense output pro-cesses.

and the error norm will be minimised if

$$b_3^* = -\sum_{j=1}^{2} \zeta_j \rho_j / \sum_{j=1}^{2} \rho_j^2.$$

The error coefficients satisfy

$$\sigma^3 \tau_1^{(3)^*} = \frac{1}{2}\sigma\left[\frac{1}{4}\sigma - \frac{1}{3}\sigma^2 + \frac{3}{16}b_3^*\right]$$

$$\sigma^3 \tau_2^{(3)^*} = \frac{1}{2}\sigma\left[-\frac{1}{3}\sigma^2 + \frac{3}{4}b_3^*\right]$$

and applying the minimisation formula yields

$$b_3^* = \frac{4\sigma(20\sigma - 3)}{153}. \tag{6.10}$$

Using the solution (6.6), the other weights are

$$b_2^* = \frac{\sigma(171 - 120\sigma)}{153}, \quad b_1^* = \frac{153 - \sigma(40\sigma - 159)}{153}.$$

Note that this optimal formula is continuous. The minimisation process must lead to this property since the third order discrete formula is consistent with the absolute minimum of $A^{(3)}(1) = 0$. The general error norm is $A^{(3)}(\sigma) = \sigma^2(1-\sigma)/2\sqrt{17}$, which is rather less than that for the simple solution $b_3^* = \frac{4}{9}\sigma$, for which $A^{(3)}(\sigma) = \sqrt{2}\sigma^2(1-\sigma)/6$.

To test the quality of the two second order interpolants and the third order continuous extension constructed for the RK3M formula, they have been applied to the orbit problem described in the last chapter. In Figure 6.1, the error of the second component of the solution ^2y is plotted against that of the first component, for each of the three dense output solutions. The normal (discrete) solution points are marked, and curves based on the dense output solutions, determined at a number of intermediate values, are plotted. Since all three solutions agree at the points given by the discrete formula, the continuity is verified. However, the lack of smoothness in the second order error is evident. Although the local errors in the RK2 dense formula do not dominate the global error, they appear to be much larger than the local error in the 3rd order case. The optimised interpolant is noticeably superior to the other second order case. Since the extra cost of finding the more accurate solution is negligible in this case, the third order method is clearly preferable.

Although the third order dense formula in Table 6.3 is uniquely determined by the FSAL specification there will be many cases of $p^* = p$ in which free parameters can be selected for an optimal formula. Simple minimisation of the principal error norm as in the $p^* = p - 1$ case may be inappropriate. If $p^* = p$ the continuity requirement will ensure

$$A^{(p^*+1)}(1) = A^{(p+1)} \neq 0,$$

the principal error norm for the discrete RK. Given that the continuous extension is constructed to avoid the necessity of extra steps with the discrete formula, the best one could expect is that

$$A^{(p^*+1)}(\sigma) = \sigma^{p+1} A^{(p+1)}, \quad 0 < \sigma < 1, \tag{6.11}$$

implying that the interpolant yields a solution as good as the main formula with a reduced step σh. In practice, the minimisation of

$$H = \int_0^1 A^{(p^*+1)}(\sigma) d\sigma \tag{6.12}$$

is achieved by parameters quite similar to those which yield a relation nearest to equation (6.11) above.

For the third order case already determined, it may be shown that

$$A^{(4)}(\sigma) = \frac{\sigma^2}{288}\sqrt{1728\sigma^4 - 7536\sigma^3 + 13132\sigma^2 - 11148\sigma + 3969}, \tag{6.13}$$

compared to an 'ideal' norm $\frac{\sqrt{145}}{288}\sigma^4$. These two functions are plotted in Figure 6.2, showing that the actual formula is not very close to 'ideal', but that its local truncation error norm is monotonic for $\sigma \in [0,1]$. This good behaviour is consistent with the results illustrated in Figure 6.1.

6.4   Higher–order formulae

Moving to fourth order formulae, a dense output formula is required to satisfy 8 equations of condition. To see how this can be achieved, consider the embedded pair RK4(3)5M from Table 5.3. Fortunately, the equations of condition originally considered for this formula were simplified using the Butcher equations (4.13, 4.14). The first of these cannot apply to the b_i^* parameters, which are dependent on the interpolation variable σ, but the second set, affecting the a_{ij}, are still effective. This reduces the number of independent equations to 6, which is still too many for a 5-stage dense output process. The addition of a sixth FSAL stage does not provide sufficient parameters in this instance since equation (4.14) requires $b_2^* = 0$, and so a seventh stage is added, providing 7 more parameters.

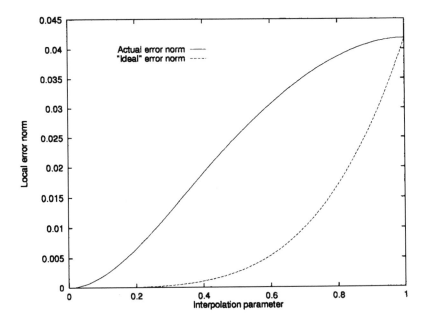

Figure 6.2: Variation of the error norm $A^{(4)}$ for the 3rd order dense output formula in Table 6.3

Unlike the third order case, there are more parameters than conditions to satisfy, and so it is necessary to use an appropriate criterion, such as the minimisation of integral (6.12), to choose them. The dense formula given in Table 6.4 is a good approximation to the optimal one for RK4(3)5M. As usual in discrete cases also, simple rational values approximating to the optimal values have been selected to 'tidy-up' the formula. The continuity properties of the interpolated solution are the same as those of the 3rd order dense formula for RK3M.

Table 6.4: The embedded RK4(3)5M formula pair with 4th order dense output extension.

c_i	a_{ij}						$\widehat{b}_i$	b_i	b_i^*
0							$\frac{13}{96}$	$\frac{23}{192}$	$-\frac{225\sigma^3-520\sigma^2+378\sigma-96}{96}$
$\frac{1}{5}$	$\frac{1}{5}$						0	0	0
$\frac{2}{5}$	0	$\frac{2}{5}$					$\frac{25}{48}$	$\frac{55}{96}$	$-\frac{25\sigma(5\sigma^2-8\sigma+2)}{48}$
$\frac{4}{5}$	$\frac{6}{5}$	$-\frac{12}{5}$	2				$\frac{25}{96}$	$\frac{35}{192}$	$-\frac{25\sigma(5\sigma^2-8\sigma+2)}{96}$
1	$-\frac{17}{8}$	5	$-\frac{5}{2}$	$\frac{5}{8}$			$\frac{1}{12}$	$\frac{1}{8}$	$-\frac{\sigma(5\sigma^2-8\sigma+2)}{12}$
1	$\frac{13}{96}$	0	$\frac{25}{48}$	$\frac{25}{96}$	$\frac{1}{12}$		0	0	$\frac{\sigma(\sigma-1)(35\sigma-11)}{24}$
$\frac{1}{5}$	$\frac{67}{480}$	0	$\frac{1}{16}$	$\frac{7}{480}$	0	$-\frac{1}{60}$	0	0	$\frac{125\sigma(\sigma-1)^2}{24}$

Actually, the formula in Table 6.4 occupies one more stage than the minimum required for a continuous RK4. In constructing the fourth order formula pair RK4(3)5M, the simplifying assumptions (4.13, 4.14) were used but, additionally, a third extra condition

$$\sum_{j=1}^{i-1} a_{ij}c_j^2 = \tfrac{1}{3}c_i^3, \quad i = 3,\ldots,s, \tag{6.14}$$

can be applied. This has the effect of eliminating the condition depending on error coefficient $\tau_3^{(4)}$ (Table 3.2) and, since it is independent of b_i, will apply equally to the derivation the interpolant formula. With only 6 stages and $b_2^* = 0$, the 5 remaining equations for the b_i^* will determine a unique solution for the dense interpolant. An example based on this model is presented by Dormand et al. (1989).

A number of fifth order continuous extensions have been derived for the popular DOPRI5 formula shown in Table 5.4. The fifth order extension by Calvo et al. (1990) is far superior to fourth order dense output formulae such as that presented by Dormand and Prince (1986). Nevertheless two extra function evaluations are necessary, and so those steps at which dense output is needed will be more expensive. The construction of a fifth order extension is much more complicated than the fourth order case but it can be carried out in a similar fashion.

An alternative method of constructing a dense output formula is based on classical interpolation. Suppose we have a discrete RK formula of order q and we seek a continuous extension of the same local order. Then it may be proved (Dormand and Prince, 1986) that this is equivalent to a Hermite interpolant $P(x)$, of degree $\geq q$, which is based on the end points of the discrete RK step and an appropriate number of internal points or derivatives which have an appropriate order of accuracy. To take $q = 5$ as an example, the interpolant must satisfy the conditions

$$
\begin{aligned}
P(x_n) &= y_n, & P'(x_n) &= f(x_n, y_n) \\
P(x_{n+1}) &= y_{n+1}, & P'(x_{n+1}) &= f(x_{n+1}, y_{n+1}) \qquad (6.15) \\
P'(x_{n+\sigma_i}) &= f(x_{n+\sigma_i}, u_{n+\sigma_i}), & i &= 1, 2,
\end{aligned}
$$

where $u_{n+\sigma_i}$, an estimate of $y(x_{n+\sigma_i})$, is obtained from a 4th order dense output interpolant which is assumed to be available. The six conditions specify a fifth degree polynomial which can be determined fairly easily. Taking the polynomial to be

$$
P(x) = \sum_{j=0}^{5} \alpha_j \sigma^j, \quad \sigma = (x - x_n)/h,
$$

the coefficients α_j may be determined by solving a system of linear equations. Otherwise a divided difference approach can be used to form the polynomial. Since two of the conditions involve derivatives of y and there are none involving the corresponding y values, a solution will not exist for all values of the parameters. However it can be shown that a suitable solution exists. The technique to be used here may be used to construct continuous extensions of increasing order (Enright et al., 1985) by a *bootstrapping* process. Thus a fourth order interpolant could be determined by computing an intermediate function evaluation derived from a cubic Hermite polynomial. A second *extrapolation* will yield the required 5th order formula. In the DOPRI5 case the derivation of a fourth order continuous extension directly from the RK equations of condition is very straightforward, and so a single extrapolation will give a fifth order dense output formula.

Since the DOPRI5 coefficients satisfy the simplifying conditions (4.14, 6.14). fourth order accuracy imposes only five conditions on the 'b_i^*' parameters, and since there are seven stages with $b_2^* = 0$, one of these parameters may be chosen freely. Actually it is an advantage to force the interpolant to be smooth and then to minimise

$$H_5 = \int_0^1 A^{(5)}(\sigma)d\sigma. \tag{6.16}$$

For C^1 continuity it is necessary to satisfy the conditions (6.4, 6.5), and so it is convenient to specify

$$b_7^* = \sigma(\sigma - 1)(\alpha\sigma + \beta), \quad \alpha + \beta = 1. \tag{6.17}$$

Solving the relevant equations of condition yields the b_i^* in terms of α, and it may be shown that $\alpha = 5/2$ yields a locally fourth order formula for which H_5 from (6.16) is near a minimum. This continuous extension is given in the left-hand column of Table 6.5. Two interpolant values, $u_{n+\sigma_i}$, $i = 1, 2$ in equation (6.15) can be obtained from

$$u_{n+\sigma_i} = y_n + \sigma_i h \sum_{j=1}^{7} b_j^*(\sigma_i) f_j,$$

and, in Runge–Kutta notation, the two extra function evaluations yielding a locally fifth order continuous extension are

$$f_{7+i} = f(x_n + c_{7+i}h, y_n + h \sum_{j=1}^{7} a_{7+i,j} f_j), \quad i = 1, 2,$$

where

$$c_{7+i} = \sigma_i, \quad a_{7+i,j} = \sigma_i b_j^*(\sigma_i).$$

Any values for σ_1, σ_2 may be chosen but optimal values for c_8, c_9 will minimize

$$H_6 = \int_0^1 A^{(6)}(\sigma)d\sigma.$$

The pair $c_8 = \frac{1}{5}, c_9 = \frac{1}{2}$ is close to optimal and has the advantage of giving relatively simple parameters. The nine $\widehat{b}^*$ of the fifth order dense formula are shown in the right-hand column of Table 6.5. The same method can be applied to still higher order formulae, although the number of extra internal points yielding extra stages must be greater. While this technique is very successful for DOPRI5, giving a simple, near optimal continuous extension in the smallest number of extra stages, its application to higher orders usually gives dense output formulae requiring more stages than the absolute minimum.

Table 6.5: Fourth and fifth order dense formulae for DOPRI5

b_i^*	a_{8i}	a_{9i}	$\widehat{b}_i^*$
$-\dfrac{435\sigma^3 - 1184\sigma^2 + 1098\sigma - 384}{384}$	$\dfrac{5207}{48000}$	$\dfrac{613}{6144}$	$\dfrac{696\sigma^4 - 2439\sigma^3 + 3104\sigma^2 - 1710\sigma + 384}{384}$
0	0	0	0
$\dfrac{500\sigma(6\sigma^2 - 14\sigma + 9)}{1113}$	$\dfrac{92}{795}$	$\dfrac{125}{318}$	$-\dfrac{500\sigma(24\sigma^3 - 51\sigma^2 + 32\sigma - 6)}{1113}$
$-\dfrac{125\sigma(9\sigma^2 - 16\sigma + 6)}{192}$	$-\dfrac{79}{960}$	$-\dfrac{125}{3072}$	$-\dfrac{125\sigma(24\sigma^3 - 51\sigma^2 + 32\sigma - 6)}{192}$
$\dfrac{729\sigma(35\sigma^2 - 64\sigma + 26)}{6784}$	$\dfrac{53217}{848000}$	$\dfrac{8019}{108544}$	$\dfrac{2187\sigma(24\sigma^3 - 51\sigma^2 + 32\sigma - 6)}{6784}$
$-\dfrac{11\sigma(3\sigma - 2)(5\sigma - 6)}{84}$	$-\dfrac{11}{300}$	$-\dfrac{11}{192}$	$-\dfrac{11\sigma(24\sigma^3 - 51\sigma^2 + 32\sigma - 6)}{84}$
$\dfrac{\sigma(\sigma - 1)(5\sigma - 3)}{2}$	$\dfrac{4}{125}$	$\dfrac{1}{32}$	$\dfrac{\sigma(\sigma - 1)(32\sigma^2 - 31\sigma + 7)}{8}$
0	0	0	$\dfrac{125\sigma(\sigma - 1)^2}{24}$
0	0	0	$\dfrac{16\sigma(\sigma - 1)^2(3\sigma - 1)}{3}$

6.5 Computational aspects of dense output

The extension of the variable step RK program, described in Chapter 5, to include dense output is very straightforward. A suitable program RkDen is listed in Appendix A. The additional steps are shown in Figure 6.3.

```
Compute a normal step
IF successful THEN
    LOOP over all solution points within the step
        Compute any new function evaluations
        (The same new evaluations are valid for any
        point within this step)
        Compute interpolation parameters
        Evaluate intermediate solution
    END LOOP
    Compute steplength for next step
    Update solution
ENDIF
```

Figure 6.3: Dense output procedure.

In Figure 6.4, the global error of the fourth order dense output solution in the gravitational two-body problem, for $x \in [2,3]$, is plotted. In the given interval there are four step points, as determined by the normal step-size control procedure. The specified tolerance was 0.0001 and 100 intermediate output points were selected. Since only three steps are involved it was necessary only to compute 3 extra function evaluations in order to yield the dense output. The smoothness of the interpolation is very remarkable when one considers that the solution component (^1y) has a value of approximately 1.

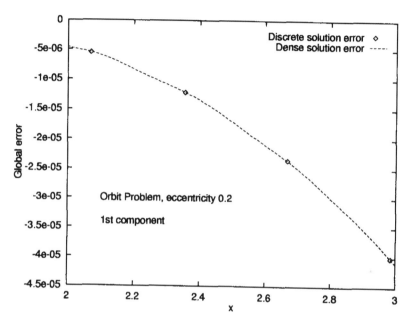

Figure 6.4: Global error of first solution component in the orbit problem with $e = 0.2$, using RK4(3)5M with 4th order dense output

The quality of continuous solution is just as high with the fifth order case using DOPRI5 and the 9 stage extension given in Table 6.5.

6.6   Inverse interpolation

A further important application of the continuous extension is the solution of the inverse interpolation problem. Instead of estimating the solution value (y) at a specified value of x, we consider computing a value x such that $y(x) = Y$. Typically the value Y would be that of a single component rather than a vector, although a least squares process could be defined to yield a solution in the latter case. Many practical problems include discontinuous derivatives which demand that a numerical method shall hit a particular solution value. Integrating past a discontinuity will give invalid results. These problems could be solved without a continuous extension but, in order to hit the required solution, an iterative technique in which a new step must be computed for each iteration would need to be employed. Such a technique would be computationally expensive in comparison with one based on a continuous extension which has polynomial form. The solution of a polynomial equation is a much easier problem.

Having determined that the required solution should be contained in a Runge–Kutta step just completed, the Newton–Raphson method can be employed to determine the x value. In this application it is convenient to express the interpolant in polynomial rather than Runge–Kutta form. Therefore we take

$$y_{n+\sigma} \;=\; y_n + \sigma h \sum_{i=1}^{s^*} b_i^* f_i$$

$$=\; \sum_{j=0}^{r} A_j \sigma^j = P(\sigma).$$

Assuming that

$$b_i^* = d_{i0} + \sigma d_{i1} + \cdots + \sigma^r d_{ir},$$

we have

$$A_0 = y_n, \quad A_j = h \sum_{i=1}^{s} d_{i,j-1} f_i, \; j = 1, \ldots, r+1. \tag{6.18}$$

Now employ Newton–Raphson to solve $P(\sigma) = Y$, assuming the scalar case. To find an initial approximation, it is easy to compute at each step

$$\Delta_n = y_{n+1} - Y,$$

and assume that the solution is contained when $\Delta_n \Delta_{n-1} < 0$. This technique does not guarantee that all solutions will be found, but the condition does verify the existence of a solution in the given step. A first approximation to the value of σ required is therefore

$$\sigma_0 = \frac{Y - y_n}{y_{n+1} - y_n},$$

and the Newton iteration can be expressed as

$$\sigma_{k+1} = \sigma_k - \frac{P(\sigma_k) - Y}{P'(\sigma_k)}, \; k = 0, 1, \ldots.$$

This seems to imply the necessity for derivative evaluations, but since only a polynomial is involved, synthetic division will yield P and P' without explicit differentiation. For a non–scalar problem, the above process can be applied for the solution based on any component value. Having determined the interpolation parameter σ, all the components of the solution can be found in the normal manner.

6.7   Problems

1. Obtain a dense output formula of local order 2 for the third order formula (4.8). Choose the free parameter to minimize the appropriate local error coefficients. Check on the continuity of your formula.

2. Add the FSAL stage to the formula from the question above, and hence obtain a dense output formula of order 3.

   Use your third order dense formula to compute $y(0.04)$, $y(0.08)$, given the initial value problem

$$y' = y^3 - y \ , \ y(0) = 0.5,$$

   and a step-size $h = 0.1$, without computing more than one step.

3. Construct a third order dense output extension to the fourth order formula of Merson(1957) given in the table below.

c_i	a_{ij}				$\widehat{b}_i$
0					$\frac{1}{6}$
$\frac{1}{3}$	$\frac{1}{3}$				0
$\frac{1}{3}$	$\frac{1}{6}$	$\frac{1}{6}$			0
$\frac{1}{2}$	$\frac{1}{8}$	0	$\frac{3}{8}$		$\frac{2}{3}$
1	$\frac{1}{2}$	0	$-\frac{3}{2}$	2	$\frac{1}{6}$

   Choose the free parameter to yield a continuous interpolant. Determine a smooth (C^1) interpolant by adding the FSAL stage to Merson's formula.

4. Modify the RkDen program in Appendix A to compute the solution to

$$y'' + 0.1(y')^2 + 0.6y = 0, \ y(0) = 1, \ y'(0) = 0,$$

   in the interval $x \in [0, 10]$, producing output at equal intervals of size 0.5 in x.

5. Use the dense output polynomial based on RK4(3)5M from Table 5.3 to compute more accurately the time of impact of the crashing starship *Enterprise*, whose equations of motion are listed in problem 8 of Chapter 5. The Newton–Raphson technique should be employed to determine the instant of zero altitude.

6. Impose the Butcher simplifying conditions (4.13, 4.14, 6.14) on the RK error coefficients up to order 4. Hence show that the equations

of condition for a fourth order formula using s stages reduce to

$$\sum_{i=1}^{s} b_i c_i^k = \frac{1}{k+1}, \quad k = 0, 1, 2, 3$$

with $b_2 = 0$. Setting $i = 3$ in the simplifying condition (4.14) and (6.14), show that $c_2 = \frac{2}{3}c_3$. Outline the method of solution of the equations when $s = 5$, and hence find the RK4 process with $c_3 = \frac{1}{2}$, $c_4 = \frac{3}{4}$.

Consider the construction of a 4th order continuous extension, utilizing the FSAL stage, to your RK4. Show that the equations to be satisfied by the six–stage dense output process are

$$\sum_{i=1}^{6} b_i^* c_i^k = \frac{\sigma^k}{k+1}, \quad k = 0, 1, 2, 3$$

$$\sum_{i=3}^{6} b_i^* a_{i2} = 0.$$

Hence determine the continuous extension. Test your formula in the starship application in an earlier problem.

7. Obtain the equations to be satisfied by a 4th order continuous extension to DOPRI5. Assume that b_7^* is given by the formula (6.17).

8. A cork of length L is on the point of being ejected from a bottle containing a fermenting liquid. The equations of motion of the cork may be written

$$\frac{dv}{dt} = \begin{cases} g(1+q)\left[\left(1 + \dfrac{x}{d}\right)^{-\gamma} + \dfrac{RT}{100} - 1 + \dfrac{qx}{L(1+q)}\right], & x < L \\ 0, & x \geq L \end{cases}$$

$$\frac{dx}{dt} = v$$

where g is the acceleration due to gravity (9.81 m.s^{-2}),
$\quad\quad\ q$ is the friction–weight ratio for the cork,
$\quad\quad\ x$ is the cork's displacement in the neck of the bottle,
$\quad\quad\ t$ is the time,
$\quad\quad\ d$ is the length of the bottle neck,
$\quad\quad\ R$ is the percentage rate at which the pressure is
$\quad\quad\quad\quad$ increasing,
$\quad$ and γ is the adiabatic constant for the gas in the bottle (1.4).
The intial condition is $x(0) = \dot{x}(0) = 0$. While $x < L$ the cork is still in the bottle but it leaves the bottle at $x = L$. Integrate the

equations of motion with DOPRI5 and tolerance 0.000001 to find the time at which the cork is ejected. Also find the velocity of ejection when

$$q = 20, \quad L = 3.75\text{cm}, \quad d = 5\text{cm}, \quad R = 4.$$

You should use a dense output formula to determine the ejection time.

Chapter 7

Stability and stiffness

7.1 Introduction

In preceding chapters, we have considered the construction of Runge-Kutta formulae with reference only to their asymptotic error properties. For many practical problems these properties do not ensure a convenient and efficient method of solution. Such problems usually comprise systems of equations with solutions containing components whose rates of change differ markedly. In many cases the property of *stability* governs the numerical process.

To illustrate this type of problem, consider a linear equation of order two modelling damped simple harmonic motion, as might be described by a vibrating spring whose motion is resisted by a force proportional to velocity. Suppose y is the displacement of a unit mass attached to the end of the spring, c is the damping constant, and k is the stiffness constant for the spring. Then the equation of motion of the mass is

$$y'' + cy' + ky = 0, \tag{7.1}$$

where the dashes indicate differentiation with respect to time x. The general solution of this equation is

$$y(x) = Ae^{\alpha_1 x} + Be^{\alpha_2 x},$$

where

$$\alpha_1, \alpha_2 = \tfrac{1}{2}\left(-c \pm \sqrt{c^2 - 4k}\right).$$

If $c^2 > 4k$, the spring is said to be strongly damped and both the exponents (α_1, α_2) are negative. For example, when $c = 26/5$, $k = 1$, and the initial values are $y(0) = y'(0) = 1$, equation (7.1) has the solution

$$y(x) = \tfrac{1}{4}\left(5e^{-x/5} - e^{-5x}\right).$$

This expression consists of two parts, usually described as the fast and slow transient terms. The fast transient term is that which decays more rapidly and, in this case, the faster term (e^{-5x}) is negligible except for small values of x. Figure 7.1 shows the solution for $x \in [0,2]$ together with the slow component. The slower component on its own provides a

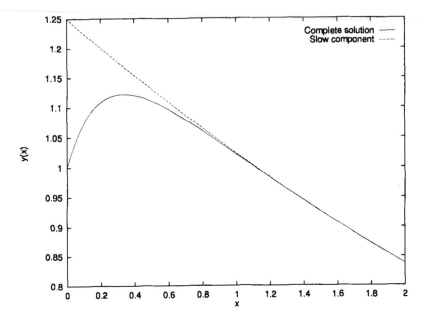

Figure 7.1: The stiff spring solution in $[0,2]$

fairly good approximation for the complete solution when $x > 1$. However, for a numerical method which makes use of derivative values, the fast component continues to influence the solution, even for values of x much larger than unity. As a consequence the step-size control depends on a quantity which is known to be negligible in the true solution. This is not to say that the steplengths will be unnecessarily small. The numerical method really does need relatively small step-sizes to satisfy what is known as a stability condition.

For the application of a standard RK formula, the second order equation (7.1) would be represented as a linear system

$$\begin{pmatrix} {}^1y' \\ {}^2y' \end{pmatrix} = \begin{pmatrix} 0 & 1 \\ -1 & -\frac{26}{5} \end{pmatrix} \begin{pmatrix} {}^1y \\ {}^2y \end{pmatrix} \qquad (7.2)$$

with initial values ${}^1y = 1$, ${}^2y = 1$.

In this chapter the concept of stability with respect to RK methods will be explored. The property of *stiffness*, in relation to systems of differential equations, will be introduced. Special RK methods for application to such systems will be considered.

7.2   Absolute stability

Although numerical methods are not normally required to process linear equations it is instructive and convenient to use them in a discussion of stability phenomena. To introduce the concept of absolute stability, consider the application of a two–stage Runge-Kutta formula to the linear scalar test problem

$$y' = \lambda y, \quad y(x_n) = y_n, \tag{7.3}$$

which has the true solution

$$y(x) = y_n e^{\lambda(x - x_n)}.$$

Applying the general two-stage RK formula (3.11) one obtains

$$
\begin{aligned}
f_1 &= \lambda y_n \\
f_2 &= \lambda(y_n + c_2 h f_1) \\
    &= \lambda(y_n + c_2 h \lambda y_n),
\end{aligned}
$$

yielding, for a single step,

$$y_{n+1} = y_n + b_1 h \lambda y_n + b_2 h \lambda y_n (1 + c_2 h \lambda).$$

Setting $h\lambda = r$ and factorising gives

$$y_{n+1} = y_n [1 + (b_1 + b_2) r + b_2 c_2 r^2],$$

where the b_i and c_i parameters are, as yet, unspecified. Further simplification is achieved by choosing the parameters so that the RK formula has order 2. Using equations (3.13) this yields

$$y_{n+1} = P(r) y_n, \qquad P(r) = 1 + r + \tfrac{1}{2} r^2, \tag{7.4}$$

where $P(r)$ is said to be the stability polynomial of the RK process. When $\lambda < 0$, the true solution of equation (7.3) is monotonic decreasing, and an acceptable numerical solution must have a similar property. Since all consistent Runge–Kutta methods are convergent, a sufficiently small step–size will guarantee this feature, but it is important to determine an appropriate steplength. Normally, this diminishing property is specified in the absolute sense, and so a condition which satisfies the inequality

$$|y_{n+1}| < |y_n|$$

is sought. From equation (7.4), this inequality requires

$$|P(r)| < 1. \qquad (7.5)$$

A method satisfying the inequality (7.5) is said to be *Absolutely Stable*. From (7.4), the absolute stability condition for any two stage RK2 can be written

$$|1 + r + \tfrac{1}{2}r^2| < 1$$

or

$$-1 < 1 + r + \tfrac{1}{2}r^2 < 1. \qquad (7.6)$$

The left–hand side of this inequality is

$$0 < 2 + r + \tfrac{1}{2}r^2,$$

which, on completing the square, yields

$$0 < \tfrac{1}{2}(r + 1)^2 + \tfrac{3}{2},$$

which is true for any real r. The right-hand side of the inequality (7.6) simplifies to

$$\tfrac{1}{2}r(r + 2) < 0,$$

which is true when $-2 < r < 0$. Hence the absolute stability condition for any two stage RK2 can be written

$$-2 < h\lambda < 0. \qquad (7.7)$$

Thus any two stage RK2 applied to equation (7.3) will produce a sequence of absolutely decreasing values when r is bounded as shown here and, if λ is specified, the condition yields a bound on the step–size h. For example, with $\lambda = -1$ absolute stability demands $h < 2$.

The foregoing introductory treatment of absolute stability is based on a scalar test problem which is much simpler than one which usually demands numerical treatment. Extension of the stability analysis to non-linear systems of differential equations proceeds first with a consideration of a linear system of dimension m,

$$y' = Ay, \quad y(x_n) = y_n,$$

where $y \in \mathbb{R}^m$ and A is an $(m \times m)$ constant matrix. Application of any two stage RK2 formula to this system gives the result, similar to the expression (7.4),

$$y_{n+1} = P(hA)y_n, \qquad P(hA) = I + hA + \tfrac{1}{2}h^2A^2,$$

where P is $(m \times m)$. The condition for absolute stability in the vector case is

$$\|y_{n+1}\| < \|y_n\| \implies \|I + hA + \tfrac{1}{2}h^2A^2\| < 1,$$

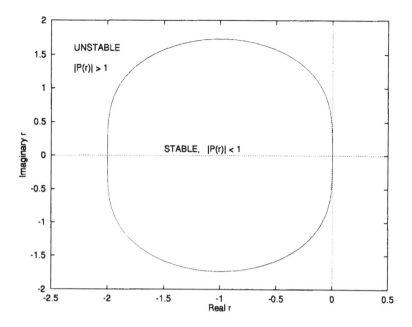

Figure 7.2: The stability region for a two-stage RK2 formula

for any norm. A suitable norm is the spectral radius of $P(hA)$, which is the largest eigenvalue in an absolute sense. Now if A has eigenvalues $\lambda_1, \ldots, \lambda_m$, then the polynomial $P(hA)$ has eigenvalues

$$1 + h\lambda_j + \tfrac{1}{2}h^2\lambda_j^2, \quad j = 1, 2, \ldots, m,$$

and so absolute stability demands

$$|1 + h\lambda_j + \tfrac{1}{2}h^2\lambda_j^2| < 1, \quad j = 1, 2, \ldots m. \tag{7.8}$$

Thus, every eigenvalue of the matrix A must satisfy the stability condition (7.7) obtained from the scalar test equation.

Since the eigenvalues of A generally are complex, the inequality (7.8) defines a region in the complex plane as shown in Figure 7.2. Actually the region within which absolute stability is assured is symmetric about the real axis, and so it is customary to display only the positive imaginary half–plane.

It should be emphasised that one requires a method to be absolutely stable only for problems with decreasing solution components. For $\lambda > 0$ the solution should increase and fortunately the figure indicates that the RK2 would yield increasing values in such a case.

A further point to note is that the stability of a numerical solution does not imply its high accuracy. Absolute stability ensures that a numerical solution will be bounded but it may differ quite significantly from the true solution. This can be demonstrated by applying an RK2 formula to $y' = -y$, $y(0) = 1$ with a step-size $h = 1$. Absolute stability requires $h < 2$ and so the numerical solution would certainly diminish with increasing steps. Ultimately, both true and RK solutions converge to zero, but transient values differ markedly.

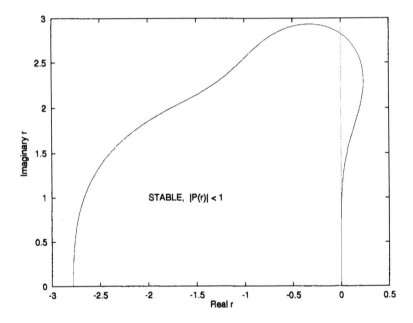

Figure 7.3: The stability region for a four-stage RK4 formula

Higher order formulae usually possess larger regions of stability than the RK2, although this is not necessarily the case. For a four stage RK4, corresponding to equation (7.4), the stability polynomial is

$$P(r) = 1 + r + \tfrac{1}{2}r^2 + \tfrac{1}{6}r^3 + \tfrac{1}{24}r^4,$$

and this gives a real negative stability limit of -2.8 as indicated in Figure 7.3.

For an RK formula of order p with s stages, it can be shown that the stability polynomial $P(r)$ is of degree s, and it may be written in terms of the RK parameters in the form

$$\begin{aligned} P(r) \;=\;& 1 + r \sum_i b_i + r^2 \sum_i b_i c_i + r^3 \sum_{ij} b_i a_{ij} c_j + r^4 \sum_{ijk} b_i a_{ij} a_{jk} c_k \\ & + \;\cdots\; + r^s \sum_{ijk\ldots vw} b_i a_{ij} a_{jk} \ldots a_{vw} c_w \end{aligned}$$

$$= \quad 1 + \sum_{i=1}^{s} W_i r^i \tag{7.9}$$

where $W_i = 1/i!$, $i = 1, 2, \ldots, p$. For the 5th order member of the DOPRI5 pair the negative real stability limit is near -3.3 as shown in Figure 7.4, while that of the 4th order member is larger at about -4.4. A curious feature of both DOPRI5 stability plots is the existence of disjoint regions of stability. To use an astronomical analogy, it is almost as though the small region at the top right were a satellite of the main one. However the small 'moon' has no practical significance.

7.3 Non-linear stability

A consideration of non–linear stability is much more difficult than the

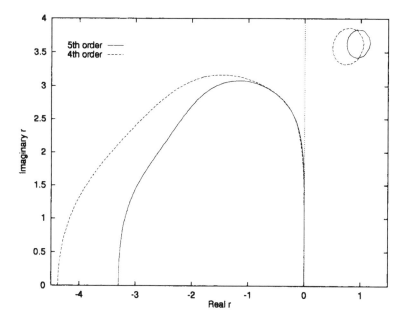

Figure 7.4: The stability regions for the 4th and 5th order members of the DOPRI5 pair

foregoing treatment of linear systems. For non–linear systems, in autonomous form $y' = f(y)$, $y \in \mathbb{R}^M$, the rôle of the matrix A above is occupied by the Jacobian matrix J of f, where

$$J = \begin{pmatrix} {}^1f_1 & {}^1f_2 & \cdots & {}^1f_M \\ {}^2f_1 & {}^2f_2 & \cdots & {}^2f_M \\ \vdots & \vdots & \vdots & \vdots \\ {}^Mf_1 & {}^Mf_2 & \cdots & {}^Mf_M \end{pmatrix}.$$

Since J is variable, the stability requirements will vary over the range of the solution of the differential equation. This variation is demonstrated by reference to a scalar non-linear problem

$$y' = \frac{1}{y} - \frac{1}{x^2}(1 + x^3), \quad y(1) = 1, \tag{7.10}$$

which has the true solution $y(x) = 1/x$. Writing this equation in autonomous form with $M = 2$, it is easy to obtain the Jacobian, which has largest eigenvalue

$$\lambda(x) = -1/y^2 = -x^2.$$

Thus the true solution decreases monotonically while the spectral radius of J increases. For an RK formula with an absolute stability requirement

$$-L < h\lambda(x) < 0,$$

it is clear that the maximum step-size h to ensure a decreasing solution will reduce as x increases. To illustrate this, Figure 7.5 shows the variation in step-size for $x \in [1, 10]$, when a three stage RK3(2) embedded pair, with a tolerance $T = 10^{-2}$, is applied to equation (7.10). The step-size control algorithm was that shown in Figure 5.3, and an initial step-size $h = 0.1$ was selected. For the first few steps the value of h increases, but then it diminishes in accordance with stability limitations. At $x = 10$ the dominant eigenvalue has magnitude 100, implying a maximum step-size of about $1/40$ since a three stage RK3 gives $L \simeq 2.5$. The vertical line segments in the figure indicate rejected steps, and the high proportion of these suggests that the step-size control mechanism is not designed for this type of problem. Recalling the derivation of the step-size control formula (5.2) this should not be very surprising. Nevertheless, the control mechanism is sufficiently robust to copy with a situation where the steplength is governed by stability. Recalling that the difference between solutions of different orders is used as the error estimate, this is likely to be large when the stability boundary for either one of the RK pair is breached, thus causing a rejection and a steplength reduction. The figure indicates that the reduction factor tends to be too large for this problem.

7.4 Stiffness

When the eigenvalues of J differ greatly the system of differential equations is said to be *Stiff*. This can present severe problems for an integrator

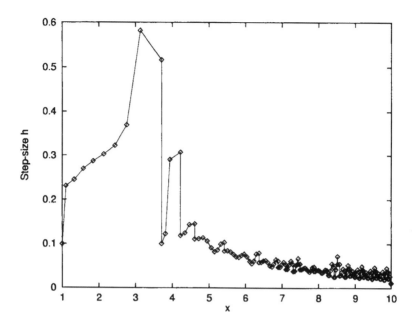

Figure 7.5: The variation of step-size in [1,10] for equation (7.10)

and much effort has been expended in the development of special methods to cope efficiently with stiffness. A definition of stiffness follows.

If the eigenvalues of J are $\lambda_1, \lambda_2, \ldots, \lambda_M$, the system is said to be stiff when

1. $\Re(\lambda_k) < 0, \quad k = 1, 2, \ldots, M$;

2. $\max |\Re(\lambda_k)| \gg \min |\Re(\lambda_k)|$.

The quotient

$$S = \frac{\max |\Re(\lambda_k)|}{\min |\Re(\lambda_k)|}$$

is called the *stiffness ratio* of the system of equations. For equation (7.2) the stiffness ratio is 25. For large values of S the absolute stability requirements of the extreme members of the set of eigenvalues differ markedly. If the same step–size is to apply to all of these it must be governed by the eigenvalue with the largest negative real part, and hence must be much smaller than would be appropriate for the small members of the set. This implies high computational cost and so the search for methods with extended regions of stability is motivated. Note that the presence of large negative eigenvalues does not necessarily indicate large derivatives for which small step-sizes would be expected on asymptotic grounds.

Let us consider a simple linear system which illustrates the problem of stiffness. The equation

$$\begin{pmatrix} ^1y' \\ ^2y' \end{pmatrix} = \frac{1}{5} \begin{pmatrix} 994 & -1998 \\ 2997 & -5999 \end{pmatrix} \begin{pmatrix} ^1y \\ ^2y \end{pmatrix} \qquad (7.11)$$

with initial values $^1y(0) = 1$, $^2y(0) = -2$, has the true solution

$$^1y(x) \;=\; 2e^{-x} - e^{-1000x}$$
$$^2y(x) \;=\; e^{-x} - 3e^{-1000x}.$$

The eigenvalues of the matrix A in this example are -1 and -1000, and so $S = 1000$. The fast transient term (e^{-1000x}) is negligible except for very small values of x, and, if one calculated the solution to modest precision, say 7 decimals, from the analytical expression, there would be no need to evaluate this term when $x > 0.02$. The solution is illustrated in Figure 7.6. Each component has a limiting value of zero. According to

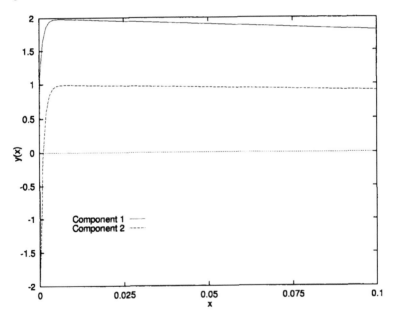

Figure 7.6: Solution of a stiff problem

the stability analysis, the step-size h needed to ensure absolute stability using a two stage RK2 must satisfy $0 < 1000h < 2$. Thus 1000 steps are necessary for $x \in [0, 2]$ and, even if the more stable RK4 were substituted, the number of steps would be reduced only by about 30%. Such an integration would actually be more costly since each of the 4th order steps has twice as many function evaluations as the RK2. If the fast transient component were absent, the same interval could have been covered with tolerable accuracy in a few tens of steps.

Unfortunately, a numerical process cannot ignore the effect of the fast transient even when its effect should be negligible. The fact that there are very large derivative values only when $x < 0.01$ does not mean that the step-size can necessarily be increased for $x > 0.01$. This can be confirmed by commencing a numerical solution at $x = 1$, when the two components of the true solution of equation (7.11), to 7 decimals, are 0.7357589 and 0.3678794. Choosing a step-size $h = 0.1$ and computing 3 steps of constant size with a simple two stage explicit RK2 from (3.15) gives the results in Table 7.1.

Table 7.1: Solution of equation (7.11) by two-stage explicit RK2

x_n	Numerical solution		Error components	
	^1y	^2y	$^1\varepsilon$	$^2\varepsilon$
1.0	0.7357589	0.3678794	0×10^0	0×10^0
1.1	0.6657450	0.3325804	2.809×10^{-6}	-2.907×10^{-4}
1.2	0.0299282	-1.4167276	-5.725×10^{-1}	-1.718×10^0
1.3	-2806.1434	-8419.7938	-2.807×10^3	-8.420×10^3

Since h is much greater than the stability limit for the problem the numerical solution produces values rapidly increasing in magnitude. The equation

$$y' = -y, \ y(1) = 0.7357589,$$

has a similar solution to that of the first component of equation (7.11) for $x \geq 1$. The same RK2 with the same step-size (0.1) produces a fairly accurate result ($\varepsilon(1.3) = 2.939 \times 10^{-4}$). The unboundedness of the numerical solution of (7.11) is due to the fast component, even though it has negligible influence on the true solution.

It should be noted that the application of an embedded RK pair to a stiff system, DOPRI5 for example, usually will give accurate results according to the selected local error criterion. However a large number of steps (c. 330) would be required to cover the range of integration [0,1] for the system (7.11). Fortunately the step–size control algorithm will automatically reduce the step–size to an appropriate value as in the non-linear case from the previous section. Nevertheless, the proportion of rejected steps would be rather high, as in the non-linear example (7.10) shown above.

7.5   Improving stability of RK methods

Given the above phenomena there is ample motivation to seek methods
which provide an extended region of stability. It will be clear that explicit
RK methods, with their stability polynomials, must be limited in this
respect. Nevertheless it is possible to determine RK parameters for the
purpose of optimising the stability characteristics rather than a local
truncation error norm. A more radical approach is to construct RK
formulae which are *implicit*; it will be seen later that such processes
can be unconditionally stable. However the interval of real stability for
explicit RK methods can be enlarged and this is considered below.

In previous chapters it was seen that the addition of stages to a Runge-
Kutta process permits an increase in its algebraic order. The development
of high order methods did not include any consideration of absolute stability properties. As an alternative approach, extra stages can be used to
extend absolute stability, without increasing the asymptotic order of the
process. A convenient method of doing this has been described by van
der Houwen (1972), who has developed a technique closely related to the
construction of minimax polynomials. Rather than considering the error

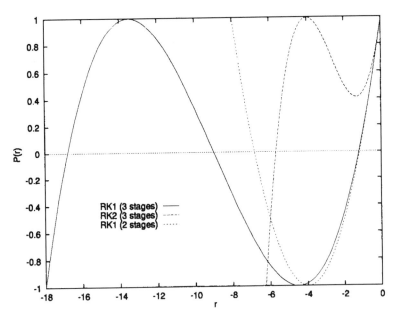

Figure 7.7: Maximal stability polynomials for 2 & 3-stage RK1 and 3-
stage RK2

coefficients of the RK process, van der Houwen places constraints on the
stability polynomial, given in equation (7.9), whose degree is the same as

the number of stages.

For a two-stage Runge-Kutta method, the stability polynomial, obtained by applying the formula to a test equation $y' = \lambda y$, has the form

$$P(r) = 1 + (b_1 + b_2)r + b_2 c_2 r^2,$$

where $r = h\lambda$, h being the step-size. Putting $b_1 + b_2 = 1$ makes the formula first order and consistent, and setting $b_2 c_2 = \frac{1}{2}$ gives an RK2. It is essential to specify a consistent formula, but the second order requirement can be substituted by a maximal stability criterion. Thus one can specify $b_2 c_2 = \beta_2 \neq \frac{1}{2}$, where β_2 will be selected to provide the largest possible absolute stability interval $[-L, 0]$.

Now absolute stability demands $|P(r)| < 1$, and a well known class of orthogonal functions, the Chebyshev polynomials, have a related property. The Chebyshev polynomials can be defined as

$$T_n(x) = \cos n\theta, \ \cos \theta = x, \ x \in [-1, 1],$$

and they may be generated from the recurrence relation

$$T_{j+1}(x) = 2x T_j(x) - T_{j-1}(x), \quad j = 1, 2, 3, \ldots.$$

The nth degree Chebyshev polynomial has $n+1$ alternating extreme values ± 1, $x \in [-1, 1]$. To match the stability specification the Chebyshev polynomial is shifted by taking

$$x = 1 + \frac{r}{\alpha},$$

then

$$x \in [-1, 1] \Longrightarrow r \in [-2\alpha, 0].$$

The Chebyshev stability polynomial recurrence for first order methods has the form

$$P_{j+1}(r) = 2\left(1 + \frac{r}{\alpha}\right) P_j(r) - P_{j-1}(r), \quad j = 1, 2, 3, \ldots$$

and the first three polynomials are

$$P_1(x) = 1 + \left(\frac{r}{\alpha}\right)$$

$$P_2(x) = 1 + 4\left(\frac{r}{\alpha}\right) + 2\left(\frac{r}{\alpha}\right)^2$$

$$P_3(x) = 1 + 9\left(\frac{r}{\alpha}\right) + 12\left(\frac{r}{\alpha}\right)^2 + 4\left(\frac{r}{\alpha}\right)^3.$$

To qualify as stability polynomials for consistent RK formulae, the polynomial of degree m must have $\alpha = m^2$, giving a real negative stability limit $L = 2m^2$. For a two–stage process, one selects P_2 with $\alpha = 4$, which yields

$$P(r) = 1 + r + \tfrac{1}{8}r^2, \tag{7.12}$$

with $|P(r)| \leq 1$, $r \in [-8, 0]$, as indicated in Figure 7.7. The derivation of this result, when only two stages are involved, is quite simple without making the Chebyshev transformation.

At this point it will be noted that the requirement for absolute stability seems to have been extended from the strict inequality (7.5). If $P(r) = 1$ the numerical solution remains bounded, but the appropriate component will not, of course, diminish. When this condition holds the process is said to be *weakly* stable.

To obtain the actual RK coefficients for the optimally stable two stage formula, set $b_2 c_2 = \tfrac{1}{8}$. Selecting $b_2 = 1$ gives $c_2 = \tfrac{1}{8}$, $\quad b_1 = 0$. This two stage RK1 is computationally twice as expensive per step as the first order Euler formula, but it will allow 4 times as large a step-size h when this is governed by stability. Clearly it is very straightforward to extend the Chebyshev technique to any number of stages.

Also the technique can be extended to higher order formulae (van der Houwen, 1972). For example, a three-stage RK2 has the stability polynomial

$$P(r) = 1 + r + \tfrac{1}{2}r^2 + \beta_3 r^3, \quad \beta_3 \neq \tfrac{1}{6}. \tag{7.13}$$

Unfortunately (7.13) is not a Chebyshev polynomial, but an optimal β_3, and the appropriate higher degree coefficients when more stages are allocated, may be computed using the Remez algorithm (Fike, 1968) applicable to the constrained minimax problem. Actually the optimal value is $\beta_3 = \tfrac{1}{16}$, and the polynomial (7.13) with this coefficient is plotted in Figure 7.7. Also illustrated in Figure 7.7 is the maximal stability polynomial, based on $P_3(x)$,

$$P(r) = 1 + r + \tfrac{4}{27}r^2 + \tfrac{4}{729}r^3, \tag{7.14}$$

for a 3 stage RK1.

To obtain suitable RK parameters for a maximally stable 3 stage formula, note from equation (7.9) that

$$\begin{aligned} b_1 + b_2 + b_3 &= 1 \\ b_2 c_2 + b_3 c_3 &= \tfrac{4}{27} \\ b_3 a_{32} c_2 &= \tfrac{4}{729}. \end{aligned} \tag{7.15}$$

Any three of these parameters may be specified and so, for simplicity, choose $a_{31} = b_1 = b_2 = 0$ giving the solution shown in Table 7.2. With a real negative stability limit of -18 this formula permits a step length 9 times that of the classical Euler (RK1) scheme at only 3 times the computational cost. The second order formula with maximal stability at the same computational cost is shown in Table 7.3.

Table 7.2: RK1 in 3 stages with maximal stability

c_i	a_{ij}		b_i
0			0
$\frac{1}{27}$	$\frac{1}{27}$		0
$\frac{4}{27}$	0	$\frac{4}{27}$	1

Table 7.3: RK2 in 3 stages with maximal stability

c_i	a_{ij}		b_i
0			0
$\frac{1}{8}$	$\frac{1}{8}$		0
$\frac{1}{2}$	0	$\frac{1}{2}$	1

For practical application of maximally stable formulae it is possible to combine embedded pairs, just as described in Chapter 5. In this case, however, both members of a pair must be chosen for their stability properties as well as their algebraic order. This is necessary because, if the difference between the two solutions is to be bounded, the step-size must be governed by the smaller of the two stability boundaries. Let us consider a first order formula using the same function evaluations as the RK2 in Table 7.3. The parameters must satisfy the equations 7.15 to give a real negative stability limit -18. In this case the values of c_2, c_3, and a_{32} are as given in Table 7.3, and so the 3 equations can be solved for the three b_i values. The resulting pair is shown in Table 7.4. This

Table 7.4: RK2(1)3S embedded pair

c_i	a_{ij}		$\widehat{b}_i$	b_i
0			0	$\frac{19}{243}$
$\frac{1}{8}$	$\frac{1}{8}$		0	$\frac{608}{729}$
$\frac{1}{2}$	0	$\frac{1}{2}$	1	$\frac{64}{729}$

formula could be used in exactly the same manner as other embedded formulae. The optimal reduction formula (5.3) will be effective provided that the stability boundary (of the RK2) is not exceeded. If a rejection occurs due to absolute instability, it is likely that the estimated error will be extremely large. Consequently, it is important to devise a step-size prediction formula which limits the reduction factor; otherwise the process may be very inefficient due to the choice of unnecessarily small steps alternating with step rejections.

The application of low order formulae such as those detailed above is not recommended for general problems for the reasons discussed in Chapter 4. However there are some types of problems, obviously those in which absolute stability is important, in which they can be valuable. The semi-discretisation technique applied to parabolic partial differential equations is one of these. A characteristic property of these systems is the very small derivative values which result in stability-governed steplengths. Slow cooling in heat conduction modelling provides many such systems.

7.6 Problems

1. Find the stability region for Euler's method

$$y_{n+1} = y_n + hf(x_n, y_n).$$

   Draw a simple diagram to illustrate this region in the complex plane.

2. Obtain by substituting the test equation $y' = \lambda y$ into the formula, the stability polynomial for a 3–stage RK3. Tabulate values of this polynomial for $r = 0, -1, -2, -3$. Hence compute the negative real stability limit correct to 2 decimal places.

3. Find the stability polynomial for the 4-stage Runge-Kutta formula represented in tabular form as

c_i		a_{ij}		b_i
0				0
c_2	a_{21}			0
c_3	0	a_{32}		0
c_4	0	0	a_{43}	1

4. Investigate the absolute stability of the Taylor series method. Find the real negative stability limits for the 5th and 6th order Taylor polynomials.

5. Show that the stability polynomial for the 5th order member of DOPRI5 in Table 5.4 is

$$P(r) = 1 + r + \tfrac{1}{2}r^2 + \tfrac{1}{6}r^3 + \tfrac{1}{24}r^4 + \tfrac{1}{120}r^5 + \tfrac{1}{600}r^6.$$

6. Find the stability polynomials for each member of the RK4(3)5M pair given in Table 5.3. Write a computer program to determine the stability regions for this pair.
   An easy technique is based on 'scanning' the complex plane. You can set $r = x_k + iy_m$, where $x_k = -k\Delta x$, $y_m = m\Delta y$, and compute $P(r)$ for every k, m and suitable Δx and Δy.

7. Investigate the stability requirements for a Runge–Kutta method applied to the non-linear equation

$$y' = -y^2, \quad y(1) = 1.$$

   Use an embedded pair to find the solution in $[1, 10]$ with a local tolerance $T = 0.001$. Compare your results with the analytical solution.

8. Obtain the Jacobian matrix for the differential equation (7.10). Hence find the eigenvalues. Apply the RK2(1)3S embedded pair to this equation and compare the step-size sequence to that plotted in Figure 7.5 for a conventional RK3(2) pair. Devise a modification to the basic step-size control procedure in Figure 5.3 which will improve the efficiency of the method.

9. Use the 4th degree Chebyshev polynomial to construct a 4-stage first order formula with real negative stability interval $[-32, 0]$. Determine the stability region in the complex plane for your formula.

10. Investigate the absolute stability of the implicit formula:

$$y_{n+1} = y_n + hy'_{n+1}.$$

(Note that the stability polynomial arising in explicit cases is replaced by a rational function in this case.)

Chapter 8

Multistep methods

8.1 Introduction

The Taylor and Runge-Kutta methods described in earlier chapters are examples of *Single step* schemes, because each new step is defined solely in terms of its initial point. No other step values are used, even though thousands of steps already may have been computed. Intuition would suggest that one should consider the possibility of making use of previous results, which will exhibit some important properties of the differential system, rather than simply discarding them. The *Multistep* technique, being based on a knowledge of a sequence of earlier numerical values for the solution, has been developed to realize this possibility.

This chapter will serve as an introduction to multistep processes. Some simple formulae will be constructed from Taylor expansions of the local truncation error, the same method in principle as that employed in the Runge–Kutta case. Unlike RK processes, which up to now have been explicit, it is normal to apply implicit multistep schemes for general problems. This leads to the idea of the Predictor–Corrector algorithm which is the most important realization of multistep methods. A number of variations on the Predictor–Corrector (P-C) will be described, and some computational aspects of the method are introduced.

8.2 The linear multistep process

Given the initial value differential system

$$y' = f(x,y), \quad y(x_n) = y_n, \quad y \in \mathbb{R}^M, \tag{8.1}$$

135

the linear multistep formula for the approximation of $y(x_n + h)$ takes the form

$$y_{n+1} = \sum_{i=0}^{p} a_i y_{n-i} + h \sum_{i=-1}^{p} b_i y'_{n-i}, \qquad (8.2)$$

where

$$
\begin{aligned}
y_{n-i} &\simeq y(x_n - ih), \\
y'_{n-i} &= f(x_n - ih, \ y_{n-i}), \\
i &= -1, 0, 1, \ldots, p.
\end{aligned}
$$

The parameters a_i, b_i and p are chosen to satisfy appropriate criteria relating to order and/or stability, in a similar manner to RK parameters.

If $p \geq 1$, then the formula (8.2) can be applied only if the sequence $\{y_{n-i}\}$, $i = 1, 2, \ldots, p$ has been computed first. Thus the multistep process is not self-starting; it requires p *starter* values, discounting the initial value (x_n, y_n), to initialize the multistep method. An obvious technique for the production of starter values would be the use of a Runge–Kutta or a Taylor series method. However, once the process (8.2) has been started, subsequent steps may be computed with the same multistep formula. For the actual implementation of the multistep formula, the parameter b_{-1} is important.

If $b_{-1} = 0$ the multistep formula is *Explicit* and implementation is straightforward, but otherwise the right–hand–side of formula (8.2) contains $y'_{n+1} = f(x_{n+1}, y_{n+1})$, making the formula *Implicit*.

In the implicit case, an iterative method may be employed to solve the non–linear system for y_{n+1}. This may seem to be a severe disadvantage, but in practice it is not so, as will be seen later.

A suitable computational procedure for the use of (8.2), assuming that the initial value is (x_0, y_0) and $p = 2$, is

1. Compute y_1, y_2 possibly using a Runge–Kutta method.

2. Use y_0, y_1, y_2 in (8.2) to compute y_3.

3. Discard y_0.

4. Use y_1, y_2, y_3 in (8.2) to compute y_4.

5. Compute a third step and so on.

Thus it is necessary to retain only three solution points during the computation. The starter values must be equally spaced since the formula (8.2) is based on a constant step-size h. It is possible to change the steplength, or to construct formulae permitting variable step–sizes, but this chapter will be restricted to the construction of some simple formulae and to their implementation in constant step Predictor–Corrector modes.

8.3   Selection of parameters

The parameters a_i, b_i in the multistep formula (8.2), rather like the parameters in RK processes, are determined so that the formula has specific properties. In the multistep case, this determination is much simpler because only total derivatives need to be considered, instead of the large number of elementary differentials which arise from Runge-Kutta expansions. As before, a power series in the step-size, h, is generated for the local truncation error of the formula, which, according to the definition (3.6), is *the amount by which the true solution fails to satisfy the numerical approximation.* For the formula (8.2), this is

$$t_{n+1} = \sum_{i=0}^{p} a_i y(x_n - ih) + h \sum_{i=-1}^{p} b_i y'(x_n - ih) - y(x_n + h) \qquad (8.3)$$

which is easily expanded in a Taylor series. Writing

$$Y^{(j)} = y^{(j)}(x_n), \quad j = 0, 1, \ldots,$$

one obtains

$$
\begin{aligned}
t_{n+1} &= \sum_{i=0}^{p} a_i \left(Y - ihY' + \tfrac{1}{2}(ih)^2 Y'' - \tfrac{1}{6}(ih)^3 Y''' + \cdots \right) \\
&+ h \sum_{i=-1}^{p} b_i \left(Y' - ihY'' + \tfrac{1}{2}(ih)^2 Y''' - \tfrac{1}{6}(ih)^3 Y^{(4)} + \cdots \right) \\
&- \left(Y + hY' + \tfrac{1}{2}h^2 Y'' + \tfrac{1}{6}h^3 Y''' + \cdots \right).
\end{aligned}
$$

Collecting terms to obtain a series of powers of h, the local truncation error is

$$
\begin{aligned}
t_{n+1} &= Y \left(\sum_{i=0}^{p} a_i - 1 \right) \\
&+ hY' \left(\sum_{i=-1}^{p} b_i - \sum_{i=0}^{p} i a_i - 1 \right) \\
&+ h^2 Y'' \left(-\sum_{i=-1}^{p} i b_i + \tfrac{1}{2} \sum_{i=0}^{p} i^2 a_i - \tfrac{1}{2} \right) \\
&+ h^3 Y''' \left(\tfrac{1}{2} \sum_{i=-1}^{p} i^2 b_i - \tfrac{1}{6} \sum_{i=0}^{p} i^3 a_i - \tfrac{1}{6} \right) \\
&+ h^4 Y^{(4)} \left(-\tfrac{1}{6} \sum_{i=-1}^{p} i^3 b_i + \tfrac{1}{24} \sum_{i=0}^{p} i^4 a_i - \tfrac{1}{24} \right) \\
&+ \cdots
\end{aligned}
$$

and assuming that $a_{-1} = 0$, the h^q term can be expressed as

$$\frac{h^q Y^{(q)}}{q!} \left[(-1)^{q-1} \left\{ \sum_{i=-1}^{p} i^{q-1}(qb_i - ia_i) \right\} - 1 \right]. \qquad (8.4)$$

Thus the local truncation error may be written as (c.f. formula(3.10)),

$$t_{n+1} = \sum_{i=0}^{\infty} h^i \varphi_{i-1}(x_n)$$

and, for a formula of order q, the conditions

$$\varphi_k = 0, \quad k = -1, 0, 1, \ldots, q-1 \qquad (8.5)$$

must be satisfied in order to yield a local truncation error of order $q+1$. These conditions should be compared with those conditions (3.21) which are necessary to specify the order of a Runge-Kutta formula. In the multistep case, only one extra condition needs to be imposed to raise the order by one, considerably simpler than the RK alternative.

8.4 A third order implicit formula

Let us consider a multistep formula with $p = 1$ and $b_{-1} \neq 0$. In this case the formula (8.2),

$$y_{n+1} = a_0 y_n + a_1 y_{n-1} + h(b_{-1} y'_{n+1} + b_0 y'_n + b_1 y'_{n-1}),$$

has 5 parameters and is implicit. Using (8.4) one obtains 4 equations of condition (8.5) for $q = 3$ which, when satisfied, will yield a third order formula. These equations are

$$\begin{aligned} a_0 + a_1 &= 1 \\ b_{-1} + b_0 + b_1 - a_1 &= 1 \\ 2b_{-1} - 2b_1 + a_1 &= 1 \\ 3b_{-1} + 3b_1 - a_1 &= 1. \end{aligned}$$

There is no unique solution in this case since there are 5 parameters, one of which may be chosen freely. Taking $a_1 = 0$, the solution is

$$a_0 = 1, \quad b_{-1} = \tfrac{5}{12}, \quad b_1 = -\tfrac{1}{12}, \quad b_0 = \tfrac{2}{3},$$

and the implicit formula is therefore

$$y_{n+1} = y_n + \tfrac{1}{12} h(5y'_{n+1} + 8y'_n - y'_{n-1}). \qquad (8.6)$$

For the moment, no justification for the choice is given. Substituting the parameters in the error coefficient (8.4) with $q = 4$ yields the principal term in the local truncation error for the formula (8.6)

$$h^4 \varphi_3 = \tfrac{1}{24} h^4 y^{(4)}(x_n).$$

Since (8.6) is implicit, the value of y_{n+1} must be obtained by an iterative method if $f(x, y)$ is non–linear in y. Assuming that starter values are already known, the formula (8.6) can be written as

$$y_{n+1} = \tfrac{5}{12} h f(x_{n+1}, y_{n+1}) + c_n, \qquad\qquad \text{' (8.7)}$$

where

$$c_n = y_n + \tfrac{1}{12} h(8 y'_n - y'_{n-1})$$

depends only on these starter values. In the scalar case the solution of equation (8.7) by simple iteration is the simplest option. An iterative process

$$z_{k+1} = g(z_k), \quad k = 0, 1, 2, \dots$$

will converge to a root of the equation $z = g(z)$ if $|g'(z)| < 1$ near the root. For the equation (8.7), this condition becomes

$$\left| \tfrac{5}{12} h \frac{\partial f}{\partial y} \right| < 1,$$

or

$$h < 2.4 \left(\left| \frac{\partial f}{\partial y} \right| \right)^{-1}. \qquad\qquad (8.8)$$

The important feature here is that the inequality (8.8) guarantees that the iteration will converge *provided that the step–size is small enough*. The Newton–Raphson procedure could be substituted in order to reduce the number of iterations. In this case convergence would be dependent only on the starting approximation for y_{n+1}.

If f is not scalar the equation (8.7) represents a non-linear system which could be tackled in a similar manner, although a Newton iteration would then require evaluation of the Jacobian of f.

To illustrate a method of implementing the multistep formula (8.6), one step in the solution of the equation

$$y' = y^3 - y, \quad y(0) = \tfrac{1}{2},$$

with $h = 0.1$, is carried out. In this case, the analytical solution $y(0.1) = 0.463032$ is used to provide a starter value, and then the multistep formula yields an estimate of $y(0.2)$. More realistically a numerical method would be used to estimate the starter value. Using the formula (8.7) one obtains

$$y_{n+1} = 0.441906 + \frac{0.5(y_{n+1}^3 - y_{n+1})}{12},$$

which is solved by simple iteration. In Table 8.1 the iterates are labelled $z_0, z_1, \ldots$, and the first approximation is taken arbitrarily to be 0.43. Since the step-size is much smaller than the critical value predicted by (8.8), convergence is rapid. Note that convergence is towards the solution

Table 8.1: Solution using an implicit method

k	x_k	y_k	y_k'
$n-1$	0	0.5	-0.375
n	0.1	0.463032	-0.363759
	0.2	$z_0 = 0.43$	
		$z_1 = 0.427302$	
		$z_2 = 0.427353$	
$n+1$	0.2	0.427352	-0.349305

of the derived non–linear equation rather than towards the true solution of the differential equation, which is $y(0.2) = 0.427355$ correct to six significant figures. In this example the initial approximation was a guess; a more systematic method for finding this would be an advantage. For this reason we consider next an explicit multistep procedure.

8.5   A third order explicit formula

Setting $b_{-1} = 0$ in the multistep formula (8.2) with $p = 1$ yields an explicit process containing 4 parameters:

$$y_{n+1} = a_0 y_n + a_1 y_{n-1} + h(b_0 y_n' + b_1 y_{n-1}').$$

These parameters are chosen, uniquely in this instance, to yield a third order formula. Using the order conditions (8.4) with $q \leq 3$ gives

$$
\begin{aligned}
a_0 + a_1 &= 1 \\
b_0 + b_1 - a_1 &= 1 \\
-2b_1 + a_1 &= 1 \\
3b_1 - a_1 &= 1.
\end{aligned}
$$

These equations have the solution $a_0 = -4$, $a_1 = 5$, $b_0 = 4$, $b_1 = 2$. A third order, explicit multistep formula is thus

$$y_{n+1} = -4y_n + 5y_{n-1} + h(4y_n' + 2y_{n-1}'), \qquad (8.9)$$

for which the principal error term (see (8.4) with $q = 4$) is

$$h^4\varphi_3 = -\tfrac{1}{6}h^4y^{(4)}(x_n).$$

This is 4 times as large as the corresponding term for the implicit formula (8.6). Applying the new formula to the same differential equation as before, using the analytical solution to provide a starter accurate to 6 decimals, yields the results in Table 8.2. This calculation is direct and

Table 8.2: Solution using an explicit method

k	x_k	y_k	y'_k
$n - 1$	0	0.5	-0.375
n	0.1	0.463032	-0.363759
$n + 1$	0.2	0.427368	-0.349312

involves no iterations. However the error is considerably larger than that of the previous iterative calculation. The two errors, which are local since only one step has been computed, are -0.000003 and 0.000013 respectively for the implicit and explicit formulae. The principal local truncation error terms of the two formulae differ by a factor -4, and so the calculation here supports the analysis, under the assumption of principal term dominance.

8.6   Predictor–Corrector schemes

In the above application of the implicit method (8.6), the first approximation to the solution was 'guessed'. Since there exists now an explicit formula of the same order of accuracy it is convenient to use this instead to *predict* a first approximation. The implicit process can then be employed to refine or *correct* the solution. The number of iterations necessary for convergence will be reduced, since the starting point will be much more accurate than before. In the example given in Table 8.3, the first 'correction' or iteration using 0.427368 to start yields $z_1 = 0.427351$, almost the same as was obtained after 4 iterations earlier. Since the iterations converge to a root of the non–linear equation (8.6) rather than the solution point of the differential equation, there is no guarantee that further iterations will improve the approximation. In this case a second iteration does give a small improvement, but by an amount less than the actual local error. Many practitioners prefer to compute just a single iteration of the Corrector formula.

The classical Predictor–Corrector scheme for the third order pair derived above may be defined as shown in Table 8.4. If it is necessary

Table 8.3: Solution using a Predictor-Corrector method

k	x_k	y_k	y'_k
$n-1$	0	0.5	-0.375
n	0.1	0.463032	-0.363759
	0.2	$z_0 = 0.427368$ $z_1 = 0.427351$	-0.349312
$n+1$	0.2	0.427352	-0.349305

Table 8.4: Predictor-Corrector algorithm

P:	$P_{n+1} = -4y_n + 5y_{n-1} + h(4y'_n + 2y'_{n-1}).$
E:	$y'_{n+1} = f(x_{n+1}, P_{n+1}).$
C:	$C_{n+1} = y_n + \frac{1}{12}h(5y'_{n+1} + 8y'_n - y'_{n-1}).$
E:	$y'_{n+1} = f(x_{n+1}, C_{n+1}),$
	If Converged $y_{n+1} = C_{n+1}$, start next step. Else Repeat C :

to iterate on the corrector k times, the procedure would be designated PE(CE)^k. Many variations on this basic algorithm are possible. An important property of the Predictor–Corrector (PECE) form is that it demands only 2 derivative evaluations per step, irrespective of the order of accuracy. For this reason it is usually seen to be a cheaper computational scheme than a Runge–Kutta process of order greater than 2. This simple comparison does not, of course, remain valid if the step–sizes of the two types of method are significantly different.

8.7   Error estimation

As with other methods for solving differential equations, the estimation of error in a Predictor–Corrector scheme is crucial for the determination of an appropriate step–size. Fortunately, in this case, where there are two estimates of the solution at each step, this is quite straightforward. Different approaches to the error estimation process depend on the choice of Predictor-Corrector formula pairs. With an extra parameter available (b_{-1}) the Corrector can be made to have a higher order than the Predictor, the difference between them forming an error estimate similar to that based on an embedded RK pair. Such a pair would, in effect, be implemented in the local extrapolation mode with the higher order result providing the initial value for the next step. The alternative strategy is to choose Predictor and Corrector formulae to be of the same order, as in the example of the last section. In this case the principal error terms of each formula can be estimated by considering the difference between the predicted and the corrected values. Since the Corrector error is then available, it is usual to perform local extrapolation.

Let us consider the error estimation process when both formulae have the same order. Suppose that the local true solution satisfying (x_n, y_n) is $u(x)$; then the local errors of the Predictor and the Corrector values are

$$P_{n+1} - u(x_{n+1}) = e_{n+1}^{(P)}$$
$$C_{n+1} - u(x_{n+1}) = e_{n+1}^{(C)},$$

and $P_{n+1} - C_{n+1} = e_{n+1}^{(P)} - e_{n+1}^{(C)}$. Now, for a method of order q, the local errors can be written as

$$e_{n+1}^{(P)} = A_P h^{q+1} y_n^{(q+1)} + O(h^{q+2})$$
$$e_{n+1}^{(C)} = A_C h^{q+1} y_n^{(q+1)} + O(h^{q+2}),$$

where A_C, A_P are independent of h. This gives

$$P_{n+1} - C_{n+1} = (A_P - A_C) h^{q+1} y_n^{(q+1)} + O(h^{q+2})$$
$$h^{q+1} y_n^{(q+1)} = \frac{P_{n+1} - C_{n+1}}{(A_P - A_C)} + O(h^{q+2}).$$

Neglecting terms of degree $q + 2$ and above, it is easy to make estimates of the local errors of the predicted and corrected values. These are

$$e_{n+1}^{(P)} \simeq \frac{A_P}{(A_P - A_C)}(P_{n+1} - C_{n+1}), \qquad (8.10)$$

$$e_{n+1}^{(C)} \simeq \frac{A_C}{(A_P - A_C)}(P_{n+1} - C_{n+1}). \qquad (8.11)$$

For the 3rd order example in the last section $A_P = -\frac{1}{6}$, $A_C = \frac{1}{24}$, and so

$$e_{n+1}^{(C)} \simeq -\frac{1}{5}(P_{n+1} - C_{n+1}), \quad e_{n+1}^{(P)} \simeq \frac{4}{5}(P_{n+1} - C_{n+1}).$$

Using the values from Table 8.3 above the difference between Predictor and Corrector is

$$P_{n+1} - C_{n+1} = 0.000016,$$

Table 8.5: Modified Predictor-Corrector algorithm

P:	$P_{n+1} = -4y_n + 5y_{n-1} + h(4y'_n + 2y'_{n-1}).$
M:	$P_{n+1}^{(m)} = P_{n+1} - \frac{4}{5}(P_n - C_n).$
E:	$y'_{n+1} = f(x_{n+1}, P_{n+1}^{(m)}).$
C:	$C_{n+1} = y_n + \frac{1}{12}h(5y'_{n+1} + 8y'_n - y'_{n-1}).$
M:	$C_{n+1}^{(m)} = C_{n+1} + \frac{1}{5}(P_{n+1} - C_{n+1}).$
E:	$y'_{n+1} = f(x_{n+1}, C_{n+1}^{(m)}), \quad y_{n+1} = C_{n+1}^{(m)}.$
	Start next step with P

and so the two error estimates are $e_{n+1}^{(C)} \simeq -0.000003$, $e_{n+1}^{(P)} \simeq 0.000013$, which agree with the actual errors to 6 decimal places.

The process of adding on the error estimate or *modifier* is sometimes called Milne's device but, to be consistent with earlier terminology, it is better described as local extrapolation. In some cases it is useful to extrapolate also the predicted value, for which an equally valid error estimate is possible. However the Predictor modifier depends on the corrected value, which is not available at the intermediate evaluation stage. A simple way of compensating for this is to take

$$P_{n+1} - C_{n+1} \simeq P_n - C_n, \quad n \geq 1,$$

and a *modified* PECE scheme is designated in Table 8.5. This has the effect of extending the number of starter values since the point (x_{n-1}, y_{n-1}) contributes to P_n and C_n. There are stability implications of such a feature and these will be considered later. The Predictor extrapolation is of no value when the Corrector is iterated, but local extrapolation of the Corrector is widely practiced.

To illustrate the effects of the different variants of Predictor–Corrector, three schemes have been applied to the logistic equation

$$y' = y(2 - y), \quad y(0) = 1, \quad h = 0.2. \tag{8.12}$$

The true solution to this problem is $y(x) = 2/(1 + e^{-2x})$ and so this has been used to calculate the global errors. The starter step uses the 4th order RK45M formula given in Table 4.2, and Table 8.6 shows values obtained in the first 5 steps for the three cases. The results were obtained from the Fortran program described in the next section. It is clear that, in this case, both extrapolations yield an improvement in the numerical solution.

Table 8.6: Predictor-Corrector results for logistic equation

i	x_i	P_i	$P_i^{(m)}$	C_i	$C_i^{(m)}$	$|y_i - y(x_i)|$
\multicolumn{7}{c}{No extrapolation}						
0	0				1	$0.000E+00$
1	0.200				1.19738	$0.684E-07$
2	0.400	1.37933		1.38019		$0.241E-03$
3	0.600	1.53490		1.53772		$0.674E-03$
4	0.800	1.66092		1.66518		$0.114E-02$
5	1.000	1.75828		1.76309		$0.149E-02$
\multicolumn{7}{c}{Extrapolate corrector}						
1	0.200				1.19738	$0.684E-07$
2	0.400	1.37933		1.38019	1.38002	$0.696E-04$
3	0.600	1.53569		1.53750	1.53714	$0.874E-04$
4	0.800	1.66297		1.66445	1.66415	$0.115E-03$
5	1.000	1.76079		1.76191	1.76169	$0.953E-04$
\multicolumn{7}{c}{Extrapolate predictor and corrector}						
1	0.200				1.19738	$0.684E-07$
2	0.400	1.37933	1.37933	1.38019	1.38002	$0.696E-04$
3	0.600	1.53569	1.53637	1.53744	1.53709	$0.384E-04$
4	0.800	1.66320	1.66460	1.66422	1.66402	$0.160E-04$
5	1.000	1.76123	1.76205	1.76165	1.76156	$0.313E-04$

8.8 A Predictor–Corrector program

Coding the Predictor–Corrector algorithm in Fortran 90 is quite straight-forward. As in earlier codes the vector operations simplify the programming and render transparent the numerical procedure.

The program is a little more complicated than the RK equivalent because a starting mechanism has to be provided. In the example PmeCme listed in full in Appendix B a fourth order RK formula is used for this purpose and the MODULE defining the differential system contains the logistic problem (8.12). A module defining the two–body gravitational system, as provided for the RKpair program, is completely compatible with the PmeCme program.

To highlight the PC process, a fragment of the program PmeCme is presented in Figure 8.1. The provision of Corrector iteration is fairly straightforward and later programs in Appendix B contain this option.

```
. . . . . . .
ALLOCATE(a(0: p), b(0: p), ac(0: p), bc(0: p + 1))
!   a, b are Predictor parameters, ac, bc for Corrector
!   p is number of starters
. . . . . . .

. . . . . . .
ALLOCATE(y(0: p, 1: neqs), v(neqs), w(neqs), pmc(neqs),&
            f(0: p+1, 1: neqs))   ! neqs is system dimension
. . . . . . . .

. . . . . . . .
w = MATMUL(a, y) + h*MATMUL(b, f(0: p, :)) ! Predictor
f(p + 1, :) = Fcn(x + h, w + pmod*pmc)     ! Evaluate
v = MATMUL(ac, y) + h*MATMUL(bc, f)        ! Correct
pmc = w - v                                ! Modifier
w = v + cmod*pmc
x = x + h
CALL Output(x, w)
f(0: p - 1, :) = f(1: p, :)                ! Update f's
y(0: p - 1, :) = y(1: p, :)                ! New starters
y(p, :) = w                                ! New step y
f(p, :) = Fcn(x, w)                        ! New y'
. . . . . . . . .
```

Figure 8.1: Program fragment for Predictor–Corrector with modifiers

8.9   Problems

1. Carry out Taylor series expansions to find the local truncation error
   of each of the multistep formulae given below.

   (a) $y_{n+1} = y_n + \frac{1}{2}h(y'_{n+1} + y'_n)$;

   (b) $y_{n+1} = y_n + \frac{1}{12}h(23y'_n - 16y'_{n-1} + 5y'_{n-2})$;

   (c) $y_{n+1} = y_{n-3} + \frac{4}{3}h(2y'_n - y'_{n-1} + 2y'_{n-2})$.

2. Find the parameters a_i, b_i, $i = -1, 0, 1$, so that the implicit formula

$$y_{n+1} = \sum_{i=0}^{1} a_i y_{n-i} + h \sum_{i=-1}^{1} b_i y'_{n-i}$$

has order 4. Given $y' = -\frac{1}{2}y^3$, $y(0) = 1$, $y(0.2) = 0.912871$, use
your formula to estimate $y(0.4)$. Employ a third order Predictor to
compute an initial approximation.

3. Find an explicit 4th order multistep formula depending on two starter values by substituting $p = 2$ in equation(8.2).

4. Obtain the appropriate modifier formulae to be used for local extrapolation, when the formula in problem 1(b) above is used as a Predictor together with the Corrector (8.6). Given the differential equation and initial value

$$y' = y^3 - y, \ y(0) = \tfrac{1}{2},$$

use a suitable RK formula to determine two starter values with $h = 0.15$. Implement the P–C pair to compute 2 further steps with $h = 0.15$:

(a) in PECE mode with local extrapolation;

(b) in $PE(CE)^k$ mode with local extrapolation.

Set the tolerance to be 10^{-7} for the iterative mode.

5. Produce a data file, compatible with PmeCme, to contain the P-C coefficients for the pair from the last problem. Test your file on the same differential equation to verify the data.

Rewrite the external MODULE units for PmeCme to solve

(a) $\quad y' = \dfrac{3x}{y} - 2xy, \quad y(0) = 2, \ x \in [0,3];$

(b) $\quad y'' = -1/y^2, \quad y(0) = 1, \ y'(0) = 0, \ x \in [0,1].$

In each case, choose a suitable step–size by considering the corrector error estimate.

6. Make alterations to the program PmeCme to enable the Corrector to be iterated k times. Compute solutions to the logistic problem (8.12) with $k = 2$ and with $k = 3$ but do not modify the Predictor value. Compare the errors in each case with those of Table 8.6. Also compare the results with those produced by a third order RK process.

Chapter 9

Multistep formulae from quadrature

9.1 Introduction

The construction of multistep formulae from Taylor expansions is somewhat easier than the derivation of RK processes. Inevitably, this method is devoted to the pursuit of algebraic order rather than any other attribute such as absolute stability. The number of starter values p determines the maximum order of a process. This is $2p + 1$ for the explicit formula and $2p + 2$ for the corresponding implicit case. Such formulae are unique and, based on criteria developed in Chapter 4 for RK methods, may seem to be optimal. Unfortunately the importance of asymptotic behaviour is not as great as for single step methods. This forces a reduction in maximal order so that other properties can be enhanced. Perhaps the situation is not radically different to the use of more RK stages than the minimum required for a given order.

A particularly convenient method of generating related formulae of different orders is based on numerical quadrature. The families of formulae so developed share certain useful characteristics which makes their computer implementation rather attractive. In particular, their coefficients arise in an iterative way, and so a change in order involves either truncation or addition of a term. Another feature depends on the relation of a multistep process to the polynomial interpolant. The variation of steplength in the solution of differential equations is just as important for multistep formulae as for single step methods. Following a change of step-size, an interpolant must be employed to compute new starter values. Interpolant-based quadrature formulae are better suited to this procedure than other multistep methods.

In this chapter we consider the systematic derivation and also the

149

implementation of the Adams processes. These are even more ancient than Runge-Kutta processes, dating back to the middle of the nineteenth century. J.C. Adams was the celestial mechanician who predicted the existence of the planet Neptune in 1845. Adams formulae remain popular to this day.

9.2   Quadrature applied to differential equations

Consider the usual ordinary differential system

$$y'(x) = f(x, y(x)), \quad y(x_0) = y_0, \quad y \in \mathbb{R}^m \qquad (9.1)$$

and the sequence $\{x_k\}$, such that

$$x_k = x_{k-1} + h, \quad k = 1, 2, \dots.$$

Integrating the differential equation (9.1) with respect to x and with appropriate limits,

$$\int_{x_n}^{x_{n+1}} y'(x)dx = \int_{x_n}^{x_{n+1}} f(x, y(x))dx,$$

gives the formula

$$y(x_{n+1}) = y(x_n) + \int_{x_n}^{x_{n+1}} f(x, y(x))dx. \qquad (9.2)$$

Since the integrand in (9.2) is generally expressed in terms of x and y, the integral cannot be evaluated directly. However it is easy to approximate $f(x, y(x))$ using a polynomial interpolant based on known values of $(x_k, y(x_k))$. The integration of this interpolant yields a quadrature formula.

A polynomial V_p of degree p can be constructed to satisfy

$$V_p(x_k) = f(x_k, y(x_k)), \quad k = n - p, n - p + 1, \dots, n - 1, n,$$

so that

$$f(x, y(x)) = V_p(x) + E_p(x), \qquad E_p(x_k) = 0, \quad k = n - p, \dots, n - 1, n,$$

where E_p is called the error of the polynomial interpolation. Thus, from the formula (9.2), one obtains

$$y(x_{n+1}) = y(x_n) + \int_{x_n}^{x_{n+1}} V_p(x)dx + \int_{x_n}^{x_{n+1}} E_p(x)dx, \qquad (9.3)$$

which is the basis of the multistep formula

$$y_{n+1} = y_n + \int_{x_n}^{x_{n+1}} v_p(x)dx, \qquad (9.4)$$

where $v_p(x_k) = f_k = f(x_k, y_k)$, $n - p \le k \le n$, interpolates the approximate function values based on the numerical solution. Using equation (9.3), and recalling its definition from Chapter 3, the local truncation error of this multistep formula will be

$$
\begin{aligned}
t_{n+1} &= y(x_n) + \int_{x_n}^{x_{n+1}} V_p(x)dx - y(x_{n+1}) \\
&= -\int_{x_n}^{x_{n+1}} E_p(x)dx.
\end{aligned}
$$

It is straightforward to write down the Lagrangian interpolant v_p and, by varying p, a family of multistep formulae can be derived from equation (9.4). Although Adams formulae are based specifically on this relation, other families can be constructed by changing the limits of the integration. Implicit formulae can be derived by extending the interpolating points. Quadrature formulae such as the ones implied above, which cover an interval beyond the extent of the data, are known as *open* formulae. Those arising in implicit processes will be of the more familiar *closed* type, such as the trapezoidal rule or Simpson's rule.

9.3   The Adams–Bashforth formulae

The Lagrange polynomial interpolant has many forms. The divided difference variant is perhaps the most useful but, in the present context, the Newton–Gregory backward difference formula is preferred. Defining the backward differences by

$$
\begin{aligned}
\nabla f_k &= f_k - f_{k-1}, \\
\nabla^{i+1} f_k &= \nabla^i f_k - \nabla^i f_{k-1}, \\
i = 1, 2, \ldots, p; \qquad k &= n, n - 1, \ldots, n - p + i,
\end{aligned}
$$

the classical formulation is based on the finite difference array shown in Table 9.1. Writing $x = x_n + sh$, the interpolating polynomial based on equally spaced data can be expressed as

$$v_p(x) = f_n + s\nabla f_n + \binom{s+1}{2}\nabla^2 f_n + \cdots + \binom{s+p-1}{p}\nabla^p f_n, \quad (9.5)$$

where

$$\binom{m}{k} = \frac{m(m-1)\cdots(m-k+1)}{k!}$$

Table 9.1: Backward difference array

x_{n-p}	f_{n-p}			
		∇f_{n-p+1}		
x_{n-p+1}	f_{n-p+1}		$\nabla^2 f_{n-p+2}$	
		∇f_{n-p+2}		$\nabla^3 f_{n-p+3}$
$\ldots$	$\ldots$		$\ldots$	
		$\ldots$		$\ldots$
$\ldots$	$\ldots$		$\ldots$	
		∇f_{n-2}		$\nabla^3 f_{n-1}$
x_{n-2}	f_{n-2}		$\nabla^2 f_{n-1}$	
		∇f_{n-1}		$\nabla^3 f_n$
x_{n-1}	f_{n-1}		$\nabla^2 f_n$	
		∇f_n		
x_n	f_n			

is a binomial coefficient. Also it may be shown (e.g. Hildebrand, 1974) that the error of this interpolant can be expressed as

$$E_p(x) = \binom{s+p}{p+1} h^{p+1} f^{(p+1)}(x_n, y(x_n)) + O(h^{p+2}). \qquad (9.6)$$

Since $s = (x - x_n)/h$, equation (9.4) becomes

$$y_{n+1} = y_n + h \int_0^1 v_p(x) ds.$$

Thus a multistep formula, called an Adams–Bashforth process, can be written in the form

$$
\begin{aligned}
y_{n+1} &= y_n + h \left\{ f_n + \nabla f_n \int_0^1 s\, ds + \cdots + \nabla^p f_n \int_0^1 \binom{s+p-1}{p} ds \right\} \\
&= y_n + h \left(\gamma_0 f_n + \gamma_1 \nabla f_n + \gamma_2 \nabla^2 f_n \cdots + \gamma_p \nabla^p f_n \right), \qquad (9.7)
\end{aligned}
$$

where

$$\gamma_0 = 1, \quad \gamma_k = \int_0^1 \binom{s+k-1}{k} ds, \quad k = 1, \ldots, p.$$

Performing these integrations in individual fashion is straightforward but unnecessary. A particularly convenient recurrence scheme (see Henrici (1962)) is generated as follows.

Consider the infinite series

$$G(x) = \sum_{k=0}^{\infty} \gamma_k x^k.$$

Substituting the expression for γ_k gives

$$G(x) = \int_0^1 \sum_{k=0}^{\infty} x^k \binom{s+k-1}{k} ds$$

$$= \int_0^1 (1-x)^{-s} ds,$$

using the binomial theorem. Integration yields

$$G(x) = -\frac{x}{(1-x)\log(1-x)},$$

or

$$-\frac{\log(1-x)}{x} G(x) = \frac{1}{1-x}$$

which, on expansion, gives

$$(1 + \tfrac{1}{2}x + \tfrac{1}{3}x^2 + \cdots)(\gamma_0 + \gamma_1 x + \gamma_2 x^2 + \cdots) = (1 + x + x^2 + \cdots).$$

Comparing coefficients on each side of this equation gives

$$
\begin{aligned}
\gamma_0 &= 1 \\
\gamma_1 + \tfrac{1}{2}\gamma_0 &= 1 \\
\gamma_2 + \tfrac{1}{2}\gamma_1 + \tfrac{1}{3}\gamma_0 &= 1 \\
\cdots &= 1 \\
\gamma_p + \tfrac{1}{2}\gamma_{p-1} + \cdots + \tfrac{1}{p+1}\gamma_0 &= 1.
\end{aligned}
\tag{9.8}
$$

The first few coefficients are given in Table 9.2. Using equation (9.6),

Table 9.2: Adams-Bashforth coefficients

k	1	2	3	4	5	6
γ_k	$\frac{1}{2}$	$\frac{5}{12}$	$\frac{3}{8}$	$\frac{251}{720}$	$\frac{95}{288}$	$\frac{19087}{60480}$

it can be seen that the local truncation error of the Adams-Bashforth formula has the principal term

$$t_{n+1} = -h^{p+2}\gamma_{p+1} y^{(p+2)}(x_n),$$

and so a third order formula member of the family will require the choice $p = 2$, and can be expressed as

$$y_{n+1} = y_n + h(f_n + \tfrac{1}{2}\nabla f_n + \tfrac{5}{12}\nabla^2 f_n).$$

If it is preferred Adams-Bashforth formulae can be written in terms of function ($y' = f$) values

$$y_{n+1} = y_n + h \sum_{k=0}^{p} \beta_k f_{n-k},$$

and a recurrence relation can be constructed. Setting the initial values

$$\beta_j = \alpha_j = 0, \quad j = 1, \dots, p, \quad \alpha_0 = 1,$$

one computes the sequence

$$\left.\begin{aligned} \alpha_j &= \alpha_j - \alpha_{j-1} \\ \beta_j &= \beta_j + \alpha_j \gamma_i \end{aligned}\right\}, \quad j = i, i-1, \dots, 1; \ i = 1, 2, \dots, p, \qquad (9.9)$$

and $\beta_0 = \sum_{i=0}^{p} \gamma_i$.

The case $p = 2$ gives

$$y_{n+1} = y_n + \tfrac{1}{12}h(23f_n - 16f_{n-1} + 5f_{n-2}), \qquad (9.10)$$

which has principal local truncation error

$$-\tfrac{3}{8}h^4 y^{(4)}(x_n).$$

Clearly this can be used as a Predictor together with formula (8.6) as a Corrector, but it does need two starter values rather than the single starter required by (8.9), which had the same order of accuracy.

   An attractive feature of the Adams–Bashforth formulae is the ease with which one can generate them. The integration scheme used here is even more direct than the Taylor expansion scheme employed earlier; the coefficients are obtained directly from definite integrals of polynomials or, more conveniently, from simple recurrence formulae (9.8).

9.4   The Adams–Moulton formulae

Families of implicit formulae for use as Correctors may be derived in a manner similar to that of the last section by adding the point (x_{n+1}, f_{n+1}) to the finite difference table. Setting $x = x_{n+1} + sh$, the appropriate polynomial interpolant of degree p is now

$$v_p^*(x) = f_{n+1} + s\nabla f_{n+1} + \binom{s+1}{2}\nabla^2 f_{n+1} + \cdots + \binom{s+p-1}{p}\nabla^p f_{n+1},$$
$$(9.11)$$

and since, with the new definition of s,

$$\int_{x_n}^{x_{n+1}} dx = h \int_{-1}^{0} ds,$$

the implicit formula is

$$y_{n+1} = y_n + h \left(\gamma_0^* f_{n+1} + \gamma_1^* \nabla f_{n+1} + \gamma_2^* \nabla^2 f_{n+1} \cdots + \gamma_p^* \nabla^p f_{n+1} \right),$$
(9.12)

where

$$\gamma_0^* = 1, \quad \gamma_k^* = \int_{-1}^{0} \binom{s+k-1}{k} ds, \quad k = 1, \ldots, p,$$

and the local truncation error has the principal term

$$t_{n+1} = -h^{p+2} \gamma_{p+1}^* y^{(p+2)}(x_n).$$

It is possible to derive recurrence relations similar to (9.8) for γ_k^* but this may be avoided by using the result

$$\gamma_k^* = \gamma_k - \gamma_{k-1}, \quad k = 1, 2, \ldots.$$
(9.13)

This relation may be justified from the properties of binomial coefficients.

As before, a third order formula will require a choice $p = 2$, but this does not involve the earliest starter value if the corresponding explicit Adams–Bashforth formula is employed as the Predictor. The Adams–Moulton coefficients up to order 6 are presented in Table 9.3.

Table 9.3: Adams-Moulton coefficients

k	1	2	3	4	5	6
γ_k^*	$-\frac{1}{2}$	$-\frac{1}{12}$	$-\frac{1}{24}$	$-\frac{19}{720}$	$-\frac{3}{160}$	$-\frac{863}{60480}$

As an example we take $p = 2$ again, which gives

$$y_{n+1} = y_n + h(f_{n+1} - \tfrac{1}{2}\nabla f_{n+1} - \tfrac{1}{12}\nabla^2 f_{n+1}),$$

and substitution of f values into the backward differences gives

$$y_{n+1} = y_n + \tfrac{1}{12}h(5f_{n+1} + 8f_n - f_{n-1}),$$

which is the same as the implicit formula (8.6) derived in the previous chapter. A notable feature of the Adams–Moulton family of implicit formulae is that, for a particular order, the principal term of the local truncation error is smaller than that of the corresponding explicit formula. As will be seen later, the implicit family also has improved absolute stability.

Good stability characteristics make Adams Predictor–Corrector pairs the most commonly–used formulae of their type. Some implementations specify the Corrector to be one order higher than the Predictor, thus making use of the same set of starter values. This scheme has computational advantages over that described earlier. Consider the Adams-Bashforth Predictor (9.7) and Adams-Moulton Corrector (9.12) with $p = 2$. The difference between corrected and predicted values can be written

$$C_{n+1} - P_{n+1} = h[f_{n+1} + \gamma_1^* \nabla f_{n+1} + \gamma_2^* \nabla^2 f_{n+1} - (f_n + \gamma_1 \nabla f_n + \gamma_2 \nabla^2 f_n)].$$

Using the relation (9.13) this simplifies to

$$
\begin{aligned}
C_{n+1} - P_{n+1} &= h(\nabla f_{n+1} + (\gamma_1 - \gamma_0)\nabla f_{n+1} - \gamma_1 \nabla f_n \\
&\quad + (\gamma_2 - \gamma_1)\nabla^2 f_{n+1} - \gamma_2 \nabla^2 f_n) \\
&= h\gamma_2 \nabla^3 f_{n+1}.
\end{aligned}
$$

Thus, in terms of backward differences, the Corrector requires one extra term, when the 'predicted' differences have been evaluated.

Alternatively the two formulae may have the same order with modifiers being used to estimate the error and also to improve the solution. Using the formulae (8.10) and (8.11), extrapolated values for the Adams–Bashforth–Moulton Predictor–Corrector pairs with the same order are

$$P - \frac{\gamma_{p+1}}{\gamma_p}(P - C) \quad \text{and} \quad C - \frac{\gamma_{p+1}^*}{\gamma_p}(P - C).$$

9.5   Other multistep formulae

The use of quadrature formulae for multistep processes can be extended beyond the Adams families which can be considered as a special case of formulae based on the integral relation

$$y(x_{n+1}) = y(x_{n-r}) + \int_{x_{n-r}}^{x_{n+1}} f(x, y(x))dx. \tag{9.14}$$

Substituting $r = 0$ in (9.14) gives the Adams formula (9.2). Most other cases are of academic interest only but the Nyström explicit family with $r = 1$ is worthy of consideration. Its implicit relations are the Milne-Simpson Correctors. Their general form is

$$P: \quad y_{n+1} = y_{n-1} + h \sum_{k=0}^{p-1} \eta_k \nabla^k f_n, \tag{9.15}$$

$$C: \quad y_{n+1} = y_{n-1} + h \sum_{k=0}^{p-1} \eta_k^* \nabla^k f_{n+1}, \tag{9.16}$$

where

$$\eta_0 = 2, \quad \eta_k = \int_{-1}^{1} \binom{s+k-1}{k} ds, \quad k = 1, \ldots, p,$$

for the pth order member, with a similar expression for the implicit coefficient. Substituting $p = 3$ (or $p = 4$) in equation (9.16) gives the
well-known Simpson's rule quadrature formula

$$y_{n+1} = y_{n-1} + \tfrac{1}{3}h(f_{n+1} + 4f_n + f_{n-1}). \tag{9.17}$$

A relation between the coefficients η_k and those for the Adams formulae
(γ_k) is easy to derive.

9.6   Varying the step–size

With RK or other single–step methods it is easy to vary the steplength,
since the increment formula is based only on a single point. For most practical problems, efficient numerical solution requires a variable steplength
since the principal local error function is often subject to large variations
in the interval of solution. Hence a constant step-size would be restricted
to a value related to a maximum of the error function, implying a large
number of steps for problems such as the eccentric 2-body problem which
was highlighted in Chapter 5. Many of these steps would be much smaller
than required by the local error function.

Multistep formulae, of the type considered so far, are much less convenient than RK processes, since any change in step-size requires the
computation of new starter information. For a steplength change from h
to θh, these new values will be at $x_{n-k}^* = x_n - k\theta h, \; k = 1, \ldots, p-1$.
Adams methods are less demanding than most other multistep processes
since they include only a single y value, the most recent, in addition to
the $p - 1$ derivative or function values. If accuracy is to be maintained
following a revision of steplength, the mechanism adopted for the new
starter information must be of a sufficiently high algebraic order. Fortunately the application of an interpolation process is relatively easy if
use is made of the available, equally spaced back values. In this respect
the multistep method seems to have an advantage over RK processes,
since such interpolation could be employed for dense output provision
when necessary. To yield the highest order of accuracy, the Hermite
interpolant, employing y and y' values, could be chosen. Given the data

$$(x_{n-k}, y_{n-k}, f_{n-k} = y_{n-k}'), \quad k = 0, 1, \ldots, p-1,$$

a polynomial of degree $2p - 1$ is available. Ignoring the effect of rounding
error, the use of a Hermite interpolant to yield new back values should
introduce errors smaller than those associated with the multistep integrator.

The foregoing assumes that the steplength is to be reduced. For an increase in step-size by interpolation it seems that a contribution from extra back values is essential. This implies a greater storage requirement but clearly there will have to be a modest upper limit on the step-size ratio.

A number of alternative schemes for changing the step-size have been examined by Krogh (1973), who recommends the use of an interpolant based only on the past derivative values. For an order p integrator, there are p derivatives available, and so a Newton-Gregory interpolant of degree $p - 1$ can be employed to approximate directly the new f values. Note that there is no need for new function evaluations, as would be required if only new back values of y were computed. Since the increment function in the Adams formula is multiplied by h, errors of order p in the f values imply a local error of $O(h^{p+1})$ when the formula is evaluated.

The actual control of the steplength in the multistep case is very similar to RK step-size control. Just as RK embedding, the Predictor-Corrector scheme leads to the estimation of local error, and the formula (5.3) could be used to predict the next step-size following a successful or an unsuccessful step. Nevertheless, changing the steplength after each step cannot be recommended in the multistep application. This would be relatively expensive and it would contribute unnecessary rounding errors. Krogh (1973) restricts changes to doubling and halving the steplength. Thus, provided that the error estimate δ_{n+1}, at step n, satisfies

$$\|\delta_{n+1}\| \in [T/2^{p+1}, T],$$

where T is the tolerance, the step-size will be unchanged. The lower bound of this interval represents the local error value which should permit the step-size to be doubled. Larger values of $\|\delta_{n+1}\|$ will result in rejection and halving of the steplength, whereas smaller values will cause the next steplength to be doubled. Krogh's method is very straightforward and aims to replace the backward differences based on step-size h

$$(\nabla f_n, \nabla^2 f_n, \ldots, \nabla^{p-1} f_n),$$

with a new set

$$(\nabla_H f_n, \nabla_H^2 f_n, \ldots, \nabla_H^{p-1} f_n),$$

for halving the step-size, or the set

$$(\nabla_D f_n, \nabla_D^2 f_n, \ldots, \nabla_D^{p-1} f_n),$$

for doubling the steplength. Clearly, this technique will be most direct when the Adams formulae are presented in terms of backward difference rather than function values. However, the conversion to and back from differences form is easily accomplished when the multistep formula is expressed as (9.10).

Suppose that the polynomial $v(x)$ interpolates the available f values for an Adams method of order p; then

$$v(x_{n-k}) = f_{n-k}, \quad x_{n-k} = x_n - kh, \quad k = 0, 1, \ldots, p - 1.$$

Then the backward differences satisfy

$$\nabla^k v(x_n) = \begin{cases} \nabla^k f_n, & k < p \\ 0, & k \geq p. \end{cases}$$

Consider first the step-size doubling process, which will assume that $v(x)$ is of degree $p - 1$. The 'doubled' backward difference is defined to be

$$\nabla_D v(x_n) = v(x_n) - v(x_{n-2}),$$

and simple manipulation gives

$$\nabla_D v(x_n) = 2\nabla v(x_n) - \nabla^2 v(x_n),$$

or, in operator form,

$$\nabla_D \equiv 2\nabla - \nabla^2.$$

A second application of this operator gives

$$\nabla_D^2 \equiv 4\nabla^2 - 4\nabla^3 + \nabla^4.$$

This process can be extended to the required order, remembering that all differences above $p - 1$ will be zero. Thus for a method of order 4 the relations are

$$\nabla_D^2 \equiv 4\nabla^2 - 4\nabla^3, \quad \nabla_D^3 \equiv 8\nabla^3.$$

The process of halving is an inverse operation and the relations

$$\begin{aligned} \nabla &\equiv 2\nabla_H - \nabla_H^2 \\ \nabla^2 &\equiv 4\nabla_H^2 - 4\nabla_H^3 \\ \nabla^3 &\equiv 8\nabla_H^3 \end{aligned}$$

are obtained. The equations arising from these are easily solved to give the 'halved' differences.

This method is employed for step-size changing in the Adams program listed in Appendix B.

Many modern computer codes based on multistep methods tend to employ variable *coefficient* formulae rather than explicit interpolations as detailed here. These multistep formulae are based on divided differences rather than the simple differences used above. Such schemes will be examined in a later chapter.

9.7   Numerical results

The program Adams from Appendix B has been applied to the orbit problem considered in Chapter 5. Selecting the same eccentricity (0.5) as
before, the system of four first order equations was solved for $x \in [0, 20]$
with a range of local error tolerances. In this case the tolerance require-
ment was applied to the difference $|P - C|$, rather than to the Corrector
error. The program permits a choice of local extrapolation and Corrector
iteration strategies. In PECE mode, each formula is applied once only,
giving two derivative (function) evaluations per step. Alternatively the
Corrector can be iterated to satisfy some convergence property. This will
usually give better accuracy, particularly at lax tolerances, but at extra
computational cost. Local extrapolation can be applied to either or both
Predictor and Corrector formulae. However there seems little point in
extrapolating the Predictor if Corrector convergence is sought.

In addition to the evaluation of different P-C strategies, it is natural
to compare the Adams P-C method with the embedded RK process which
was described in Chapter 5. Since the two types of methods are radically
different, their true computational costs in CPU time must be compared.
Three curves are plotted in Figure 9.1. In the 5th order PECE case, both

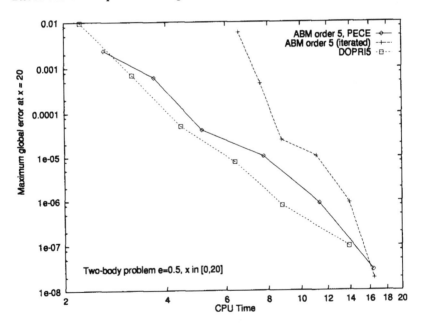

Figure 9.1: Efficiency curves for 5th order Adams multistep and DOPRI5

formulae are extrapolated, whereas in the iterated case, only the Corrector is extrapolated. Consequently, one may regard the multistep pairing

as yielding order 6 rather than 5. Over the range of global error presented, the RK5 (DOPRI5 from Table 5.4) seems to be the most efficient. Perhaps predictably, the iterated P-C method is relatively expensive at lax tolerances and the steeper slope of its efficiency curve indicates a higher order than 5. Some readers may be surprised at the efficiency of the RK method which needs 6 function evaluations per step compared with 2 for the PECE process. In fact the PECE process requires 386 steps with $T = 10^{-5}$; to give a similar global error, with $T = 10^{-6}$, DOPRI5 needs only 111 steps, thus compensating for the extra cost per step. With $T = 10^{-5}$, the step control procedure works very efficiently, with only 6 steps being rejected and only 8 step doublings. The range of step-sizes was [0.025, 0.10].

9.8   Problems

1. Given the differential equation and starter values

   $$y' = x + y, \ y(0) = 1, \ y(0.1) = 1.110342, \ y(0.2) = 1.242805,$$

   use the 3rd order Adams P-C pair to compute two steps of size $h = 0.1$.

2. By direct integration, obtain a recurrence relation for the Adams-Moulton coefficients, similar to that given for the Adams-Bashforth case in equation (9.8).

3. Use the integral relation

   $$y(x_{n+1}) = y(x_{n-r}) + \int_{x_{n-r}}^{x_{n+1}} f(x, y(x))dx$$

   to derive a 4th order explicit formula when $r = 3$. Determine the modifier to be used for local extrapolation when your formula is combined with the 4th order Milne-Simpson corrector (9.17).

   Use Taylor series to determine the local truncation error of the Milne–Simpson implicit formula.

4. Use the properties of binomial coefficients to show that the relation connecting explicit and implicit Adams coefficients

   $$\gamma_k - \gamma_{k-1} = \gamma_k^*, \ k > 0,$$

   is satisfied. Also show that this result leads to

   $$\gamma_k = \sum_{i=0}^{k} \gamma_i^*.$$

5. Show that the relations from the previous problem lead to the property

   $$\sum_{k=0}^{p-1} (\gamma_k^* \nabla^k f_{n+1} - \gamma_k \nabla^k f_n) = \gamma_{p-1} \nabla^p f_{n+1}.$$

   Hint: try substitution in the formula with $p = 3$ first.

6. The Adams P-C pair of order p can be written in difference form as

   $$P_{n+1} = y_n + h \sum_{k=0}^{p-1} \gamma_k \nabla^k f_n,$$

   $$C_{n+1} = y_n + h \sum_{k=0}^{p-1} \gamma_k^* \nabla^k f_{n+1},$$

where $f_{n+1} = f(x_{n+1}, P_{n+1})$. Use the property from the previous problem to show that

$$C_{n+1} = P_{n+1} + h\gamma_{p-1}\nabla^p f_{n+1}.$$

Also obtain the local extrapolation formula for the corrector in this case.

7. Find integral expressions for η_k, η_k^*, the coefficients in the Nyström and Milne-Simpson formulae. Show that

$$\eta_k = \gamma_k + \gamma_k^*, \quad k \geq 0,$$

and hence tabulate the coefficients up to order 6. Also show that

$$\eta_k - \eta_{k-1} = \eta_k^*,$$

and hence tabulate up to the same order the Milne-Simpson coefficients. Determine the local extrapolation formulae corresponding to these P-C pairs.

8. Given the equation and starter values

$$\begin{aligned} y' &= y^3 - y, \; y(0) = 0.5, \\ y(0.1) &= 0.463032, \; y(0.2) = 0.427355, \; y(0.3) = 0.393251, \end{aligned}$$

use the 4th order Adams P-C in difference form to estimate $y(0.4)$. By employing the Krogh method for halving and doubling, compute also approximations to $y(0.35)$ and $y(0.5)$.

9. Apply the program **Adams** to the two-body problem with eccentricity $e = 0.7$. Investigate the error behaviour when the tolerance is varied, and when the order is varied. Modify the program to record the maximum and the minimum step-sizes.

   Also modify the program to use directly the backward difference versions of the Adams formulae. Verify your changes by comparison with the earlier calculations.

10. Use **MODULE** units from §B.1 for **Adams** to solve the logistic equation

$$y' = y(2 - y), \quad y(0) = 0.1, \quad x \in [0, 5].$$

11. The equations of motion of a spaceprobe moving in the Earth–Moon system are

$$\begin{aligned} \ddot{x} &= 2\dot{y} + x - \frac{E(x + M)}{r_1^3} - \frac{M(x - E)}{r_2^3} \\ \ddot{y} &= -2\dot{x} + y - \frac{Ey}{r_1^3} - \frac{My}{r_2^3}, \end{aligned}$$

where

$$r_1 = \sqrt{(x+M)^2 + y^2}, \quad r_2 = \sqrt{(x-E)^2 + y^2},$$
$$M = 1/82.45, \quad E = 1 - M,$$

with initial conditions

$$x(0) = 1.2, \quad \dot{x}(0) = 0, \quad y(0) = 0,$$
$$\dot{y}(0) = -1.049357509830320.$$

These equations are based on a rotating coordinate system whose period is 2π, in which the Earth is fixed at $(-M, 0)$ and the Moon at $(E, 0)$. The spaceprobe has negligible mass. Write a suitable MODULE defining this system for Adams, and solve it for

$$t \in [0, 6.192169331319640],$$

using a local error tolerance $T = 10^{-5}$. Plot a graph depicting the orbit in the rotating coordinate system.

This problem has a periodic solution and the spaceprobe should return to its starting coordinates at the end of the given interval. Since the system is unstable, small errors will destroy the periodicity and so it is necessary to use high precision in the numerical solution.

12. Solve the system of equations from the previous problem with revised initial values

$$x(0) = 1.15, \quad \dot{y}(0) = 0.0086882909,$$

with $M = 0.012277471$, for $t \in [0, 29.4602]$. In the rotating coordinate system the orbit for this problem is shown in Figure 9.2. Also plot the orbit in the non-rotating coordinate system together with the orbits (circular) of the Earth and the Moon.

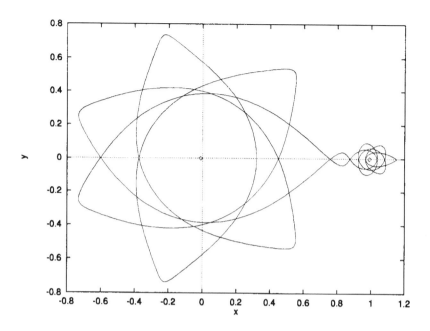

Figure 9.2: Solution of Problem 12 (rotating system)

Chapter 10

Stability of multistep methods

10.1 Introduction

The subject of multistep stability is much more complicated than the corresponding Runge–Kutta topic. With RK methods their consistency guarantees convergent processes, but this is not the case in the multistep situation, in which rounding error at any step of the calculation must be damped out by subsequent steps. Consequently it may not be possible to improve the accuracy of a numerical solution by selecting a reduced step-size.

Since the RK increment formula reduces to a first order difference equation when the scalar linear test problem, $y' = \lambda y$, is solved, the absolute stability properties are easy to determine. One can determine a step-size which ensures that the numerical solution y_n, when $\lambda < 0$, satisfies

$$\lim_{n \to \infty} y_n = 0,$$

the absolute stability property. In the same circumstances, multistep formulae yield higher order difference equations and, as a consequence, more solutions are possible.

While single-step methods sometimes are required to be absolutely stable, they exhibit no instability when applied to differential equations with positive transients. However, multistep processes can contain characteristic values which destroy all numerical solutions. This unfortunate property is associated with the amplification of rounding errors present in starter values. In this chapter we introduce the concept of *zero-stability* which is essential for any process to be convergent, whatever the application. Other important stability criteria for multistep processes, including absolute stability, are presented and discussed. In particular, the influence of the P-C mode of operation on stability is highlighted.

10.2 Some numerical experiments

To illustrate the phenomenon of zero–instability the third order Hermite Predictor (8.9),

$$y_{n+1} = -4y_n + 5y_{n-1} + h(4y'_n + 2y'_{n-1}),$$

so-called because it is equivalent to extrapolation with a two-point Hermite formula, is applied to the linear problem

$$y' = y, \quad y(0) = 1, \quad x \in [0, 0.2], \qquad (10.1)$$

with two step-sizes $h = 0.02$ and 0.01. The results are presented in Table 10.1. The single starter value in this case was obtained from a double precision computation of (e^x), and so was accurate to about 15 significant figures. Since $y' > 0$ this problem does not require an absolutely stable formula. With the larger of the two step-sizes the solution soon becomes poor, and the alternating sign of the error is obvious. Close examination will show that errors at successive steps differ by a factor approximately -5. Halving the step-size might be expected to reduce the error, at any point, by a factor about 8, since the formula is known to have algebraic order 3. The results do not confirm this. In fact, over the same interval of integration, the error growth with $h = 0.01$ is catastrophic. Halving the step-size has had the opposite of the desired effect. Note the same factor between successive errors. Also recall that a Predictor-Corrector solution employing formula (8.9) was illustrated in Table 8.6. The results gave no hint of the behaviour highlighted in Table 10.1.

To understand the rapidly growing oscillating error, it is necessary to consider the difference equation arising from the application of the Hermite Predictor to $y' = y$. This equation is of order two, being

$$y_{n+1} + 4(1 - h)y_n - (5 + 2h)y_{n-1} = 0, \qquad (10.2)$$

and, assuming solutions of the form $y_n = \theta^n$, direct substitution gives a quadratic characteristic equation

$$\theta^2 + 4(1 - h)\theta - (5 + 2h) = 0.$$

The roots of this quadratic are

$$\theta_0, \theta_1 = 2h - 2 \pm \sqrt{9 - 6h + 4h^2}, \qquad (10.3)$$

Table 10.1: Errors of a zero-unstable formula (Hermite Predictor)

x_n	$h = 0.02$ $y_n - y(x_n)$	$h = 0.01$ $y_n - y(x_n)$
0.00	$0.000E+00$	$0.000E+00$
0.01		$0.000E+00$
0.02	$0.000E+00$	$-0.168E-08$
0.03		$0.496E-08$
0.04	$-0.271E-07$	$-0.298E-07$
0.05		$0.141E-06$
0.06	$0.786E-07$	$-0.710E-06$
0.07		$0.352E-05$
0.08	$-0.473E-06$	$-0.175E-04$
0.09		$0.869E-04$
0.10	$0.222E-05$	$-0.432E-03$
0.11		$0.215E-02$
0.12	$-0.111E-04$	$-0.107E-01$
0.13		$0.530E-01$
0.14	$0.547E-04$	$-0.264E+00$
0.15		$0.131E+01$
0.16	$-0.271E-03$	$-0.651E+01$
0.17		$0.324E+02$
0.18	$0.134E-02$	$-0.161E+03$
0.19		$0.799E+03$
0.20	$-0.660E-02$	$-0.397E+04$

and so the solution of the difference equation, giving the numerical solution at any step, can be written

$$y_n = C_0 \theta_0^n + C_1 \theta_1^n,$$

where C_0, C_1 are determined from the starter values y_0 and y_1. Now application of the binomial theorem to the characteristic roots (10.3) gives

$$\theta_0 = e^h + O(h^4), \quad \theta_1 = -5 + O(h).$$

The first of these provides an approximation to the true solution, as does the corresponding result as given by equation (7.4) in a Runge-Kutta application. In this case the other solution, termed parasitic, provides a component which increases more rapidly, and which will eventually dominate the value of y_n, *even when h is zero*. This must occur, since it is not possible to eliminate rounding error. The phenomenon is called *zero-instability*, or sometimes *strong instability*.

10.3   Zero–stability

Actually it is possible to analyse zero-stability for a multistep formula by applying it to the trivial problem

$$y' = 0, \ y(0) = g.$$

This will become clear by setting $h = 0$ in equation (10.2), which is then the difference equation corresponding to an integration with zero step-size, or to the trivial case. The solution of this equation is obviously $y(x) = g$, but its importance here is that it can be considered to complement any non–trivial equation and, in the multistep application, its starter values will be corrupted by small rounding errors depending on the precision of the arithmetic in use. Substituting $y' = 0$ in the multistep formula

$$y_{n+1} = \sum_{i=0}^{p} a_i y_{n-i} + h \sum_{i=-1}^{p} b_i y'_{n-i} \qquad (10.4)$$

gives

$$y_{n+1} - a_0 y_n - a_1 y_{n-1} - \cdots - a_p y_{n-p} = 0, \qquad (10.5)$$

a linear, constant coefficient difference equation of order $p + 1$. Of course *exactly* equal starter values will yield an *exact* solution to equation (10.5) since for any linear multistep formula, the consistency condition $\sum a_i = 1$ must be satisfied (see §8.3). The characteristic polynomial for the difference equation (10.5) is

$$\rho(\theta) = \theta^{p+1} - a_0\theta^p - a_1\theta^{p-1} - \cdots - a_p, \qquad (10.6)$$

giving the solution

$$y_n = \sum_{i=0}^{p} c_i \theta_i^n, \qquad (10.7)$$

assuming no repeated zeros. The polynomial $\rho(\theta)$ is usually called the *first* characteristic polynomial of the multistep formula (10.4). If one of the roots θ_r, say, of $\rho(\theta) = 0$ is repeated, then it may be shown that θ_r^n and $n\theta_r^n$ are linearly independent solutions of the difference equation.

For any consistent multistep formula, one of the roots of $\rho(\theta) = 0$ may be determined immediately since $\sum a_i = 1$. This is easily seen by substituting $\theta = 1$ in equation (10.6), yielding

$$\rho(1) = 1 - a_0 - a_1 - \cdots - a_p.$$

This root, usually denoted by θ_0, is called the *principal* root of the first characteristic equation of the multistep process.

A convergent formula will satisfy

$$\lim_{h \to 0} y_n = g, \ n = 1, 2, \ldots,$$

but it is easily seen that the solution (10.7) implies that y_n is unbounded if any distinct root θ_i has modulus greater than unity, or if a repeated root has unit modulus. Thus, convergence requires

$$\theta_0 = 1, \quad |\theta_i| \le 1, \quad i = 1, 2, \ldots, p,$$

and any multistep method satisfying this property is zero–stable.

It is instructive to examine the zero–stability of some of the formulae encountered earlier. First, note that all Adams formulae, of explicit and implicit types, reduce to $y_{n+1} = y_n$ on substitution of $y' = 0$. Thus the first characteristic polynomial effectively is reduced to

$$\rho(\theta) = \theta - 1,$$

yielding the only non-zero, and principal root, $\theta = 1$. Clearly this satisfies the zero–stability criterion. The reader is invited to repeat the test of the last section using the third order Adams-Bashforth formula in place of the Hermite Predictor (8.9).

The first characteristic polynomial for (8.9) is

$$\rho(\theta) = \theta^2 + 4\theta - 5 = (\theta + 5)(\theta - 1),$$

and has zeros 1 and -5, values (c.f. (10.3)) which fail to satisfy the zero–stability property. In this case the solution of $y' = 0$ will be

$$y_n = c_0(1)^n + c_1(-5)^n,$$

which is clearly unbounded for non–zero c_1. From the point of view of a practical computation, the rounding errors of the starter values will eventually be amplified without limit. The problem (10.1) has been used already to demonstrate the non-convergence of formula (8.9).

Given that the Hermite Predictor is zero–unstable, consider the derivation of a formula with $p = 1$, by the method of Chapter 8, which deliberately avoids this undesirable property. The explicit formula can be written as

$$y_{n+1} = a_0 y_n + a_1 y_{n-1} + h(b_0 y'_n + b_1 y'_{n-1})$$

and using the equations of condition (8.4) with $q \le 3$ one obtains, for a third order scheme,

$$
\begin{aligned}
a_0 + a_1 &= 1 \\
b_0 + b_1 - a_1 &= 1 \\
-2b_1 + a_1 &= 1 \\
3b_1 - a_1 &= 1.
\end{aligned}
$$

which is actually a member of the Nyström explicit family.

The importance of zero–stability is clear from the results in Table 10.2 which derive from the application of the zero-stable formula (10.8) to the equation (10.1). These are directly comparable to the data in Table 10.1.

A number of features are very clear. First, it is obvious that the third order formula gives better accuracy than (10.8) for a few steps before the effect of zero–instability becomes large. Thereafter the second order formula is more accurate, with a smoothly increasing error which does not reverse sign at each step. For the zero–stable formula, halving the step reduces the error at corresponding points by an approximate factor 4, exactly as would be predicted for a convergent formula of order 2.

At this stage, it is natural to question the wisdom of introducing the subject of explicit multistep formulae in Chapter 8 with a candidate that seems now to be quite useless. In spite of its failure in the current context, the Hermite formula was quite successful as one half of a Predictor-Corrector pair applied to the logistic equation in §8.7. There is no contradiction between this success and the results obtained more recently. In the first application, the explicit formula (8.9) was paired with an implicit Adams formula, which we know now to be zero-stable, and thus one needs to analyse the stability properties of the *combined* P-C formulae. Considering the trivial test problem, the effect of the Predictor method is not transmitted to the Corrector, which therefore determines on its own the zero-stability of the combination.

10.4 Weak stability theory

Zero or strong stability must be a property of any multistep process in any application. Other factors which are associated with non-convergence of multistep processes are generally placed under the umbrella of weak stability properties. These properties are often conditional, as was the absolute stability of RK methods.

Consider the application of the multistep formula (8.2) to the linear scalar test problem

$$y' = \lambda y, \ y(x_0) = y_0,$$

to which Runge-Kutta methods were applied in Chapter 7. Substituting for y' in the multistep formula (10.4) yields

$$\sum_{i=-1}^{p} (a_i + h\lambda b_i)y_{n-i} = 0, \tag{10.9}$$

where $a_{-1} = -1$. This is a difference equation of order $p + 1$ whose solution may be written

$$y_n = \sum_{i=0}^{p} C_i\theta_i^n, \tag{10.10}$$

where θ_j are the roots, assumed distinct, of the characteristic polynomial equation

$$P(\theta) = \sum_{i=-1}^{p} (a_i + h\lambda b_i)\theta^{p-i} = 0. \qquad (10.11)$$

P is called the *Stability* polynomial of the multistep formula and setting $\lambda = 0$ or $h = 0$ in (10.11) yields the first characteristic polynomial ρ (10.6). Now $\rho(1) = 0$, and so $P(\theta)$ must have a zero θ_0 such that

$$\lim_{h \to 0} \theta_0 = 1.$$

To investigate this root, which is termed the *principal* root of P, it is assumed that

$$\theta_0 = 1 + d_1 h + d_2 h^2 + \cdots$$

and substituting in the polynomial (10.11) gives

$$\begin{aligned}
P(\theta_0) &= \sum_{i=-1}^{p} (a_i + h\lambda b_i)(1 + d_1 h + d_2 h^2 + \cdots)^{p-i} \\
&= \sum_{i=-1}^{p} [a_i + h(\lambda b_i + a_i d_1 (p - i)) + O(h^2)].
\end{aligned}$$

Now consistency demands $\sum_{i=-1}^{p} a_i = 0$, and so

$$P(\theta_0) = \sum_{j=1}^{\infty} r_j h^j,$$

where the r_j depend on a_i and b_i. If θ_0 is to be a zero for general h, then $r_j = 0$ for all j. Setting $r_1 = 0$ gives

$$\sum_{i=-1}^{p} (\lambda b_i + a_i d_1 (p - i)) = 0,$$

yielding

$$d_1 = \frac{\lambda \sum\limits_{i=-1}^{p} b_i}{\sum\limits_{i=-1}^{p} (i - p) a_i} = \frac{\lambda \sum\limits_{i=-1}^{p} b_i}{1 + \sum\limits_{i=0}^{p} i a_i}.$$

Using equation (8.4), and assuming that the multistep formula (10.4) is consistent, it is easy to see that $d_1 = \lambda$.

Further analysis shows that for formulae of order at least q,

$$d_j = \frac{\lambda^j}{j!}, \quad j = 1, 2, \ldots, q$$

and so $\theta_0 = e^{h\lambda} + O(h^{q+1})$.

Thus the principal root of the stability polynomial corresponds to the true solution, $y(x) = y_0 e^{\lambda(x-x_0)}$. Since all the other roots $\theta_1, \ldots, \theta_p$ also will contribute to the solution (10.10), it is essential that the principal root is dominant. If this is not true then the method cannot be convergent. This leads to a definition of relative stability for a multistep method:

- If θ_0 is the principal zero of the stability polynomial of a multistep formula with p starter values, the formula is *Relatively Stable* if

$$|\theta_0| > |\theta_j|, \quad j = 1, 2, \ldots, p. \tag{10.12}$$

When $\lambda < 0$, an appropriate numerical solution demands absolute stability as introduced in Chapter 7, which leads to the definition:

- If θ_j, $j = 0, 1, 2, \ldots, p$ are the zeros of the stability polynomial of a multistep formula with p starter values, the formula is *Absolutely Stable* if

$$|\theta_j| < 1, \quad j = 0, 1, 2, \ldots, p. \tag{10.13}$$

The two stability properties here are often satisfied by specification of a sufficiently small step–size. In this respect they are different to the zero–stability property which cannot be affected by reducing the steplength.

It should be stated that the concept of relative stability is not relevant to single step methods since no spurious solutions arise with such schemes.

10.5 Stability properties of some formulae

As with RK formulae, absolute stability regions in the complex plane relating to multistep methods can be plotted. However, the investigation of relative stability needs a consideration of the separate roots. This is easily achieved for low order formulae, and here we examine the 2nd order Adams–Bashforth Predictor formula (AB2)

$$y_{n+1} = y_n + \tfrac{1}{2}h(3y'_n - y'_{n-1}). \tag{10.14}$$

Applying AB2 to the scalar test equation, $y' = \lambda y$, where λ is real, gives

$$y_{n+1} - y_n\left(1 + \tfrac{3}{2}h\lambda\right) + \tfrac{1}{2}h\lambda y_{n-1} = 0,$$

and substituting $H = \frac{1}{2}h\lambda$, the stability polynomial can be written as

$$P(\theta) = \theta^2 - \theta(1 + 3H) + H.$$

This has zeros

$$\theta = \frac{1}{2}(1 + 3H \pm \sqrt{1 + 2H + 9H^2}). \tag{10.15}$$

Carrying out a binomial expansion, it is easy to see that the '+' sign gives the principal root θ_0, which leads to an approximation of the true solution.

Relative stability requires $|\theta_0| > |\theta_1|$, there being only two roots of $P(\theta) = 0$ in this case. Since

$$w = \sqrt{1 + 2H + 9H^2} = \sqrt{(3H + \frac{1}{3})^2 + \frac{8}{9}}$$

is essentially real and positive, the relative stability requirement becomes

$$|1 + 3H + w| > |1 + 3H - w|,$$

which will be satisfied if $1 + 3H > 0$ or, in terms of the usual variables,

$$h\lambda > -\frac{2}{3}.$$

Relative stability is assured for all positive step-sizes and positive λ, but negative values of λ place an upper bound on h.

The simple form of the characteristic roots permits a similar analysis of absolute stability for AB2. The condition (10.13) splits into two conditions in this case. Taking the principal root first, this yields the inequality

$$|1 + 3H + w| < 2, \tag{10.16}$$

which must be treated separately for the positive and negative cases. Recall that absolute stability is concerned chiefly with those instances where $H < 0$.

1. For a positive root

$$1 + 3H + w \;<\; 2$$
$$w \;<\; 1 - 3H.$$

For $H < 0$, both sides of this inequality are positive. Squaring gives

$$w^2 \;<\; (1 - 3H)^2$$
$$1 + 2H + 9H^2 \;<\; 1 - 6H + 9H^2$$
$$H \;<\; 0,$$

and the inequality (10.16) is satisfied.

2. When the root is negative

$$-2 \; < \; 1 + 3H + w$$
$$-3 - 3H \; < \; w,$$

which is obviously true if $H > -1$ since $w > 0$. For $H < -1$ it is easy to verify by squaring that the inequality still holds.

For the other parasitic root absolute stability demands

$$|1 + 3H - w| < 2, \tag{10.17}$$

and, as before, two cases must be considered.

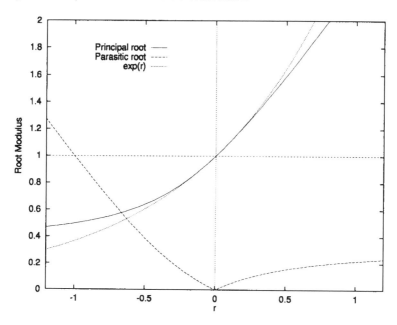

Figure 10.1: Characteristic roots for AB2 $(r = h\lambda)$

1. For a positive root

$$1 + 3H - w \; < \; 2$$
$$3H - 1 \; < \; w,$$

which is satisfied for all negative H.

2. When the root is negative

$$-2 \; < \; 1 + 3H - w$$
$$w \; < \; 3 + 3H.$$

Both sides of this inequality are positive if $H > -1$, and so squaring yields

$$1 + 2H + 9H^2 \; < \; 9 + 18H + 9H^2$$
$$-\frac{1}{2} \; < \; H$$
$$-1 \; < \; h\lambda.$$

Combining the results, the negative real interval of absolute stability for the second order Adams–Bashforth Predictor (10.14) is evidently

$$h\lambda \in [-1, 0],$$

and so it is possible to choose a sufficiently small step to guarantee this property.

In Figure 10.1 is plotted the characteristic roots of AB2 in the interval of interest. The dominance of the principal root when $h\lambda > \frac{2}{3}$ is clear as is the absolute stability limit of -1. At this critical value, it is the parasitic root which attains the bound. To cater for more general systems of differential equations, the complex behaviour of the roots must be analysed. The absolute stability region in the complex plane for AB2 is plotted in Figure 10.2. The region is not dissimilar in shape to the

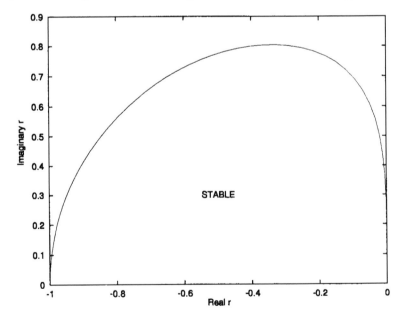

Figure 10.2: Stability Region for AB2 $(r = h\lambda)$

corresponding region for a two–stage RK2 but the real limit in this case is only half that of the RK2.

The midpoint Predictor (MP2), derived earlier (10.8) on the basis of zero-stability, may be analysed in the same way as AB2. Its characteristic roots are

$$h\lambda \pm \sqrt{1 + (h\lambda)^2},$$

and these ensure relative stability for positive $h\lambda$, but also that no region of absolute stability exists. The successful application of MP2 to problem (10.1), as shown in Table 10.2, confirms the relative stability just predicted, but it is instructive to view the application of the formula to a problem which demands absolute stability as well. The example taken is

$$y' = -y, \ y(0) = 1, \ x \in [0,6], \ h = 0.25.$$

The problem is solved also with AB2 which is absloutely stable for stepsizes smaller than unity. The two numerical solutions are compared with the true solution in Figure 10.3. As in the earlier illustration of zero-instability, the unstable formula exhibits growing oscillations as steps proceed. The effect grows less quickly than the instability in Table 10.1, because the offending root

$$\theta_1 = \frac{-1 - \sqrt{17}}{4}$$

has a smaller modulus in this case. For the stable formula the error diminishes since the numerical solution has a limiting value of zero, as does the true solution.

10.6 Stability of Predictor–Corrector pairs

A disappointing feature of multistep processes is the deterioration of stability properties with increasing order. Thus the expected step-size increase normally associated with increasing order is threatened by the shrinking region of stability. The Adams-Bashforth formulae exhibit very well this feature, and Figure 10.4 illustrates the shrinkage from order 2 to order 6.

To highlight the difficulty, consider the application of AB5 to the linear equation $y' = -y$, previously solved by AB2 and MP2. Since AB5 has a real stability limit -0.16, the step-size of $h = 0.25$ used for the second order formulae would be *too large for this 5th order process*. In effect it implies that a solution with modest accuracy is unattainable with the higher order formula. Obviously it is possible to solve the problem more accurately with a reduced step-size ($h < 0.16$), using the 5th order formula.

Proceeding to implicit formulae, the Adams-Moulton Correctors are much more stable than their explicit relatives. Nevertheless they also

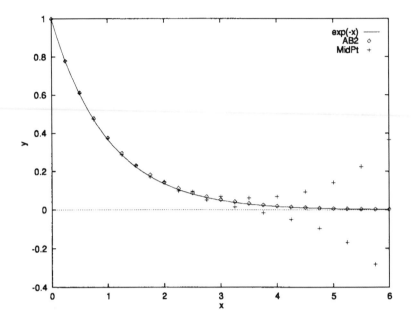

Figure 10.3: Solution of $y' = -y$ with AB2 and MP2

have stability regions which shrink with increasing order. The analysis of their stability proceeds along the same lines as that for AB methods. The first order Adams-Moulton formula (AM1) is usually called the implicit Euler method and yields a first order difference equation when applied to the scalar linear test problem. Therefore

$$y_{n+1} = y_n + hy'_{n+1} \implies P(\theta) = (1 - h\lambda)\theta - 1, \qquad (10.18)$$

and so the single root is

$$\theta = (1 - h\lambda)^{-1},$$

which is less than unity for all $h\lambda < 0$. The second order member of the Adams-Moulton set (AM2),

$$y_{n+1} = y_n + \tfrac{1}{2}h(y'_{n+1} + y'_n), \qquad (10.19)$$

also possesses a first degree stability polynomial with the single zero

$$\theta = \frac{1 + \tfrac{1}{2}h\lambda}{1 - \tfrac{1}{2}h\lambda},$$

which gives absolute stability at any point in the negative half of the complex plane. Such methods as AM1 and AM2 are said to be A-stable,

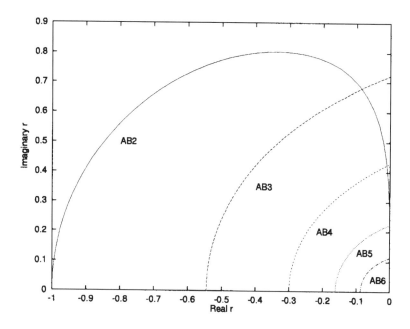

Figure 10.4: Adams-Bashforth stability regions, $r = h\lambda$

and these are most valuable in stiff applications, where a conventional process may require a very small step-size. Moving to third order with AM3 gives a bounded stability region, and further increases in order are accompanied by the shrinkage similar to Figure 10.4. In Figure 10.5 the real stability limits of Adams methods up to order 8 are plotted.

Multistep processes invariably involve composite schemes, and usually one of the well-established Predictor-Corrector methods. The stability properties of such methods depend on the precise strategy followed. If one chooses to iterate to convergence the Corrector formula, then the stability properties will be those of the implicit process. This result, of course, depends on the success of the iteration, which provides another limitation on the step-size. Other composite strategies are not necessarily more stable than the lone explicit process. A detailed study of the stability of many different P-C schemes is described by Lambert (1991), but a few examples are considered below.

Consider the ABM2 Predictor-Corrector combination in PECE mode. Applying the AB2 to the scalar test equation $y' = \lambda y$ yields the predicted value

$$p_{n+1} = y_n + \tfrac{1}{2}r(3y_n - y_{n-1}), \ \ r = h\lambda,$$

and so the derivative value

$$y'_{n+1} = \lambda p_{n+1}$$

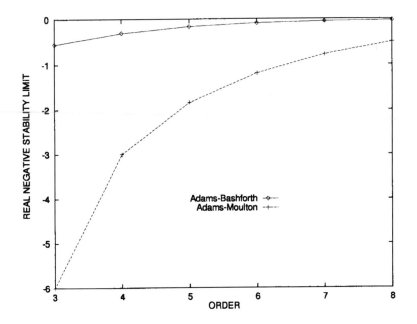

Figure 10.5: Adams real stability limits

is substituted in AM2 (10.19). After simplification, this gives

$$y_{n+1} - y_n(1 + r + \tfrac{3}{4}r^2) + \tfrac{1}{4}r^2 y_{n-1} = 0, \qquad (10.20)$$

a second order difference equation. The stability polynomial is a quadratic and so the characteristic roots are easily determined, leading to intervals of absolute and relative stability of $[-2, 0]$ and $[-\tfrac{2}{3}, \infty]$ respectively. This is an improvement over the explicit process alone.

Table 10.3: Real stability of second order Adams P-C methods

Mode	Absolute	Relative
PE	$[-1, 0]$	$[-0.67, \infty]$
PECE	$[-2, 0]$	$[-0.67, \infty]$
PE(CE)^2	$[-1.5, 0]$	$[-1.1, \infty]$
PECME	$[-2.4, 0]$	$[-0.81, \infty]$
CE	$[-\infty, 0]$	Not appl.

Most practitioners prefer to use local extrapolation with P-C methods and so (10.20) needs modification to reflect PECME mode. From §9.4, the local extrapolation for ABM2 is

$$y_{n+1} = c_{n+1} + \frac{1}{6}(p_{n+1} - c_{n+1}),$$

where $c_{n+1} = y_{n+1}$, from equation (10.20). Substitution yields a difference equation of the same order as before, with coefficients of the same degree (2) in r. The stability polynomial is

$$P(\theta) = \theta^2 - (1 + \tfrac{13}{12}r + \tfrac{5}{8}r^2)\theta + r(\tfrac{1}{12} + \tfrac{5}{24}r).$$

The characteristic roots for this and the unextrapolated case are plotted in Figure 10.6.

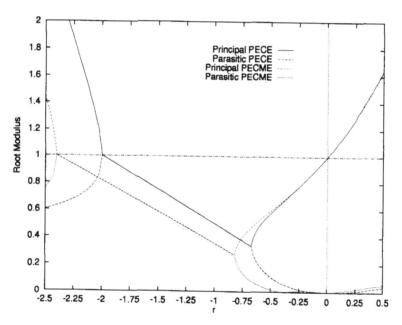

Figure 10.6: Characteristic roots for ABM2 in two modes, $r = h\lambda$

Both relative and absolute stability show useful improvements over the unextrapolated process.

The effect of computing a second iteration of the Corrector $(\mathrm{PE(CE)}^2)$ is to raise the degree in r of the stability polynomial coefficients, and for ABM2 this gives

$$P(\theta) = \theta^2 - (1 + r + \tfrac{1}{2}r^2 + \tfrac{3}{8}r^3)\theta + \tfrac{1}{8}r^3.$$

The outcome is an improvement in relative stability but a reduction in the real absolute limit. Some properties are summarized in Table 10.3.

The results obtained for the 2nd order case are quite typical and so the use of local extrapolation is now to be recommended on two grounds.

Also it is quite possible for a Predictor-Corrector pair to achieve the absolute stability property even when one or both of the formula pair cannot do so individually. An example of this, the third order pair given in Table 8.4, already tested on a simple problem in Chapter 8, has a zero-unstable Predictor, but the combination works well. The same Predictor can be paired with the 4th order Milne-Simpson Corrector (9.17), which has no real absolute stability.

A similar analysis to that above could be applied to a pair using the PMECME mode. However this introduces values from the previous step, thus increasing the degree of the stability polynomial. As a consequence this technique will generally be less stable than the unextrapolated mode.

10.7   Problems

1. Investigate the zero-stability and the order of the explicit formula

$$y_{n+1} = y_n + y_{n-1} - y_{n-2} + 2h(y'_n - 3y'_{n-1}).$$

2. Show that the 3rd order formula

$$
\begin{aligned}
y_{n+1} = {} & -4(3\beta + 1)y_n + (12\beta + 5)y_{n-1} \\
& + h[(7\beta + 4)y'_n + (4\beta + 2)y'_{n-1} + \beta y'_{n-2}]
\end{aligned}
$$

is zero-stable when $-\frac{1}{2} < \beta \le -\frac{1}{3}$. Also show that the fourth order β option for this formula does not lie in this interval.

3. Obtain the first characteristic polynomial for the implicit formula

$$y_{n+1} = (1 - \beta)y_n + \beta y_{n-1} + \tfrac{1}{4}h[(\beta + 3)y'_{n+1} + (3\beta + 1)y'_{n-1}]$$

and hence find a range of β for which it is zero–stable. What is the order of this method for general β? Obtain the principal term of the local truncation error and hence suggest an optimal choice of β.

4. Show that all members of the Nyström family of multistep formulae (§9.5) are zero-stable.

5. Analyse the absolute and relative stability properties of the mid–point Predictor

$$y_{n+1} = y_{n-1} + 2hy'_n.$$

Show that the two characteristic roots of the stability polynomial may be expressed as

$$\theta_0 = e^{h\lambda} + O(h^3), \quad \theta_1 = -e^{-h\lambda} + O(h^3).$$

Sketch a graph of the characteristic roots over a suitable interval of $h\lambda$.

6. From an analytical consideration of the roots of the stability polynomial, show that the 3rd order Adams–Moulton Corrector

$$y_{n+1} = y_n + \tfrac{1}{12}h(5y'_{n+1} + 8y'_n - y'_{n-1})$$

is absolutely stable if $-6 < h\lambda < 0$, and relatively stable if $h\lambda > -1.5$.

7. The Milne–Simpson implicit formula is

$$y_{n+1} = y_{n-1} + \tfrac{1}{3}h(y'_{n+1} + 4y'_n + y'_{n-1}).$$

Obtain its stability polynomial (based on $y' = \lambda y$),

$$P(\theta) = (1 - \tfrac{1}{3}h\lambda)\theta^2 - \tfrac{4}{3}h\lambda\theta - (1 + \tfrac{1}{3}h\lambda),$$

and show that the two zeros, θ_0 and θ_1, approximate $e^{h\lambda}$ and $-e^{-h\lambda/3}$ respectively, for small values of $h\lambda$. Comment on the stability properties of the formula.

8. Obtain a formula for the stability polynomial relating to a general Predictor–Corrector pair. Assume standard notation.

9. Show that the PECE combination of the Hermite Predictor

$$y_{n+1} = -4y_n + 5y_{n-1} + 2h(2y'_n + y'_{n-1})$$

and the 4th order Milne-Simpson implicit formula is absolutely stable if $h\lambda \in (-1, 0)$.

10. Find the stability polynomial for the Hermite–AM3 PECE Predictor-Corrector pair given in Table 8.4. Verify that the real stability limit is $h\lambda = -\tfrac{2}{5}$. Also show that the modified stability polynomial, when the PECME local extrapolated scheme is adopted, is

$$P(\theta) = \theta^2 - \tfrac{4}{3}r^2\theta - (1 + 2r + \tfrac{2}{3}r^2), \quad r = h\lambda$$

which is identical to that for the Hermite-Milne pair from the last question.

11. The multistep formula

$$y_{n+1} = \sum_{i=0}^{p} a_i y_{n-i} + h \sum_{i=-1}^{p} b_i y'_{n-i}$$

has the stability polynomial

$$P(\theta) = \sum_{i=-1}^{p} (a_i + h\lambda b_i)\theta^{p-i},$$

where $a_{-1} = -1$. The principal root θ_0 of the polynomial satisfies

$$\lim_{h \to 0} \theta_0 = 1.$$

Show that, for a formula of order 2 or greater,

$$\theta_0 = 1 + h\lambda + (h\lambda)^2/2 + O(h^3).$$

12. Show that the stability polynomial of a linear multistep formula may be written in the form

$$P(\theta) = h\lambda\xi(\theta) - \rho(\theta),$$

where ρ is the first characteristic polynomial, and

$$\xi(\theta) = \sum_{j=-1}^{p} b_j \theta^{p-j}.$$

On the boundary of the absolute stability region in the complex plane, one of the roots, say θ_k, of $P(\theta) = 0$ has modulus unity, and the other roots have smaller moduli. In polar form the root can be expressed as

$$\theta_k = e^{i\phi}, \quad i = \sqrt{-1}, \quad \phi \in [0, 2\pi].$$

Direct substitution in the stability polynomial now yields $h\lambda$ for a specified ϕ; thus

$$h\lambda = h\lambda(\phi) = \frac{\rho(e^{i\phi})}{\xi(e^{i\phi})},$$

and a sequence of points on the stability boundary is obtained by varying ϕ in the given interval. This method of determining the stability region is called the *boundary locus* technique.

Write a computer program to implement the boundary locus method and use it to plot the regions for Adams-Moulton formulae of orders 3 to 5. Adapt your method to cope with PECE mode pairs.

Chapter 11

Methods for Stiff systems

11.1 Introduction

The importance of highly stable formulae for application to Stiff systems has been considered in Chapter 7, in which some special explicit Runge-Kutta methods were constructed. A particularly desirable property, that of A–Stability, was identified in Chapter 10, where it was found that the backward Euler and trapezoidal integrators provided absolute stability anywhere in the negative real half of the complex plane. These two simple formulae are of implicit type and the general superiority of implicit processes over their explicit multistep counterparts has been confirmed already. Unfortunately, the Adams-Moulton methods of increasing order still feature shrinking stability regions, making them quite unsuitable for Stiff problems. In this chapter, another important class of multistep methods, the backward differentiation formulae (BDF) with unlimited real absolute stability, will be introduced.

Previous development of Runge–Kutta methods has concentrated on explicit formulae. However, it is clear that implicit definitions of the intermediate stages can be allowed. The resulting formulae are computationally much less convenient than explicit RK methods, but it is possible to achieve A–Stability with such processes. Some examples will be presented below.

Implicit multistep and Runge-Kutta processes require solutions of systems of non-linear algebraic equations at each step. In the RK case, the equations arise at each stage of the process. The iterative solution of these equations comprises the major computational cost in implementing all implicit schemes for Stiff systems. For these methods to be more efficient than those discussed in earlier chapters, the high cost of a single step must be compensated by a sufficient increase in step-size. This is often achieved by A–Stable processes applied to Stiff equations.

11.2 Differentiation formulae

As before we consider the differential equation

$$y' = f(x, y), \ y(x_n) = y_n, \ y \in \mathbb{R}^M,$$

for which a number of back values

$$(x_{n-k}, y_{n-k}), \ k = 1, 2, \ldots, p - 1$$

have been determined. In contrast to Chapter 9, where the Adams multistep formulae were based on the integration of polynomials interpolating back values of the function f, the backward differentiation formulae (BDF) are constructed from a numerical differentiation formula. An interpolant of degree p is based on y values, including (x_{n+1}, y_{n+1}), rather than f values. In backward difference form the interpolant is written

$$Y_p(x) = y_{n+1} + s \nabla y_{n+1} + \binom{s+1}{2} \nabla^2 y_{n+1} + \cdots + \binom{s+p-1}{p} \nabla^p y_{n+1},$$
$$(11.1)$$

where $s = (x - x_{n+1})/h$. Assuming that this interpolant approximates $y(x)$, $x \in [x_{n-p+1}, x_{n+1}]$, its derivative $Y_p'(x_{n+1})$ is an approximation to $y'(x_n + h)$. On changing the variable from x to s, the differential system to be solved is replaced by

$$\frac{d}{ds}\{Y_p\}_{s=0} = hy'_{n+1}.$$

Substituting equation (11.1), one obtains

$$\sum_{k=1}^{p} \tfrac{1}{k} \nabla^k y_{n+1} = hy'_{n+1}, \qquad (11.2)$$

and setting $p = 2$ in this gives the implicit multistep formula

$$y_{n+1} = \tfrac{4}{3} y_n - \tfrac{1}{3} y_{n-1} + \tfrac{2}{3} hy'_{n+1}. \qquad (11.3)$$

The local truncation error of this formula can be obtained directly from the error of the interpolant (11.1), as was done for the quadrature based Adams methods, or from a Taylor series expansion. For the BDF2 (11.3) the principal term is

$$\tfrac{2}{9} h^3 y'''(x_n).$$

However the stability properties of formula (11.3) are more interesting

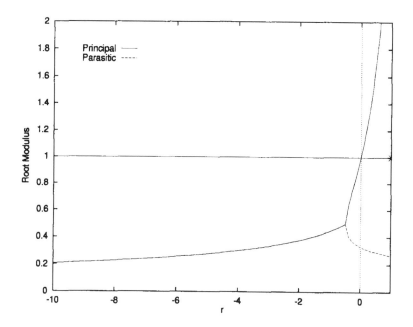

Figure 11.1: Characteristic roots for 2nd order BD formula, $r = h\lambda$

than its algebraic order. The first characteristic polynomial is

$$\rho(\theta) = \theta^2 - \tfrac{4}{3}\theta + \tfrac{1}{3} = (\theta - 1)(\theta - \tfrac{1}{3}),$$

and so (11.3) is zero-stable. To investigate absolute stability, the usual linear scalar test equation, $y' = \lambda y$, is applied, giving the stability polynomial

$$P(\theta) = (1 - \tfrac{2}{3}h\lambda)\theta^2 - \tfrac{4}{3}\theta + \tfrac{1}{3},$$

which has zeros

$$\theta = \frac{2 \pm \sqrt{1 + 2h\lambda}}{3 - 2h\lambda}.$$

It may be shown that absolute stability, requiring $|\theta| < 1$, is achieved for all negative real $h\lambda$. The characteristic roots for this method are plotted in Figure 11.1. Furthermore, absolute stability is guaranteed for all $h\lambda$ in the left half of the complex plane and so the second order BDF is A–Stable.

Higher order BDFs are readily obtained from the equation (11.2) and, if the pth order member of the family is expressed as

$$y_{n+1} = \sum_{k=1}^{p} a_k y_{n-k+1} + b h y'_{n+1},$$

the coefficients may be generated by a relation similar to the one used
(9.9) for the Adams formulae. Setting the initial values

$$\beta_j = \alpha_j = 0, \quad j = 1,\ldots,p, \quad \alpha_0 = 1, \quad b = \left(\sum_{k=1}^{p} \frac{1}{k}\right)^{-1},$$

one computes the sequence

$$\left.\begin{array}{l} \alpha_j = \alpha_j - \alpha_{j-1} \\ \beta_j = \beta_j + \frac{1}{k}\alpha_j \end{array}\right\}, \quad j = k, k-1,\ldots,1; \;\; k = 1, 2,\ldots,p, \qquad (11.4)$$

and $a_k = b\beta_k$, $k = 1,\ldots,p$. The third order formula is

$$y_{n+1} = \frac{18}{11}y_n - \frac{9}{11}y_{n-1} + \frac{2}{11}y_{n-2} + \frac{6}{11}hy'_{n+1}. \qquad (11.5)$$

This and higher order members of the family are not A–Stable but, for

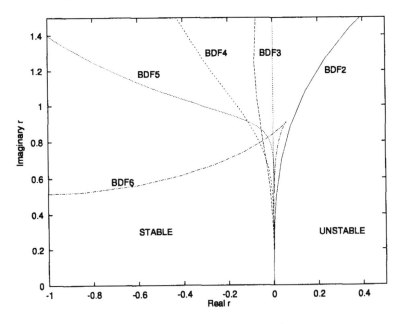

Figure 11.2: Stability boundaries near the origin for zero-stable BDFs,
$r = h\lambda$

$p \le 6$, the whole of the negative real axis is included in the stability
regions. The formulae fail the zero–stability test for $p \ge 7$.

Figure 11.3 illustrates the regions of stability and instability for the
zero-stable BDF family. In contrast to the Adams formulae, the closed
regions contain instability rather than stability. Fortunately, these are
chiefly contained in the positive half-plane. Thus the BDF2 is unstable

in a region containing that part of the real axis between $h\lambda = 0$ and $h\lambda = 4$, and stable everywhere else. This constitutes A–Stability. The BDF3 region of instability is larger and makes a small incursion into the negative real half plane as shown in Figure 11.2, thus violating full A–Stability. Higher order BDF have still larger regions of instability with greater incursions into the negative half-plane but, for systems of differential equations possessing Jacobians with real eigenvalues, BDF methods up to order 6 will yield stable solutions.

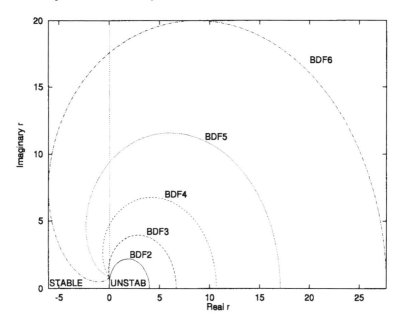

Figure 11.3: Stability regions for zero-stable BDFs, $r = h\lambda$

11.3   Implementation of BDF schemes

Since the BDF are implicit of the same form as the Corrector formulae encountered earlier, they may be applied in a similar fashion. However, in practice they are applied to systems of equations for which the step–size for a conventional Corrector would be very small. Now the simple iteration applied to a Corrector, as described in §8.4 for a scalar case, converges if

$$h < \left\{ \left| b \frac{\partial f}{\partial y} \right| \right\}^{-1} ,$$

but in Stiff systems some of the eigenvalues of the Jacobian matrix are likely to be large, implying a small step–size when this method is applied. Of course this will not violate the stability requirements but, if an A–Stable process has been selected, it defeats the main objective of increasing the steplength.

The use of a conventional Predictor is not to be recommended in the Stiff context. Such a formula will depend on derivative evaluations at previous steps and is likely to be subject to the ill–conditioning errors which are a feature of functions yielding fast transient behaviour. To avoid this drawback, a simple extrapolation formula, based on previous y values alone but similar to the polynomial (11.1), is often selected as a predictor. Thus a pth order predictor is obtained by substituting $s = 1$ in

$$z_p(x) = y_n + s\nabla y_n + \binom{s+1}{2}\nabla^2 y_n \cdots + \binom{s+p-1}{p}\nabla^p y_n,$$

giving

$$P_{n+1} = z_p(x_{n+1}) = y_n + \nabla y_n + \nabla^2 y_n \cdots + \nabla^p y_n,$$

which has a principal local truncation error term

$$-h^{p+1}y^{(p+1)}(x_n).$$

Simple iteration on the BDF Corrector of the same order will yield

$$z^{(k)} = hbf(x_{n+1}, z^{(k-1)}) + c, \ k = 1, 2, \ldots,$$

where $c = \sum_{i=1}^{p} a_i y_{n-i+1}$, and $z^{(0)} = P_{n+1}$. As indicated above, this will converge only for small h when the Jacobian matrix of f, $J(f)$, has large eigenvalues. To allow larger step–sizes, Newton's method, which will converge for small enough h, should be used. The disadvantage of this choice is the need to evaluate the elements of the Jacobian $J(f)$. However, the cost of this is often more than offset by the increase in steplength. The non–linear equation to be solved is

$$F(z) = z - hbf(x, z) - c = 0, \quad z, \ f \in \mathbb{R}^M. \tag{11.6}$$

The Newton iteration can be expressed as

$$z^{(k)} = z^{(k-1)} + \Delta z^{(k-1)}, \quad k = 1, 2, \ldots,$$

where

$$J(F)\Delta z^{(k)} = -F(z^{(k)}), \quad J(F) = I - hbJ(f),$$

and I is the identity matrix of dimension M. When f is linear, a single iteration suffices.

To illustrate this technique consider the non–linear system of equations

$$\begin{aligned} {}^1y' &= \frac{1}{{}^1y} - x^2 - \frac{2}{x^3} \\ {}^2y' &= \frac{{}^1y}{{}^2y^2} - \frac{1}{x} - \frac{1}{2x^{3/2}}, \end{aligned} \qquad (11.7)$$

with initial condition ${}^1y(1) = 1$, and ${}^2y(1) = 1$. This system has true solution ${}^1y = 1/x^2$, ${}^2y = 1/\sqrt{x}$, and so the Jacobian matrix, in terms of the independent variable, may be expressed as

$$J(f) = \begin{pmatrix} -x^4 & 0 \\ x & -2/\sqrt{x} \end{pmatrix}.$$

In this example the eigenvalues of J are its diagonal elements, and so the system (11.7) stiffens as x increases, even though the actual derivatives diminish. For any method with a finite real negative stability limit, such as an Adams multistep or a Runge-Kutta process, the step-size will have to be reduced as x increases. The largest eigenvalue at $x = 10$ has modulus 10^4, implying a maximum steplength $L \times 10^{-4}$ for an integrator possessing a real negative stability interval $[-L, 0]$. The actual derivatives of (11.7) near $x = 10$ are very small. The example here is an artificial one which has been constructed for the purpose of illustration, but systems with varying stiffness are frequently encountered in practice.

The RK program RKden given in Appendix A has been applied to the system (11.7) with $x \in [1, 10]$. The DOPRI5 coefficients were selected, and with a tolerance 10^{-6} the error results in Table 11.1 were obtained. The solution values were based on the dense output formula and the numbers of rejected steps are not included in the totals. As might be expected from the tolerance specified, the solution is very accurate. However, the cost is very high, with a total of 6092 steps being evaluated. The steplength varies as predicted, with only 14 steps needed for the interval $[1, 2]$, increasing to 2477 steps for the last interval $[9, 10]$. The step-size at $x = 10$ is very close to the above prediction based on the stability limit. Increasing or decreasing the tolerance has little effect on the total number of steps required since the step-size is governed largely by stability properties in this case. Thus it is not possible to reduce the computational cost by increasing the tolerance. The expectation of tolerance proportionality, observed in non-stiff problems, is not valid here. Application of an Adams multistep procedure, with poorer stability than DOPRI5, would be even more expensive in steps but not necessarily in overall cost.

An interesting comparison is obtained by applying the same algorithm

Table 11.1: DOPRI5 applied to the Non-linear Stiff system (11.7)

x	Steplength	$^1\varepsilon$	$^2\varepsilon$	Steps
1	1.00×10^{-1}	0.00×10^0	0.00×10^0	0
2	5.10×10^{-2}	3.97×10^{-7}	-2.94×10^{-7}	14
3	2.45×10^{-2}	4.39×10^{-7}	-9.38×10^{-8}	43
4	1.20×10^{-2}	8.73×10^{-7}	-4.09×10^{-8}	101
5	5.25×10^{-3}	4.91×10^{-7}	-1.48×10^{-8}	232
6	2.55×10^{-3}	4.41×10^{-7}	-7.09×10^{-9}	514
7	1.38×10^{-3}	1.10×10^{-6}	-6.20×10^{-9}	1060
8	8.77×10^{-4}	7.43×10^{-7}	-3.98×10^{-9}	2025
9	4.96×10^{-4}	4.77×10^{-7}	-3.54×10^{-9}	3615
10	3.45×10^{-4}	4.46×10^{-7}	-4.23×10^{-9}	6092

to the system

$$
\begin{aligned}
^1y' &= -\frac{2\,^1y}{x} \\
^2y' &= -\frac{^1y}{2\,^2y},
\end{aligned}
\tag{11.8}
$$

with initial condition $^1y(1) = 1$, and $^2y(1) = 1$, which has an identical true solution with system (11.7). The RKden integration of this non-Stiff system takes only 20 steps, yielding similar accuracy to that for the earlier Stiff problem.

The advantage of a BDF for problems such as (11.7) is that the step-size is not limited by stability requirements, and so less accurate solutions may be obtained by increasing the steplength. However, there is an upper limit on the step-size of BDF integration; this is governed by the Newton iteration on equation (11.6). In Table 11.2 are given the results of a constant step solution of (11.7) using the 3rd order BDF (11.5). A step-size of $h = 0.1$ yields an adequate solution in only 88 steps. In this case the starter values were computed by a 4th order Runge-Kutta process. Larger step-sizes are possible and, as would be expected, these give less accurate results. With $h = 0.1$ the Newton algorithm converges in 2 or 3 iterations when the tolerance is set to 10^{-6}, but the method fails to converge when $h = 0.5$ is selected. The solution is much less expensive than the explicit RKden process applied to this Stiff system.

Table 11.2: BDF3 applied to the Stiff system (11.7)

x	Steplength	$^1\varepsilon$	$^2\varepsilon$	Steps
1	1.00×10^{-1}	0.00×10^{0}	0.00×10^{0}	0
2	1.00×10^{-1}	6.82×10^{-5}	5.34×10^{-4}	8
3	1.00×10^{-1}	6.82×10^{-7}	1.78×10^{-4}	18
4	1.00×10^{-1}	3.48×10^{-8}	6.53×10^{-5}	28
5	1.00×10^{-1}	3.57×10^{-9}	2.68×10^{-5}	38
6	1.00×10^{-1}	5.62×10^{-10}	1.20×10^{-5}	48
7	1.00×10^{-1}	1.18×10^{-10}	5.74×10^{-6}	58
8	1.00×10^{-1}	3.06×10^{-11}	2.91×10^{-6}	68
9	1.00×10^{-1}	9.33×10^{-12}	1.55×10^{-6}	78
10	1.00×10^{-1}	3.23×10^{-12}	8.58×10^{-7}	88

11.4 A BDF program

The Backward Differentiation program used for the results in Table 11.2 is listed in Appendix C. As in previous codes, the system of equations is defined in a special MODULE. In this case an extra FUNCTION for the Jacobian evaluation is provided, but the module can be used without change by the earlier Runge–Kutta and multistep programs.

The core of the BDF program is the Newton iteration which yields the BDF Corrector solution and this is shown in Figure 11.4. At each iteration a system of linear equations based on the Jacobian matrix of f must be solved. Any linear equation solver could be employed but the subroutine references in this example are based on an LU factorisation scheme. The factorisation process is separated from the forward and backward substitution phase, the relevant subroutines LU and FBSubst being listed in Appendix C. There is no advantage in this separation for the example given but a more sophisticated strategy may improve efficiency by not evaluating the Jacobian after each iteration. Obviously, one does not then perform the matrix factorisation until a new matrix is applied. Thus a modified Newton scheme could require only one LU factorisation per integration step. In fact, for many problems, it may be unnecessary even to evaluate the Jacobian at every step.

Since the Predictor extrapolation of the same order as the BDF requires one more back value than the Corrector, the first step is started at one order less than subsequent steps. Although this could retard or prevent the convergence of the first step iteration, it seems preferable to the computation of an extra step by the Runge-Kutta starter, which will

```
. . . . . . . . .
ALLOCATE(amat(neqs, neqs), yp(neqs), order(neqs),c(neqs),&
    dy(neqs), ident(neqs,neqs)) ! neqs is system dimension
    . . . . . . . . . . .
    DO While(error > tol)           ! Commence Newton iteration
        amat = ident - h*b*Jac(x, yp)         ! Use Jacobian
        dy = -yp + h*b*Fcn(x, yp) + c         ! f evaluation
        errfun = MAXVAL(Abs(dy))
        CALL LU(amat, neq, order)          ! Solve Newton equns
        CALL FBSubst(amat, dy, neq, order)   ! by LU factoring
        error = MAXVAL(Abs(dy)) + errfun     ! error statistic
        yp = yp + dy;   its = its + 1;       ! Update solution
        . . . . . . . . . .
    END DO
    . . . . . . . . . .
```

Figure 11.4: Program fragment for BDF Newton iteration

be expensive in some cases. The example given is not stiff for small x, and therefore the use of the RK is appropriate and quite cheap. An alternative, and perhaps superior, option for starting would be the backward Euler process. To compensate for the low algebraic order, the chosen step-size could be suitably reduced. In the RK case given, the code aims to 'hit' the required starter values, taking as many steps as required. The number could be fairly large for a problem with large Jacobian matrix eigenvalues near the initial point. Such a case is the well–known stiff system (Robertson, 1967)

$$
\begin{aligned}
{}^1y' &= -\alpha\,{}^1y + \beta({}^2y\,{}^3y) \\
{}^2y' &= \alpha\,{}^1y - \beta({}^2y\,{}^3y) - \gamma({}^2y)^2 \\
{}^3y' &= \gamma({}^2y)^2, \\
{}^1y(0) &= 1, \;\; {}^2y(0) = {}^3y(0) = 0,
\end{aligned}
\tag{11.9}
$$

where the components of y represent the concentrations of three compounds involved in a chemical reaction. Typical values of the parameters are $\alpha = 0.04$, $\beta = 10^4$, $\gamma = 3 \times 10^7$. In this case the Jacobian matrix has small eigenvalues at $x = 0$, but the spectral radius is large for nonzero x. The program BDF can be applied successfully to this problem but the starting phase by Runge-Kutta does require a large number of steps before the stable BDF can be initiated. Since there is a fast transient involved, the application of a small steplength near $x = 0$ is essential for an accurate solution.

The BDF program shown here is not particularly efficient or very robust. Excessive Jacobian evaluation has been indicated already and there is no variation of steplength. The usual measure of local error, involving the difference between Predictor and Corrector solutions, could provide step-size control, and the Krogh algorithm described in Chapter 9 could be employed to halve or to double the steplength. In this case the difference operators are applied to the solution values rather than derivatives as in the Adams case. Another factor affecting the step-size choice is the number of Newton iterations required by a step. For too large a step-size this may be substantial or perhaps no convergence takes place. In the second case a step-size reduction is essential. With its fast transient near $x = 0$, the Robertson problem (11.9) requires some step-size variation for a cheap and accurate solution.

11.5 Implicit Runge-Kutta methods

The importance of unrestricted absolute stability, at least in the real sense, is evident from the results of the preceding sections. Although Runge-Kutta methods generally have better stability characteristics than multistep methods, the explicit methods considered so far cannot provide A–stability or even match the higher order BDFs in this respect. While it is possible to add stages, while restricting the algebraic order, to extend stability, one has to consider implicit RK processes to obtain unbounded regions of absolute stability. Fortunately the analytical foundation of such processes is just the same as that of the explicit methods already described.

Using the Butcher tabular notation, an s–stage implicit RK formula is displayed in Table 11.3.

Table 11.3: Modified Butcher table for an implicit Runge-Kutta formula

c_i	a_{ij}				b_i
c_1	a_{11}	a_{12}	$\cdots$	a_{1s}	b_1
c_2	a_{21}	a_{22}	$\cdots$	a_{2s}	b_2
c_3	a_{31}	a_{32}	$\cdots$	a_{3s}	b_3
$\vdots$	$\vdots$	$\vdots$	$\ddots$	$\vdots$	$\vdots$
c_s	a_{s1}	a_{s2}	$\cdots$	a_{ss}	b_s

For this type of process $a_{ij} \neq 0$ for $j \geq i$, and so the s–stage formula has $s(s + 1)$ parameters, twice the number in the explicit case. The implementation of such a formula is much less convenient than an explicit RK since, at each stage i, the f_i value depends on the other stage f's. Generally the f's are obtained by solving a set of non-linear equations. The implicit RK has an increment formula exactly like that of the explict method (2.12)

$$y_{n+1} = y_n + h_n \sum_{i=1}^{s} b_i f_i,$$

where the function values are given by

$$f_i = f(x_n + c_i h_n, \ y_n + h_n \sum_{j=1}^{s} a_{ij} f_j) \qquad (11.10)$$

$$i = 1, 2, \ldots, s.$$

For a scalar f, the system (11.10) comprises s equations in the unknowns f_i, $i = 1, \ldots, s$. If $y \in \mathbb{R}^M$, then the dimension of the problem is sM. The solution of such a system, particularly in the non-linear case, is likely to be much more expensive than the evaluation of an explicit formula. The usual comparisons of function evaluations between RK and multistep methods, as discussed in Chapter 9, can be made in this case also. The justification for using such methods can only be a significant increase in step-size. Actually, the situation is not quite as bad as it first seems, as will be discovered shortly.

Let us consider an implicit Runge-Kutta formula with 2 stages. The equations of condition to be satisfied by the RK parameters are obtained by setting the usual error coefficients, as given in Table 3.1, to zero. Since $s = 2$, and assuming the row sum condition,

$$a_{11} + a_{12} = c_1, \quad a_{21} + a_{22} = c_2,$$

Table 11.3 gives 6 independent parameters. Consequently we seek a third order formula and the four equations of condition are

$$\begin{aligned} b_1 + b_2 &= 1 \\ b_1 c_1 + b_2 c_2 &= \tfrac{1}{2} \\ b_1 c_1^2 + b_2 c_2^2 &= \tfrac{1}{3} \\ b_1(a_{11}c_1 + a_{12}c_2) + \\ b_2(a_{21}c_1 + a_{22}c_2) &= \tfrac{1}{6}. \end{aligned}$$

There are two free parameters and so, to yield an easy solution, we choose $a_{11} = a_{12} = 0$. Then $c_1 = 0$, and the formula

c_i	a_{ij}		b_i
0	0	0	$\frac{1}{4}$
$\frac{2}{3}$	$\frac{1}{3}$	$\frac{1}{3}$	$\frac{3}{4}$

$$(11.11)$$

is readily derived. The choice of free parameters has ensured that the first stage is explicit but the second function evaluation, which can be written as

$$f_2 = f(x_n + \tfrac{2}{3}h, y_n + \tfrac{1}{3}hf_1 + \tfrac{1}{3}hf_2),$$

generally will be non–linear in f_2. Solution by Newton's method may be straightforward but this will imply extra function evaluations. Note the similarity of this formula to the optimum 2–stage RK2 in §4.2, which has the same c_2.

Let us examine the absolute stability of the implicit RK3 (11.11) by applying it to the usual scalar test equation $y' = \lambda y$, $y(x_n) = y_n$. This gives

$$
\begin{aligned}
f_1 &= \lambda y_n \\
f_2 &= \lambda(y_n + \tfrac{1}{3}h\lambda y_n + \tfrac{1}{3}hf_2).
\end{aligned}
$$

Writing $r = h\lambda$, the second function evaluation yields

$$hf_2 = ry_n\left(\frac{1+\tfrac{1}{3}r}{1-\tfrac{1}{3}r}\right),$$

and so the increment function gives

$$y_{n+1} = y_n\left(\frac{1+\tfrac{2}{3}r+\tfrac{1}{6}r^2}{1-\tfrac{1}{3}r}\right).$$

This contains a rational function rather than the stability polynomial deriving from the explicit Runge-Kutta. The condition for absolute stability is

$$\left|\frac{1+\tfrac{2}{3}r+\tfrac{1}{6}r^2}{1-\tfrac{1}{3}r}\right| < 1,$$

and it is easy to show that this is satisfied by

$$-6 < h\lambda < 0,$$

a significant improvement on the explicit 3rd order methods so far encountered.

It is possible to improve stability even further. Consider the one–stage formula

$$y_{n+1} = y_n + b_1 hf_1, \quad f_1 = f(x_n + c_1 h, y_n + ha_{11}f_1).$$

This is a second order process if $b_1 = 1$, $b_1 c_1 = \tfrac{1}{2}$, and so $c_1 = a_{11} = \tfrac{1}{2}$. Application to the standard test equation gives

$$y_{n+1} = \left(\frac{1+\tfrac{1}{2}h\lambda}{1-\tfrac{1}{2}h\lambda}\right) y_n,$$

a result identical to the recursion from the trapezoidal formula (10.19). Thus the one–stage implicit RK2 is A-Stable.

Implicit RK methods for which $a_{ij} = 0$, for $j > i$, are called semi–implicit formulae, the class of which (11.11) and the one-stage RK2 are members. These can be made A–Stable. For practical purposes they are simpler to implement than the fully implicit formulae, since each stage consists of the determination of only a single f. This requires the simultaneous solution of M rather than sM algebraic equations. A still more convenient form is the *diagonally implicit RK* process (DIRK) in which

$$a_{11} = a_{22} = \ldots = a_{ss}.$$

A member of this class is the A–Stable, third order, 2–stage DIRK (Alexander, 1977) given in Table 11.4. Applying this to the linear test

Table 11.4: An A-stable DIRK formula of order 3

c_i	a_{ij}		b_i
$\frac{3+\sqrt{3}}{6}$	$\frac{3+\sqrt{3}}{6}$	0	$\frac{1}{2}$
$\frac{3-\sqrt{3}}{6}$	$-\frac{\sqrt{3}}{3}$	$\frac{3+\sqrt{3}}{6}$	$\frac{1}{2}$

problem from Chapter 7 (see Figure 7.6) with constant step–size 0.1, as used earlier, the results in Table 11.5 are obtained. Using an A–stable formula has eliminated the earlier catastrophic growth in the solution. Of course the amount of computational effort for each step is considerably greater than in the explicit RK integrator, even when applying the technique to a linear system. For non–linear equations, each stage would require the solution of a non–linear system of algebraic equations.

Although these results are extremely encouraging, it must be emphasized that the DIRK has been applied in an interval in which the effect of the fast transient should be negligible. If the formula is applied to the same system with the same step–size for $x \in [0, 1]$, the numerical solution oscillates in a stable manner but produces very inaccurate results. In order to give small errors in the transient interval, a small step–size is essential and so it may be more efficient to apply an explicit RK in such an interval. An ideal automatic differential equation solver would employ both forms of RK integrators.

Table 11.5: Linear test problem (7.11) solution with DIRK formula

x_n	${}^1y$	${}^2y$	${}^1\varepsilon$	${}^1\varepsilon$
1.0	0.7357589	0.3678794	0	0
1.1	0.6657363	0.3328658	-5.834×10^{-6}	-5.271×10^{-6}
1.2	0.6023790	0.3011877	-9.371×10^{-6}	-6.474×10^{-6}
1.3	0.5450505	0.2725234	-1.310×10^{-5}	-8.400×10^{-6}
1.4	0.4931774	0.2465843	-1.653×10^{-5}	-1.264×10^{-5}
1.5	0.4462442	0.2231235	-1.609×10^{-5}	-6.643×10^{-6}
1.6	0.4037739	0.2018847	-1.904×10^{-5}	-1.174×10^{-5}
1.7	0.3653475	0.1826740	-1.951×10^{-5}	-9.473×10^{-6}
1.8	0.3305767	0.1652866	-2.106×10^{-5}	-1.223×10^{-5}
1.9	0.2991156	0.1495557	-2.158×10^{-5}	-1.284×10^{-5}
2.0	0.2706498	0.1353244	-2.074×10^{-5}	-1.088×10^{-5}

11.6 A semi–implicit RK program

The implementation of implicit Runge-Kutta processes is quite similar to that of backward differentiation formulae in that some means of evaluating the elements of the Jacobian matrix of f is essential. In this work it is assumed that the elements of the Jacobian can be obtained analytically, but a numerical scheme for their evaluation could be substituted if preferred. One advantage of the RK scheme is that it is self-starting.

It is convenient to express the semi-implicit RK formula as

$$y_{n+1} = y_n + h \sum_{i=1}^{s} b_i f(x_n + c_i h, Y_i),$$

$$Y_i = y_n + h \sum_{j=1}^{i} a_{ij} f(x_n + c_j h, Y_j), \quad i = 1, \ldots, s, \quad (11.12)$$

and then, at each stage i, it is necessary to solve $G(Y_i) = 0$, where

$$G(Y_i) = Y_i - W_i - h a_{ii} f(x_n + c_i h, Y_i), \quad (11.13)$$

and

$$W_i = y_n + h \sum_{j=1}^{i-1} a_{ij} f(x_n + c_j h, Y_j).$$

The relevant Jacobian matrix for this system at stage i may be written

$$J(G_i) = I - h a_{ii} J(f_i), \quad (11.14)$$

where, assuming $y \in \mathbb{R}^M$, I is the identity matrix of dimension M.

A fragment of code demonstrating the core of a simple semi-implicit RK program is shown in Figure 11.5. This algorithm depends on the

```
. . . . . . . . .
DO WHILE(x < xend)                        ! LOOP OVER STEPS
  DO i = 1, s                             ! LOOP OVER STAGES
    z = x + c(i)*h
    w = y + h*MATMUL(a(i, 1: i-1), f(1: i-1, :)) !Earlier
    yy = y; iter = 0                      ! stages
    f(i, :) = Fcn(z, yy)
    DO WHILE(MAXVAL(ABS(dy)) > tol)       ! NEWTON ITERATION
      amat = ident - h*a(i, i)*Jac(z, yy)     ! Jacobian
      CALL LU(amat, neq, order)          ! FACTORISE MATRIX
      dy = w + h*a(i, i)*f(i, :) - yy
      CALL FBSubst(amat, dy, neq, order)   ! SOLVE SYSTEM
      yy = yy + dy; iter = iter + 1      ! UPDATE SOLUTION
      f(i, :) = Fcn(z, yy)                  ! RE-EVALUATE f
      . . . . . . . . . . . . .
    END DO                              ! End Newton iteration
  END DO
  y = y + h*MATMUL(b, f)                      ! Updated SOLUTION
  x = x + h
END DO
. . . . . . . . . .
```

Figure 11.5: Program fragment for semi-implicit RK scheme

same linear equation solver as used in the BDF program. However the Newton iteration is needed at each stage, implying an increase in computational cost over the multistep process. Fortunately a modified Newton procedure, in which the Jacobian is held constant over the iterations, is appropriate.

A further cost-saving advantage of diagonally-implicit formulae is the constancy of a_{ii}, permitting the Jacobian evaluation and the matrix factorisation to be reduced to once per step. This is implemented in the program Dirk3 in Appendix C which applies the 3rd order two-stage DIRK (11.4) to a system of non-linear equations defined in a MODULE, as in the backward differentiation program BDF. As discussed earlier, the use of a constant step-size is not very efficient and so Dirk3 varies this according to the number of Newton iterations required. A large number of iterations implies a poor first approximation and hence an overlarge steplength. Conversely, rapid convergence suggests an increase may be possible. This strategy involves no direct error estimate but it proves to be fairly robust.

To test the Dirk3 program it has been applied to the Robertson system (11.9) which starts with a fast transient phase requiring small steps for good accuracy. This affects the second component most strongly and Figure 11.6 shows the solution obtained for ^2y from system (11.9), using a starting step-size $h = 1.0 \times 10^{-5}$ to give a good solution for very small x. The total numbers of function f and Jacobian $J(f)$ evaluations for the solution with $x \in [0, 100]$ were 467 and 68 respectively. At the end of the integration the step-size had grown to 10.5. To compare this result

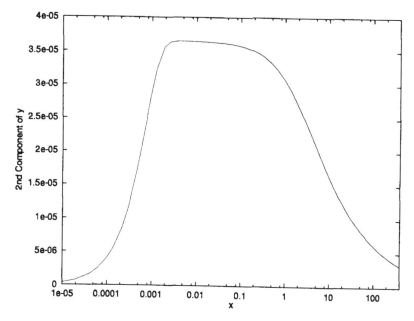

Figure 11.6: Solution for ^2y from the Robertson Stiff system

with that from a conventional RK embedded pair, the RKden program from Appendix A was applied to the same problem. Using DOPRI5 with tolerance 10^{-6}, a total of 807979 function evaluations was required! Of course the RKden algorithm is not designed to cope with a problem whose step-size is governed by stability and its performance in this context could be improved, perhaps by 20 to 30%. This comparison certainly proves the value of A-Stable processes.

The efficiency of Dirk3 is by no means optimal. Assuming that the Jacobian matrix (11.14) of f varies only slowly, it is not necessary to compute it even at each step. In many cases it is sufficient to compute it every 10 or 20 steps.

Error estimation with a DIRK is not as simple as in the explicit case, or indeed, as in the BDF process. Of course it is possible to compute a lower order solution, but not with an embedded formula. One method of evaluating local error is by classical extrapolation. This requires a second solution at the specified x based on a different (usually halved) steplength.

Another uncertainty in the DIRK process is the initial approximation used for the Newton iteration. Convergence is not guaranteed since the first approximation must be sufficiently close to the root. A poor initial value can give convergence to an incorrect solution.

11.7   Problems

1. Use a Taylor series expansion to derive the principal local truncation error terms for the 2nd and 3rd order Backward Differentiation formulae.

2. Show that the roots of the stability polynomial of the BDF2 (11.3), for $h\lambda < -\frac{1}{2}$, have modulus

$$|\theta| = \frac{1}{\sqrt{3 - 2h\lambda}}.$$

3. Use (a) the general formula (11.2), and (b) the recursion formula (11.4), to construct third and fourth order BDFs. Show that these are zero–stable. Employ the BDF program given in Appendix C to compute the solution to problem (11.7) with the 4th order BDF and step-size $h = 0.1$. Compare the results with those relating to the third order formula in Table 11.1.

4. Assuming that the backward difference interpolant (11.1) of degree p, used to approximate $y(x)$, has an error which can be written in the form

$$E_p(x) = \binom{s+p}{p+1} h^{p+1} y^{(p+1)}(x_n) + O(h^{p+2}),$$

show that the principal truncation error of the pth order BDF is

$$\left((p+1) \sum_{k=1}^{p} (1/k) \right)^{-1}.$$

Hence obtain formulae for the Corrector error estimates when BDFs of orders 2 to 4 are implemented with suitable extrapolation formulae as Predictors. Test the accuracy of these estimates by reference to the problem (11.7).

5. Modify the BDF program to solve the stiff non-linear system (11.9) with $x \in [0, 100]$ and a step-size $h = 0.01$.

6. Improve the efficiency of the BDF program by reducing the number of Jacobian evaluations as suggested in §11.6. Also incorporate a step-size control mechanism based on local error estimation. Test your program on problem (11.9) with $x \in [0, 100]$.

7. Find the Jacobian matrix of the Robertson system (11.9) and find the eigenvalues at $x = 0$.

8. By considering the appropriate inequality, show that the real interval of absolute stability for the semi-implicit RK formula (11.11) is $h\lambda \in [-6, 0]$.

9. Find the stability function R, defined by

$$y_{n+1} = R(h\lambda)y_n,$$

for a general two-stage fully implicit Runge-Kutta process. Substitute in this the parameters of the DIRK3 formula from Table 11.4. Hence show that

$$|R(h\lambda)| < 1, \quad \forall \, h\lambda < 0$$

in this case.

10. Determine the stability function for a general 3-stage DIRK formula.

11. Compute two steps of size $h = 1$ when the DIRK3 in Table 11.4 is applied to the linear equation $y' = -5y$, $y(0) = 1$.

12. Show that an implicit RK process for $y' = f(x, y)$ can be expressed as

$$y_{n+1} = y_n + h \sum_{i=1}^{s} b_i f(x_n + c_i h, Y_i),$$

$$Y_i = y_n + h \sum_{j=1}^{s} a_{ij} f(x_n + c_j h, Y_j), \quad i = 1, \ldots, s.$$

Write out the equations which need to be solved at each stage if a 2-stage formula is applied to

(a) the linear system

$$\begin{pmatrix} {}^1f \\ {}^2f \end{pmatrix} = \frac{1}{5} \begin{pmatrix} 994 & -1998 \\ 2997 & -5999 \end{pmatrix} \begin{pmatrix} {}^1y \\ {}^2y \end{pmatrix};$$

(b) the non-linear equations

$$ {}^1f = \frac{1}{{}^1y} - x^2 - \frac{2}{x^3} $$

$$ {}^2f = \frac{{}^1y}{2y^2} - \frac{1}{x} - \frac{1}{2x^{3/2}}. $$

Also give the relevant equations when the RK becomes semi-implicit.

13. Find α so that the 2-stage DIRK formula, given in tabular form

$$
\begin{array}{c|cc|c}
c_i & \multicolumn{2}{c|}{a_{ij}} & b_i \\
\hline
\alpha & \alpha & 0 & 1 - \alpha \\
1 & 1 - \alpha & \alpha & \alpha
\end{array}
$$

has order 2. Find the stability function R for this formula and show that

$$\lim_{r \to \infty} R(r) = 0.$$

14. Apply the DIRK3 program from Appendix C to the system (11.7). Use an initial step-size $h = 0.1$.

Chapter 12

Variable coefficient multistep methods

12.1 Introduction

An inconvenient feature of the multistep schemes so far considered is their reliance on equally spaced starter values. The advantages of a variable steplength have been adequately stated in earlier chapters and, although step–size variation with Adams formulae is fairly straightforward by re-computing backward differences, it is by no means as easy as in the Runge–Kutta case. One remedy for this situation is the construction of variable coefficient multistep formulae. These arise from the integration of divided difference interpolants based on unequally spaced data points. The principle is no different from that employed in the derivation of Adams formulae from backward difference interpolants as described in Chapter 9. As the name suggests, these variable coefficient schemes are more complicated than conventional methods. Rather like dense output formulae, the integrator coefficients depend on the desired output point. Nevertheless, variable coefficient techniques offer extra flexibility and consequently they have become very popular. In this chapter we describe a simple approach to the application of a variable coefficient procedure which is closely related to the Adams methods. A more sophisticated process, on which a widely-used computer code is based, is also described.

12.2   Variable coefficient integrators

As in Chapter 9 we consider the solution of the ordinary differential system

$$y'(x) = f(x, y(x)), \quad y \in \mathbb{R}^m.$$

From §9.2, the multistep procedure is based on the quadrature

$$\int_{x_n}^{x_{n+1}} y'(x)dx = \int_{x_n}^{x_{n+1}} f(x, y(x))dx,$$

yielding

$$y(x_{n+1}) = y(x_n) + \int_{x_n}^{x_{n+1}} f(x, y(x))dx.$$

The integrand can be approximated by a polynomial interpolant v_k based on known values of $(x_i, y(x_i))$. Of course, in practice the polynomial is constructed from the previous solution or starter values, which leads to a multistep formula

$$y_{n+1} = y_n + \int_{x_n}^{x_{n+1}} v_k(x)dx, \tag{12.1}$$

where

$$v_k(x_i) = f_i = f(x_i, y_i), \quad i = n, n-1, \ldots, n-k.$$

In the equal-interval case v_k was expressed as a Newton-Gregory backward difference interpolant but, more generally, one can specify a divided difference form of the interpolant which does not rely on the equal spacing of the abscissae x_i.

Divided differences can be defined recursively and, with four data points, the diagonal entries are

$$f[i, i-1] \quad = \quad \frac{f_i - f_{i-1}}{x_i - x_{i-1}},$$

$$f[i, i-1, i-2] \quad = \quad \frac{f[i, i-1] - f[i-1, i-2]}{x_i - x_{i-2}},$$

$$f[i, i-1, i-2, i-3] \quad = \quad \frac{f[i, i-1, i-2] - f[i-1, i-2, i-3]}{x_i - x_{i-3}}.$$

It is normal to tabulate them as shown in the Table 12.1, and with this notation the polynomial v_k can be written

$$\begin{aligned}
v_k(x) \quad = \quad & f[n] + (x - x_n)f[n, n-1] \\
+ \quad & (x - x_n)(x - x_{n-1})f[n, n-1, n-2] + \cdots \\
+ \quad & \prod_{j=0}^{k-1}(x - x_{n-j})f[n, n-1, \ldots, n-k], \tag{12.2}
\end{aligned}$$

Table 12.1: A divided difference table

x_0	f_0				
		$f[1,0]$			
x_1	f_1		$f[2,1,0]$		
		$f[2,1]$		$f[3,2,1,0]$	
$\vdots$	$\vdots$		$\vdots$		$\vdots$
			$\vdots$	$\vdots$	
$\vdots$	$\vdots$		$\vdots$		$\vdots$
		$f[7,6]$		$f[8,7,6,5]$	
x_7	f_7		$f[8,7,6]$		$f[9,8,7,6,5]$
		$f[8,7]$		$f[9,8,7,6]$	
x_8	f_8		$f[9,8,7]$		
		$f[9,8]$			
x_9	f_9				

where $f[n] \equiv f_n$. An explicit formula, suitable as a Predictor, is determined by substituting (12.2) in the equation (12.1). Integration yields a formula

$$y_{n+1}^p = y_n + \sum_{i=0}^{k} c_i f[n, \ldots, n-i], \qquad (12.3)$$

where the c_i are given by

$$c_0 = \int_{x_n}^{x_{n+1}} dx, \qquad c_i = \int_{x_n}^{x_{n+1}} \prod_{j=0}^{i-1}(x - x_{n-j}) dx.$$

The first few *variable* coefficients c_i are

$$c_0 = h_n, \quad c_1 = \tfrac{1}{2}h_n^2, \quad c_2 = \tfrac{1}{2}h_n^2[\tfrac{2}{3}h_n + h_{n-1}], \qquad (12.4)$$
$$c_3 = \tfrac{1}{2}h_n^2[\tfrac{1}{2}h_n^2 + \tfrac{2}{3}h_n(2h_{n-1} + h_{n-2}) + h_{n-1}(h_{n-1} + h_{n-2})],$$

where $h_i = x_{i+1} - x_i$. Setting $h_i = h$, $i = n, \ldots, n-k$, in the formula (12.3) will give an Adams-Bashforth Predictor of order $k+1$.

   In a similar way it is possible to construct an implicit or Corrector formula by use of an interpolant which satisfies the predicted derivative value (x_{n+1}, f_{n+1}), where $f_{n+1} = f(x_{n+1}, y_{n+1}^p)$. The interpolant becomes

$$v_k^*(x) = f[n+1] + (x - x_{n+1})f[n+1, n]$$
$$+ (x - x_{n+1})(x - x_n)f[n+1, n, n-1] + \ldots$$

$$+ \prod_{j=0}^{k-1} (x - x_{n+1-j}) f[n+1, n \ldots, n-k+1].$$

Replacing v_k in equation (12.1) by v_k^*, integration yields the Corrector formula

$$y_{n+1}^c = y_n + \sum_{i=0}^{k} c_i^* f[n+1, \ldots, n+1-i], \tag{12.5}$$

where the first few c_i^* are

$$c_0^* = h_n, \quad c_1^* = -\tfrac{1}{2}h_n^2, \quad c_2^* = -\tfrac{1}{6}h_n^3,$$
$$c_3^* = -\tfrac{1}{12}h_n^3[h_n + 2h_{n-1}]. \tag{12.6}$$

This Corrector reduces to an Adams-Moulton formula of order $k+1$ when the steplength is constant.

It is clear that the expressions for c_i, c_i^* will become quite lengthy for high order formulae. The development of an efficient method for computing them will be important if their evaluation is not to be a significant fraction of the total cost of solving a problem. One way of avoiding much of this cost is to restrict the number of steplength changes. It will be recalled that a variable step Adams process in Chapter 9 proved to be quite efficient without frequent step changes.

12.3   Practical implementation

In the starting phase of an integration, the advantages of the above variable coefficient formulae over Adams formulae are very clear. Since the step-size can be changed from step to step without forming explicit interpolants, the increase in step-size appropriate to a gain in order can be accommodated. Therefore one can start the solution with a first order Predictor and a very small step-size, increase the order to two for the second step and to order 3 for the third step, with increased step-sizes, and so on. Of course the coefficients c_i, c_i^* need to be computed at each step. The order may be permitted to increase until a maximum value, probably limited by stability requirements, is attained.

Rather than compute modifiers for the Corrector and Predictor formulae, the Corrector stage will be regarded as a local extrapolation. Thus a Corrector of order $K+1$ will be formed following a Kth order Predictor. This scheme is motivated by the properties of the divided difference interpolant, but it would be equally suitable for a constant step multistep process. Suppose the polynomial $v_k(x)$ of degree k, which interpolates the data (x_i, f_i), $i = n, n-1, \ldots, n-k$, is the basis of a $(k+1)$th order Predictor formula (12.3). With an extra predicted point (x_{n+1}, f_{n+1}), a polynomial of degree $k+1$, incorporating the higher divided difference

$f[n+1, n, n-1, \ldots, n-k]$, may be computed. Since a given set of points defines an interpolant uniquely, the order of coding in a divided difference tableau does not affect the value of the highest difference. Thus the new polynomial can be expressed as

$$v_{k+1}^*(x) = v_k(x) + \prod_{j=0}^{k}(x - x_{n-j})f[n+1, n, n-1, \ldots, n-k]. \quad (12.7)$$

Using this as the basis for the Corrector formula of order $k+2$ it is easily seen that

$$y_{n+1}^c = y_{n+1}^p + c_{k+1}f[n+1, n, n-1, \ldots, n-k], \quad (12.8)$$

and so the c_i^* defined in (12.6) are redundant. For step-size control purposes, as well as order variation, it is useful also to provide a Corrector of the same order as that of the Predictor. The difference of the two Correctors will then serve as a local error estimate. From equation (12.7), replacing k by $k-1$, the basis of the $k+1$th order Corrector is

$$v_k^*(x) = v_{k-1}(x) + \prod_{j=0}^{k-1}(x - x_{n-j})f[n+1, n, \ldots, n-k+1],$$

but formula (12.2) can be written as

$$v_k(x) = v_{k-1}(x) - \prod_{j=0}^{k-1}(x - x_{n-j})f[n, n-1, \ldots, n-k].$$

Eliminating v_{k-1} yields the convenient form for $v_k^*(x)$,

$$\begin{aligned}
v_k^*(x) &= v_k(x) + \\
&\quad \prod_{j=0}^{k-1}(x - x_{n-j})\{f[n+1, \ldots, n-k+1] - f[n, \ldots, n-k]\} \\
&= v_k(x) + \prod_{j=0}^{k-1}(x - x_{n-j})(x_{n+1} - x_{n-k})f[n+1, n, \ldots, n-k],
\end{aligned}$$

and integration leads to the Corrector formula of order $k+1$

$$y_{n+1}^{c1} = y_{n+1}^p + (x_{n+1} - x_{n-k})c_k f[n+1, n, \ldots, n-k], \quad (12.9)$$

and hence a local error estimate is

$$\begin{aligned}
e_{n+1} &= y_{n+1}^{c1} - y_{n+1}^c \\
e_{n+1} &= \{(x_{n+1} - x_{n-k})c_k - c_{k+1}\}f[n+1, n, \ldots, n-k]. \quad (12.10)
\end{aligned}$$

12.4   Step-size estimation

Since the order as well as the step-size can be varied with the method derived above, it is possible to start the solution of an initial value problem with a first order formula if a small-enough steplength is specified. The choice of this initial steplength is important. Too large a value will not give sufficient accuracy and thus will lead to rejection, while too small a step will imply a wasteful sequence of operations in the starting phase of the integration. These comments are relevant to any method, variable order or not. Shampine and Gordon (1975) have made a detailed study of the problem of step-size control for their variable coefficient multistep code (STEP) and they propose, as 'a rule of thumb', that the local error of a first order method should be assumed to be h times that of the method of order zero. If (x_0, y_0) is the initial point, this rule predicts a local error for a first order process

$$e_1 \sim h^2 \|f(x_0, y_0)\|.$$

If the local error tolerance is T, then one requires

$$|e_1| < T$$

and so a starting step of length

$$h \sim \left[\frac{T}{\|f(x_0, y_0)\|} \right]^{\frac{1}{2}}$$

is predicted. However, Shampine and Gordon prefer to be conservative and actually choose an initial steplength to be

$$h_0 = \frac{1}{4} \left[\frac{0.5T}{\|f(x_0, y_0)\|} \right]^{\frac{1}{2}}, \tag{12.11}$$

provided $\|f(x_0, y_0)\| \neq 0$.

As the number of steps increases, the number of available starter values, and hence the potential order of the method, can increase. The same authors recommend raising the order and doubling the step-size after each successful step during the start phase of an integration. This phase is terminated when a failure, with $|e_{n+1}| > T$, occurs, or when the maximum permissible order is attained. When the method is operating at order k the predicted maximum step-size for the *next* step will be

$$h_{n+1} = h_n \left[\frac{0.5T}{\|e_{n+1}\|} \right]^{\frac{1}{k+1}},$$

which is identical to the formula (5.2) used in RK embedding with $\theta = 0.5$, and the local error estimate e_{n+1} is given by the formula (12.10).

For reasons already discussed with respect to the variable step Adams implementation, the step-size may not actually be raised when the step prediction formula yields $h_{n+1} > h_n$.

To illustrate the starting phase of a variable multistep procedure, consider the equation

$$y' = x + y, \quad y(0) = 1, \tag{12.12}$$

to be solved with the rather unlikely absolute error tolerance $T = 3.2 \times 10^{-3}$. The formula (12.11) gives $h_0 = 0.01$ in this instance, and the first three steps are enumerated below.

1. The first order Predictor formula is

$$y_1^p = y_0 + c_0 f[0] = 1.01.$$

This allows the first difference estimate $f[1, 0] = 2$, and a Corrector of order 2 based on this is given by

$$y_1^c = y_1^p + c_1 f[1, 0] = 1.0101.$$

Comparing with a first order Corrector, a local error estimate from (12.10) with $n = 0$ is

$$e_1 = [(x_1 - x_0)c_0 - c_1]f[1, 0] = 1.0 \times 10^{-4}.$$

Since this is smaller than the tolerance T, the second order corrected solution is acceptable and a second step, size $h_1 = 2h_0 = 0.02$, is attempted with a second order Predictor, after recomputing $f[1, 0]$ from the corrected solution.

2. The second order Predictor is

$$y_2^p = y_1 + c_0 f[1] + c_1 f[1, 0] = 1.030904 \text{ (6 decimals).}$$

The next divided difference $(f[2, 1, 0] = 1.006667)$ is now estimated and the 3rd order Corrector is applied. This yields

$$y_2^c = y_2^p + c_2 f[2, 1, 0] = 1.030909.$$

Comparison with a 2nd order Corrector gives the error estimate

$$e_2 = [(x_2 - x_0)c_1 - c_2]f[2, 1, 0] = 1.342 \times 10^{-6} < T,$$

and so, after recalculating $f[2, 1, 0]$, a third step of length $h_2 = 0.04$ is computed.

3. The 3rd order Predictor is

$$y_3^p = y_2 + c_0 f[2] + c_1 f[2,1] + c_2 f[2,1,0] = 1.075016 \text{ (6 decimals)}.$$

Using the predicted solution, $f[3,2,1,0] = 0.313173$, a 4th order Corrector gives

$$y_3^c = y_3^p + c_3 f[3,2,1,0] = 1.075016 \text{ (6 decimals)},$$

and once again the estimated error is smaller than the tolerance T.

The divided difference tableau with corrected solutions is shown in Table 12.2. The above step-size/order control procedure is very conservative

Table 12.2: A divided difference table showing the starting phase

x_i	y_i	f_i	$f[,]$	$f[,,]$	$f[,,,]$
0	1	1			
			2.01		
0.01	1.0101	1.0201		1.015	
			2.04045		0.315476
0.03	1.030909	1.060909		1.037083	
			2.102675		
0.07	1.075016	1.145016			

with error estimates being much smaller than the tolerance. With continued stepping, eventually a steplength appropriate to the tolerance will be determined.

The scheme described here is quite adequate if the formulae are restricted to low orders. However, in order to take full advantage of the features of multistep methods, one must proceed to high orders of accuracy. In such cases the direct evaluation of the c_i, according to the formulae (12.4), would be very tedious and so a more sophisticated computational scheme is required.

12.5   A modified approach

For their STEP integrator code, Shampine and Gordon (1975) adopted an approach in which were defined modified divided differences, reducing to backward differences for constant steplengths. If circumstances permit, a constant steplength will reduce significantly the computational overhead and so the steplength is increased only when a doubling is possible. This is the same strategy as used in the variable Adams process described in Chapter 9, and clearly it yields groups of steps of constant length. The process is described in detail in the book by Shampine and Gordon (1975) which contains full listings of computer programs. A summary of this scheme with suitably modified notation now follows.

Suppose $h_i = x_{i+1} - x_i$, $i = 0, 1, 2, \ldots$, and $s = (x - x_n)/h_n$. We introduce the notation

$$\psi_i(n) = \sum_{j=0}^{i} h_{n-j}, \quad i \geq 0;$$

$$\alpha_i(n) = h_n/\psi_i(n), \quad i \geq 0;$$

$$\beta_0(n) = 1; \quad \beta_i(n) = \prod_{j=0}^{i-1} \frac{\psi_j(n)}{\psi_j(n-1)}, \quad i > 0;$$

$$\phi_0(n) = f[n];$$

$$\phi_i(n) = \prod_{j=0}^{i-1} \psi_j(n-1) f[n, n-1, \ldots, n-i], \quad i > 0.$$

The ϕ_i are called modified divided differences and they reduce to backward differences when step-sizes are constant. Consider the term of degree i in the polynomial v_k (12.2); this is

$$\prod_{j=0}^{i-1}(x - x_{n-j}) f[n, n-1, \ldots, n-i]$$

$$= sh_n.(sh_n + h_{n-1}) \ldots (sh_n + h_{n-1} + \cdots + h_{n-i+1}) \frac{\phi_i(n)}{\prod_{j=0}^{i-1} \psi_j(n-1)}$$

$$= \frac{sh_n}{\psi_0(n)} \frac{sh_n + h_{n-1}}{\psi_1(n)} \cdots \frac{sh_n + h_{n-1} + \cdots + h_{n-i+1}}{\psi_{i-1}(n)} \beta_i(n)\phi_i(n)$$

$$= C_i(s)\phi_i^*(n)$$

where $\phi_i^*(n) = \beta_i(n)\phi_i(n)$, $C_0(s) = 1$, $C_1(s) = s$, and

$$C_i(s) = s \prod_{j=1}^{i-1} \frac{sh_n + \psi_{j-1}(n-1)}{\psi_j(n)}, \quad i \geq 2.$$

Integration yields the Predictor of order k, similar to the earlier formula (12.3),

$$y_{n+1}^p = y_n + h_n \sum_{i=0}^{k-1} \phi_i^*(n) \int_0^1 C_i(s)ds. \qquad (12.13)$$

The terms introduced here can easily be updated when a new step is started from the following relations

$$\psi_0(n) = h_n, \qquad \psi_i(n) = \psi_{i-1}(n-1) + h_n,$$

$$\beta_0(n) = 1, \qquad \beta_i(n) = \beta_{i-1}(n). \frac{\psi_{i-1}(n)}{\psi_{i-1}(n-1)}.$$

Also the coefficients C_i may be generated recursively from

$$C_i(s) = \left[\alpha_{i-1}(n)s + \frac{\psi_{i-2}(n-1)}{\psi_{i-1}(n)} \right] C_{i-1}(s), \quad i \geq 2.$$

Shampine and Gordon show that the Predictor can be expressed as

$$y_{n+1}^p = y_n + h_n \sum_{i=0}^{k-1} g_{i,1} \phi_i^*(n) \tag{12.14}$$

where the coefficients $g_{i,j}$ satisfy

$$g_{i,j} = \begin{cases} 1/j, & i = 0, \\ 1/j(j+1), & i = 1, \\ g_{i-1,j} - \alpha_{i-1}(n)g_{i-1,j+1}, & i \geq 2. \end{cases} \tag{12.15}$$

Note that the formula is recursively defined for $i > 1$. It is instructive to check the equal interval case for which

$$\beta_i = 1, \quad \alpha_i = 1/(i+1), \quad i = 0, 1, 2, \ldots.$$

The predictor formula contains only $g_{i,1}$ but the recursion requires $g_{i,j}$ with values of j up to $i+2$. A few coefficients are shown in Table 12.3, in which the first column contains the coefficients $g_{i,1}$ as required by the Predictor (12.14). Comparison with Table 9.2 shows that these are identical with the Adams-Bashforth coefficients. A consideration of the ϕ^* shows that, in this case, the Predictor may be written

$$y_{n+1}^p = y_n + h \sum_{i=0}^{k-1} \gamma_i \nabla^i f_n.$$

The Corrector of order $k+1$ is obtained by adding one term as in equation (12.8), where the p superscript indicates that the new divided difference is dependent on the predicted solution. Thus $\phi_0^p = f(x_{n+1}, y_{n+1}^p)$, and the order $k + 1$ Corrector becomes

$$y_{n+1}^c = y_{n+1}^p + h_n g_{k,1} \phi_k^p(n+1). \tag{12.16}$$

Table 12.3: Variable multistep coefficients $g_{i,j}$ when the step-size is constant

		j	
i	1	2	3
0	1	$\frac{1}{2}$	$\frac{1}{3}$
1	$\frac{1}{2}$	$\frac{1}{6}$	$\frac{1}{12}$
2	$\frac{5}{12}$	$\frac{1}{8}$	
3	$\frac{3}{8}$		
$\vdots$	$\vdots$		

Similarly the Corrector (12.9) of order k, used for error estimation, has the form

$$y_{n+1}^{c1} = y_{n+1}^{p} + h_n g_{k-1,1} \phi_k^p(n+1).$$

For practical computation of the modified divided differences, note that

$$\phi_{i+1}(n+1) = \prod_{j=0}^{i} \psi_j(n) f[n+1, n, \ldots, n-i],$$

in which

$$f[n+1, n, \ldots, n-i] =$$
$$\frac{f[n+1, n, \ldots, n-i+1] - f[n, n-1, \ldots, n-i]}{\psi_i(n)}.$$

This leads to the recursion

$$\phi_{i+1}(n+1) = \phi_i(n+1) - \beta_i(n)\phi_i(n) = \phi_i(n+1) - \phi_i^*(n), \qquad (12.17)$$

with a similar relation for the ϕ^p terms used in the Corrector formulae.

To demonstrate the modified approach it is useful to apply the Shampine and Gordon scheme to the differential equation (12.12) which was the subject of the less sophisticated approach outlined earlier. The quantities obtained in the first three steps of the modified process are given in Table 12.4. Since the new process is mathematically the same as before, the numerical solution at the third step is

$$y(0.07) \simeq y_3 = 1.075016 \quad (6 \text{ decimals}).$$

Proceeding to a fourth step of size $h_3 = 0.08$, the values of $\psi_i(3)$ and

Table 12.4: The first three steps in the solution of $y' = x + y$, $y(0) = 1$

n	x_n	y_n	i	$\psi_i(n)$	$\alpha_i(n)$	$\beta_i(n)$	$\phi_i(n)$	$\phi_i^*(n)$
0	0	1	0	0.01	1	1	1	1
1	0.01	1.0101	0	0.02	1	1	1.0201	1.0201
			1	0.03	$\frac{2}{3}$	2	0.0201	0.0402
2	0.03	1.030909	0	0.04	1	1	1.060909	1.060909
			1	0.06	$\frac{2}{3}$	2	0.040809	0.081618
			2	0.07	$\frac{4}{7}$	4	0.000609	0.002436

Table 12.5: Variable multistep coefficients $g_{i,j}$ with step doubling

i	\multicolumn{4}{c}{j}
	1	2	3	4
0	1	$\frac{1}{2}$	$\frac{1}{3}$	$\frac{1}{4}$
1	$\frac{1}{2}$	$\frac{1}{6}$	$\frac{1}{12}$	$\frac{1}{20}$
2	$\frac{7}{18}$	$\frac{1}{9}$	$\frac{1}{20}$	
3	$\frac{41}{126}$	$\frac{26}{315}$		
4	$\frac{2659}{9450}$			
$\vdots$	$\vdots$			

$\alpha_i(3)$ are determined first. From them the variable coefficients $g_{i,1}$ for the 4th order Predictor are obtained from the recurrence relations (12.15), and these are given in Table 12.5. Next, the modified divided differences are evaluated from (12.17)

$$\phi_0(3) = f(x_3, y_3), \qquad \phi_i(3) = \phi_{i-1}(3) - \phi_{i-1}^*(2), \quad i = 1, 2, 3.$$

The Predictor (12.14) depends on $\phi_i^*(3)$, which are easily derived given β_i and the above values. These yield

$$y_4^p = 1.173667$$

to 6 decimal accuracy, from which ϕ_0^p is computed. The 5th order Corrector is given by

$$y_4^c = y_4^p + h_3 g_{4,1} \phi_4^p(4),$$

where

$$\phi_i^p(4) = \phi_{i-1}(4) - \phi_{i-1}^*(3), \quad i = 1, 2, 3, 4.$$

Finally we obtain

$$y(0.15) \simeq y_4^c = 1.173668, \quad e_4 = h_3(g_{3,1} - g_{4,1})\phi_4^p(4) = 2 \times 10^{-7} < T.$$

Thus the step is acceptable and for the next one a fifth order Predictor with step-size 0.16 would be selected.

The problem of producing intermediate output, for values of x which are not normal step points, is easily solved. Having passed the specified output point X with the normal step, so that $x_n < X < x_{n+1}$, it is possible to compute a new step of length $H = X - x_n$, without recomputing divided differences, and to hit the desired point with a predictor. However it is necessary to calculate new g values to use in the Predictor (12.14), and also one must compute new modified divided differences if a corrector stage is employed. The process is analogous to that of dense output for Runge-Kutta methods.

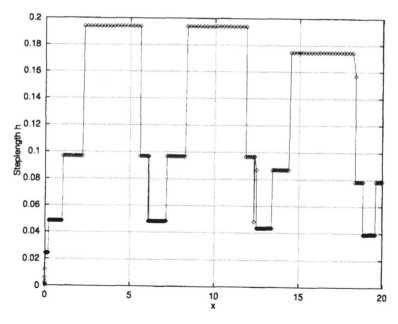

Figure 12.1: Variation of step-size in STEP90 for the orbit problem with $e = 0.5$

The ideas presented above are the foundation of a multistep procedure (STEP), originally coded in FORTRAN66 by Shampine and Gordon (1975). This has been recoded in Fortran 90 (STEP90) and the program is lightly documented and listed in Appendix B.

12.6   An application of STEP90

To demonstrate some features of the STEP90 program, it has been applied to the 2-body orbit problem as described in Chapter 5, in which the variable step-size RK algorithm was considered. The new package contains a MODULE and subroutines to implement the various features of the method. A user must write a main program unit and provide a MODULE with the name System to define the system of equations to be solved. The form of the MODULE is identical with those applicable to the earlier programs in this book and an example is provided with the STEP90 code in Appendix B.

The listing below shows a suitable main program which will give output at specified points. Other outputs are indicated in the appendix listing. In some cases it may be essential to gain direct access to other parts of STEP90. This is easily achieved if the statement USE Data is included in the user program. In this instance care must be taken not to confuse variable names.

```
PROGRAM Driver   ! Simple example of main program to drive
                 ! the STEP90 package. The equations to be
                 ! solved are defined in the MODULE System
                 ! which is USEd also by the STEP90 DATA
                 ! MODULE. Based on the STEP package of
                 ! Shampine & Gordon (1975).
!------------------------------------------J.R.Dormand 8/95
USE System       ! MODULE System defines the equations
IMPLICIT NONE
  INTEGER :: flag, i, neqs, nout
  REAL (KIND = 2), ALLOCATABLE :: y(:), yd(:)
  REAL (KIND = 2):: t, relerr, abserr, tout, tinc
  neqs = neq
  ALLOCATE(y(neqs), yd(neqs))      ! Dimension dynamic arrays
  CALL Initial(t, y)               ! Set initial values
  flag = 1
  PRINT*, 'Enter relerr and abserr';  READ*, relerr, abserr
  tout = 0.0d0
  PRINT*, 'Output increment, No of outs'; READ*, tinc, nout
  DO i = 1, nout
     tout = tout + tinc
     DO
        ! Call STEP90 subroutine to advance solution to tout
        CALL De(neqs, y, yd, t, tout, relerr, abserr, flag)
        IF(flag == 6) STOP
        IF(flag == 2) EXIT          ! tout reached
```

```
      END DO
      CALL Output(t, y)
   END DO
END PROGRAM Driver
```

Figure 12.1 shows the variation of step-size in the orbit integration, with $e = 0.5$, by the variable coefficient method using a pure absolute local error tolerance $T = 10^{-5}$. This figure can be usefully compared with Figure 5.4. Unlike the RK solution, the STEP procedure prefers not to vary the steplength too often, and so equal length steps are grouped.

Another interesting feature, not encountered with earlier processes, is the variation of order plotted in Figure 12.2. Starting with first order, the

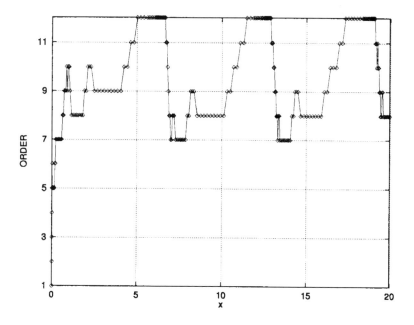

Figure 12.2: Variation of order in orbit problem with $e = 0.5$

code increases order very rapidly, as in the simple example shown earlier. However this does not remain constant at the highest level (12) which occurs in the ranges where step-size reduction is most notable in the comparable RK integration. The selection of order depends on the error estimates of multistep formulae of different orders at each step. From the differences computed to yield order k, all lower order solution estimates are available. If the trend of decreasing error with order is maintained the order may be increased. When it appears that the local error may be decreased by selecting a lower order, the order is reduced. Since the

stability regions of multistep formulae shrink as order increases, the idea
of improving accuracy by lowering order is not surprising.

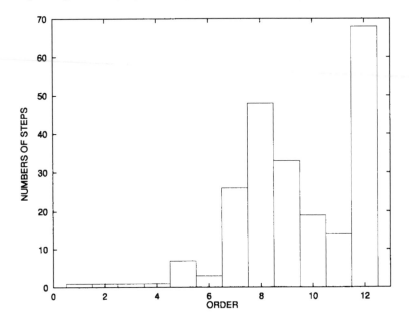

Figure 12.3: Distribution of orders in orbit problem with $e = 0.5$

Rejected steps are indicated in Figure 12.1 by vertical line segments.
Few steps are rejected.

The distribution of order (Figure 12.3) indicates the dominance of the
12th order phase. However this is not typical. With a mixed tolerance in
which the relative and absolute local error estimates are limited to 10^{-5}
the dominant order is reduced to 9. Of course the relative tolerance will
be more stringent for values of x near πn, where the components of the
solution attain extreme values, than is the absolute component. It will
be clear that the variation of order is highly problem dependent.

The actual step-sizes are quite similar to those obtained in the DO-
PRI5 integration with the same tolerance. As noted in Chapter 9, this
is typical in a multistep–RK comparison. Even with a large increase in
order the multistep process requires relatively small step-sizes. However
this reduction, as compared with Runge-Kutta, is compensated by the
need for only two function evaluations at each step.

12.7 Problems

1. Verify by direct integration of the appropriate divided difference interpolant the coefficients c_i from (12.3) up to $i = 3$.

2. Use a variable coefficient Predictor-Corrector formula to estimate $y(0.5)$ given the starter values

x	y
0	1
0.1	1.104987
0.25	1.280696

   and the differential equation $y' = y \cos x$. Compute Corrector values of different orders and estimate the maximum length for a second step if the tolerance is 10^{-5}. Check your approximation using the analytical solution $(y(x) = \exp(\sin x))$.

3. Use the recurrence relation (12.15) to compute the coefficients $g_{i,1}, i = 0, 1, 2, 3, 4$, for the initial step-doubling, order increasing phase of variable step multistep operation (see Table 12.5).

4. Verify the formula (12.17) for the evaluation of modified divided differences.

5. Compute a fifth step of length 0.16 for the differential equation (12.12) and determine an estimate of the Corrector error. Also estimate the solution at $x = 0.2$ by recomputing the multistep coefficients $g_{i,j}$.

6. Carry out 3 integration steps for the problem $y' = -\frac{1}{2}y^3$, $y(0) = 1$, with a tolerance (error/step) $T = 10^{-2}$ using the modified divided difference scheme. Use the starting step-size as given by the formula (12.10). Without computing a special step, estimate the solution at the intermediate value $x = 0.1$.

7. The equations of motion of n planets in the solar system may be expressed in heliocentric form as

$$\ddot{r}_i = -\frac{k^2(M_\odot + M_i)r_i}{|r_i|^3} - \sum_{j \neq i}^{n} k^2 M_j \left(\frac{r_i - r_j}{|r_{ij}|^3} + \frac{r_j}{|r_j|^3} \right),$$

   where $i = 1, 2, \ldots, n$, $|r_{ij}| = |r_i - r_j|$, and r_i is the position vector of the ith planet. Taking the solar mass $M_\odot$ as unity and the Gaussian gravitational constant $k = 0.01720209895$, the units of length

and time are Astronomical Units (AU) and days. The planetary
reciprocal masses $1/M_i$ are

Planet	Reciprocal mass
Mercury	6023600
Venus	408523.5
Earth + Moon	328900.55
Mars	3098710
Jupiter	1047.350
Saturn	3498.0
Uranus	22960
Neptune	19314
Pluto	130000000

The data for epoch Julian Day 2440400.5, equivalent to 0.00 hrs
on 28th June 1969, taken from the *Explanatory supplement to the
astronomical almanack* (Seidelmann, 1992), is given in Table 12.6
below. Carry out an integration using STEP90 to tabulate the rectangular coordinates of the planets, at intervals of 5 days, for 90
days.

Table 12.6: Planetary heliocentric coordinates at JD2440400.5

Planet	Heliocentric coordinates r_i	Heliocentric velocity $\dot{r}_i$
Mercury	$3.57260212546963E - 1$	$3.36784520455775E - 3$
	$-9.15490552856159E - 2$	$2.48893428375858E - 2$
	$-8.59810041345356E - 2$	$1.29440715971588E - 2$
Venus	$6.08249437766441E - 1$	$1.09524199354744E - 2$
	$-3.49132444047697E - 1$	$1.56125069115477E - 2$
	$-1.95544325580217E - 1$	$6.32887643692262E - 3$
Earth	$1.16014917044544E - 1$	$1.68116200395885E - 2$
+	$-9.26605558053098E - 1$	$1.74313126183694E - 3$
Moon	$-4.01806265117824E - 1$	$755975079765192E - 4$
Mars	$-1.14688565462040E - 1$	$1.44820048365775E - 2$
	$-1.32836653338579E + 0$	$2.37285174568730E - 4$
	$-6.06155187469280E - 1$	$-2.83748756861611E - 4$
Jupiter	$-5.38420864140637E + 0$	$1.09236745067075E - 3$
	$-8.31249997353602E - 1$	$-6.52329390316976E - 3$
	$-2.25098029260032E - 1$	$-2.82301211072311E - 3$
Saturn	$7.88988942673227E + 0$	$-3.21720514122007E - 3$
	$4.59570992672261E + 0$	$4.33063208949070E - 3$
	$1.55842916634453E + 0$	$1.92641681926973E - 3$
Uranus	$-1.82698911379855E + 1$	$2.21544461295879E - 4$
	$-1.16273304991353E + 0$	$-3.76765491663647E - 3$
	$-2.50376504345852E - 1$	$-1.65324389089726E - 3$
Neptune	$-1.60595043341729E + 1$	$2.64312595263412E - 3$
	$-2.39429413060150E + 1$	$-1.50348686458462E - 3$
	$-9.40042772957514E + 0$	$-6.81268556592018E - 4$
Pluto	$-3.04879969725404E + 1$	$3.22541768798400E - 4$
	$-8.73216536230233E - 1$	$-3.14875996554192E - 3$
	$8.91135208725031E + 0$	$-1.08018551253387E - 3$

Chapter 13

Global error estimation

13.1 Introduction

From a practitioner's point of view it is essential that a numerical process shall yield reliable solutions to differential equations. The application of step–size control based on local error estimates has gone a long way towards achieving this objective. Many test problems do yield global errors proportional to specified local error tolerance and so, in most practical cases, we may be confident of the extension of this property. Nevertheless, in some situations, one would prefer to have a direct estimate of the global error of the numerical solution. Most numerical analysts regard this as a difficult and/or a computationally expensive procedure but, with modern methods, this is not necessarily true. This chapter will be concerned with the development of such methods dealing with global error.

Many methods have been constructed to provide global error estimation. A typical procedure, often employed when local error control is practised, is called *tolerance reduction*. This depends on the assumption of tolerance proportionality being correct. Having solved the differential equations over the desired interval, a new solution is obtained using a reduced or increased tolerance. The differences in the solution, taken at comparable points, can be used to estimate global error. Since varying step–sizes do not normally give comparable output points unless dense output is employed, the technique is sometimes applied rather crudely. Nevertheless, the technique is quite effective for non-Stiff systems.

A more systematic approach is the classical *Richardson extrapolation* scheme, which requires a second solution with halved or with doubled step-sizes. This method is fairly reliable but costly.

Perhaps the best process for global error computation is based on a parallel solution of a related system of differential equations. These are constructed to have a solution satisfied by the actual global error of the main system of equations. Much of the present chapter will deal with this powerful technique, known as *solving for the correction*.

Assuming that a reliable global error estimate exists, one is faced with a dilemma similar to that in the local scenario. Does one extrapolate or not? The numerical analyst is rather cautious in considering this question, but many practitioners would argue that it would be unwise to 'throw away' the best solution available. The final topic in this chapter deals with global embedding, which provides a convenient method of global extrapolation.

13.2 Classical extrapolation

The classical extrapolation method is well-known in several numerical processes. Its application in the differential equations context is essentially the same as in the quadrature problem. Consider the usual system of first order equations with an initial value, $y' = f(x, y)$, $y(a)$ given. Apply any pth order method with N steps, $x \in [a, b]$, of sizes

$$h_0, h_1, h_2, \ldots, h_{N-1}; \quad h_i = \theta_i h, \quad 0 < \theta_i \leq 1,$$

yielding the approximations $y(x_i; h)$ to the true solution $y(x_i)$. The global error of the numerical solution at $x = b$ is expressed as

$$y(b; h) - y(b) = \varepsilon(b; h) = Ah^p + O(h^{p+1}).$$

To employ the Richardson extrapolation procedure the system is re-integrated, usually with halved step–sizes

$$\tfrac{1}{2}h_0, \tfrac{1}{2}h_0, \quad \tfrac{1}{2}h_1, \tfrac{1}{2}h_1, \ldots, \tfrac{1}{2}h_{N-1}, \tfrac{1}{2}h_{N-1}.$$

Then the global error of the second solution, at the same point $x = b$, is

$$\varepsilon(b; \tfrac{1}{2}h) = A(\tfrac{1}{2}h)^p + O(h^{p+1}).$$

Neglecting higher powers of h, the difference in the two solutions is

$$y(b; h) - y(b; \tfrac{1}{2}h) = Ah^p \left(1 - \frac{1}{2^p} \right),$$

giving an approximation for the global error

$$\varepsilon(b; h) = \frac{2^p \{ y(b; h) - y(b; \tfrac{1}{2}h) \}}{2^p - 1}, \tag{13.1}$$

or, for the more accurate of the two solutions,

$$\varepsilon(b; \tfrac{1}{2}h) = \frac{y(b; h) - y(b; \tfrac{1}{2}h)}{2^p - 1}.$$

Although this method is straightforward and quite reliable, it is costly. The first solution took N steps whereas the error estimation, also involving a more accurate solution, has taken a further $2N$ steps. For constant step-sizes the second integration could employ a doubled rather than halved steplength. This would be rather less costly but it is possible only when the variation of steplength is restricted. Also the quality of error estimation, in a relative sense, is much worse with the step doubling.

13.3   Solving for the correction

The method to be outlined here was proposed by Skeel (1986) and by Peterson (1986), but it is closely related to the technique formulated by Zadunaisky (1976), and developed by Dormand, Duckers and Prince (1984) and Dormand and Prince (1985). Like the Zadunaisky method, the scheme is plausible and simple in practice. Its implementation with Runge-Kutta methods has been developed by Dormand, Lockyer, McGorrigan and Prince (1989).

Consider the usual initial value problem,

$$y' = f(x, y), \quad y(x_0) = y_0, \quad y \in \mathbb{R}^M,$$

to which a Runge-Kutta method

$$y_{n+1} = y_n + h_n \Phi(x_n, y_n, h_n), \quad n = 0, 1, \ldots, N$$

is applied in some standard fashion. By the usual definition the global error is expressed as $\varepsilon_n = y_n - y(x_n)$ and it is easy to construct a differential equation whose true solution is satisfied by the global error at the integration points. Define the function $\varepsilon(x)$ to be

$$\varepsilon(x) = P(x) - y(x), \quad \text{where } P(x_n) = y_n, \quad n = 0, 1, \ldots, N. \qquad (13.2)$$

Thus P interpolates the numerical solution values in some, as yet unspecified, manner. Then $\varepsilon(x_n) = P(x_n) - y(x_n) = \varepsilon_n$, and differentiating equation (13.2) gives

$$\varepsilon' = P' - y' = P' - f(x, y(x)),$$

yielding the differential system

$$\varepsilon'(x) = P'(x) - f(x, P(x) - \varepsilon(x)). \qquad (13.3)$$

Given an initial value $\varepsilon(x_0) = \varepsilon_{h0} = 0$, the equation (13.3) can be solved numerically using the same step sequence as in the main problem to yield $\{\varepsilon_{hn}\}$, $n = 1, \ldots, N$. These values will be estimates of the global error of the y solution.

The choice of the interpolant P is crucial. Since typically N will be large, a piecewise interpolant is suggested. Normally a polynomial is chosen to interpolate a block of steps. If the block–size is 2 then a Lagrange interpolant will be of degree 2 as indicated in the Figure 13.1. The interpolant for the first block may be expressed in terms of divided differences as

$$P(x) = y_0 + (x - x_0)y[0, 1] + (x - x_0)(x - x_1)y[0, 1, 2],$$

where the differences are shown in Table 13.1. Alternatively, one may

Table 13.1: A divided difference table for 2nd degree interpolant

x_0	y_0		
		$y[0, 1]$	
x_1	y_1		$y[0, 1, 2]$
		$y[1, 2]$	
x_2	y_2		

choose Hermite interpolation so that, in each block,

$$P(x_i) = y_i, \quad P'(x_i) = f(x_i, y_i),$$

ensuring that P will be C^1 continuous. These methods, based on conventional interpolants, can be successful if certain conditions are satisfied. The conditions involve the blocksize m of the piecewise interpolant as well as the properties of the Runge-Kutta formula being applied. However, the most convenient form of P is identical to the dense output interpolant, $P(x) = y_{n+\sigma}$, of the RK formula (6.2) being used for the main integration. It would be convenient to use the same RK formula to solve the system (13.3) as applied to the main problem, but this is not essential. An important consideration is the order requirement for the solution of equations (13.3). Assuming that a qth order method is used to integrate the main problem the global error will satisfy $\varepsilon \sim O(h^q)$ and so if the same process is used for error equation (13.3) the numerical solution would have the same order of accuracy. Defining the global error of solution of equation (13.3) to be E_n at $x = x_n$, this determines

$$E_n = \varepsilon_{hn} - \varepsilon(x_n) \sim O(h^q),$$

which does not guarantee that, for small enough h, $E_n < \varepsilon_n$. This essential condition requires

$$E_n = \varepsilon_{hn} - \varepsilon_n \sim O(h^r), \quad r > q.$$

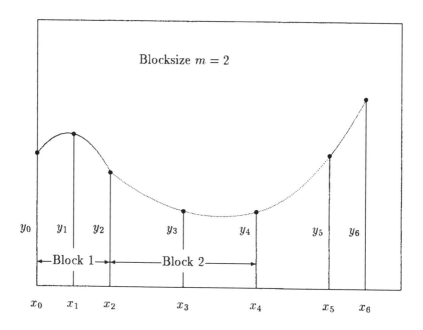

Figure 13.1: Piecewise interpolation of degree 2

When this condition is satisfied the global error estimation is said to be
valid. As in the case of classical extrapolation, it seems inevitable that
the global error is more costly to estimate than the numerical solution
itself. However the method here demands a *higher* order formula which
does not normally triple the cost as does the extrapolation technique with
step halving.

Actually the situation is even more favourable. When a detailed analysis of the application of an RK method to equation (13.3) is performed
(Dormand, Lockyer, McGorrigan and Prince, 1989), it emerges that there
is a deficiency in the RK error coefficients of the global error problem.
The exact form of the error depends on the type of interpolation employed, and it turns out that, for a sufficiently large blocksize when using
Lagrangian interpolants for P, valid global error estimation is always possible. If P is based on Hermite approximation or on the dense output
extension, it is necessary to construct a special estimator formula but
this needs to satisfy fewer conditions than a general integrator. In many
cases this will ensure that the RK *estimator* to be applied to equation
(13.3) will have less stages than the main RK formula. This economy
will be largely offset by extra stages which must be added to those of the
main formula to construct the continuous extension which gives $P(x)$.
For example, a locally 5th order dense output formula (see Table 6.5) for

RK5(4)7FM requires two extra function evaluations, f_8 and f_9.

Consider the application of a Runge-Kutta process of order q to the global error equation (13.3), termed the *correction*, which may be expressed in the form

$$\varepsilon' = \bar{f}(x, \varepsilon), \quad \varepsilon(0) = 0. \tag{13.4}$$

Defining the *defect* function as

$$d_h(x) = P'(x) - f(x, P(x)),$$

and also

$$f_h(x, P) = f(x, P) + d_h(x) = P'(x),$$

yields the relation

$$\bar{f}(x, \varepsilon) = f_h(x, P) - f(x, y). \tag{13.5}$$

Similar to equation (3.19), the local truncation error arising from the correction (13.4) will be

$$t_{\varepsilon n+1} = \sum_{i=1}^{\infty} h^i \left\{ \sum_{j=1}^{n_i} \tau_j^{(i)} \bar{F}_j^{(i)}(x_n, \varepsilon(x_n)) \right\}, \tag{13.6}$$

where $\bar{F}_j^{(i)}$ are the elementary differentials of $\bar{f}$. Using equation (13.5) one can write

$$\bar{F}_j^{(i)}(x_n, \varepsilon(x_n)) = F_{hj}^{(i)}(x_n, P(x_n)) - F_j^{(i)}(x_n, y(x_n)).$$

If the polynomial P is based on the solution from a method of order q, then

$$P(x_n) = y(x_n) + \varepsilon_n = y(x_n) + O(h^q),$$

and so the relevant elementary differential can be expressed as

$$\bar{F}_j^{(i)}(x_n, \varepsilon(x_n)) = F_{hj}^{(i)}(x_n, y(x_n)) - F_j^{(i)}(x_n, y(x_n)) + O(h^q).$$

In the scalar case it is easy to assess low order elementary differentials as in Chapter 3. Using the definition of the defect function, the following expressions are obtained.

$$
\begin{aligned}
F_{h1}^{(1)} - F_1^{(1)} &= f_h - f = d_h \\
F_{h1}^{(2)} - F_1^{(2)} &= f_{hx} + f_h f_{hy} - (f_x + f f_y) \\
&= f_x + d_h' + (f + d_h) f_y - (f_x + f f_y) \\
&= d_h' + f_y d_h \\
F_{h1}^{(3)} - F_1^{(3)} &= d_h'' + 2 f_{xy} d_h + f_{yy}(2 f d_h + d_h^2) \\
F_{h2}^{(3)} - F_2^{(3)} &= d_h' f_y + f_y^2 d_h.
\end{aligned}
$$

Consequently the local truncation error when a Runge-Kutta process is applied to the correction (13.4) becomes

$$
\begin{aligned}
t_{\varepsilon n+1} \;=\;& h\{\tau_1^{(1)}(d_h + O(h^q))\} \\
+\;& h^2\{\tau_1^{(2)}(d_h' + f_y d_h + O(h^q))\} \\
+\;& h^3\{\tau_1^{(3)}[d_h'' + 2f_{xy}d_h + f_{yy}(2f d_h + d_h^2) + O(h^q)] \\
+\;& \tau_2^{(3)}[d_h'f_y + f_y^2 d_h + O(h^q)]\} + \dots \qquad (13.7)
\end{aligned}
$$

As the notation implies, the defect function d_h is h-dependent. Thus the asymptotic properties of the expression above will depend on the type of interpolant chosen for $P(x)$. If $P(x)$ is of degree q interpolating values of $y(x)$, then its error can be expressed in the form

$$
P(x) - y(x) = W(\sigma)h^{q+1}, \quad \sigma = (x - x_n)/h.
$$

Differentiation yields

$$
P'(x) - y'(x) = \frac{dW}{d\sigma}h^q,
$$

and, assuming P is taken to be the dense output interpolant of degree and order q, it may be shown that the defect and its derivatives satisfy

$$
d_h^{(j)}(x) \sim \left\{ \begin{array}{ll} O(h^{q-j}), & j < q \\ O(1), & j \geq q \end{array} \right..
$$

Using this result in equation (13.7), it is possible to determine which orders contain the various error coefficients. For example in equation (13.7), d_h' will be typically $O(h^{q-1})$ while d_h'' will be $O(h^{q-2})$ provided $q \geq 2$, thus ensuring the error coefficient $\tau_2^{(3)}$ is moved one order higher than the other third order coefficient. Taking $q = 3$ and grouping the error coefficients of (13.7) with the appropriate powers of h gives

$$
\begin{aligned}
t_{\varepsilon n+1} \;\sim\;& \tau_1^{(1)}h^4 + \tau_1^{(2)}h^4 + \tau_1^{(3)}h^4 + \tau_2^{(3)}h^5 + \\
& \tau_1^{(4)}h^4 + \tau_2^{(4)}h^6 + \tau_3^{(4)}h^5 + \tau_4^{(4)}h^6 + O(h^5).
\end{aligned}
$$

For a valid estimator when $q = 3$ and $t_{n+1} \sim O(h^4)$, the requirement is $t_{\varepsilon n+1} \sim O(h^5)$, which is guaranteed when $\tau_1^{(i)} = 0$, $i = 1, 2, 3, 4$. This result extends to general order q, and conditions which will ensure minimal global error estimation by an estimator accompanying an RKq formula, with continuous extension of the same order, presented by Dormand, Gilmore, and Prince (1994), are

$$
\tau_1^{(i)} = 0, \quad i = 1, 2, \dots, q + 1.
$$

The error conditions featured in the above sequence are usually called *quadrature conditions* because a formula whose parameters eliminate these will be applicable, to the given order, to the quadrature problem $y' = f(x)$. To quote an example, the RK3Q derived in Chapter 2, given in tabular form as

$$
\begin{array}{c|cc|c}
c_i & a_{ij} & & b_i \\
\hline
0 & & & \frac{1}{6} \\
\frac{1}{2} & \frac{1}{2} & & \frac{2}{3} \\
1 & -1 & 2 & \frac{1}{6}
\end{array}
\qquad (13.8)
$$

has a principal error coefficient $\tau_1^{(4)} = 0$, and so it will yield *4th order* accuracy when applied to the correction (13.3) when this refers to a 3rd order RK, with $P(x)$ being taken as the locally–third–order continuous extension.

More accurate global error estimates can be obtained by selecting an estimator which ensures $r - q = 2$. With this condition the error computed is consistent with the first two terms of the asymptotic expansion of the global error. For the 3rd order formula here, a two term estimator must additionally satisfy

$$
\tau_2^{(3)} = \tau_3^{(4)} = \tau_1^{(5)} = \tau_5^{(5)} = 0.
$$

Such a formula, satisfying the seven equations of condition, is shown in Table 13.2. Note that this formula, although only of order 3 for a gen-

Table 13.2: The two-term estimator RK3E5 for RK3 with a 3rd order dense formula

$$
\begin{array}{c|ccc|c}
c_i & & a_{ij} & & b_i \\
\hline
0 & & & & \frac{1}{12} \\
\frac{5-\sqrt{5}}{10} & \frac{5-\sqrt{5}}{10} & & & \frac{5}{12} \\
\frac{5+\sqrt{5}}{10} & 0 & \frac{5+\sqrt{5}}{10} & & \frac{5}{12} \\
1 & \frac{-3+\sqrt{5}}{2} & 0 & \frac{5-\sqrt{5}}{2} & \frac{1}{12}
\end{array}
$$

eral problem, provides 5th order solutions for the correction. Similarly RK3E5 can be paired with a 4th order integrator provided that a continuous extension of order ≥ 3 is available. One-term estimation would be obtained in this case. As orders increase the relative cost of solving the correction diminishes if appropriate interpolants are available. For example, a five-stage estimator formula (Dormand, Gilmore, and Prince, 1994) can yield two-term estimation when allied with DOPRI5.

In practice, the correction procedure is straightforward and effective. Some computational examples will be presented below.

13.4   An example of classical extrapolation

A suitable example to demonstrate classical extrapolation is based on the non-linear equation

$$y' = -\tfrac{1}{2}y^3, \; y(0) = 1, \tag{13.9}$$

which has the true solution $y(x) = (x+1)^{-\frac{1}{2}}$. Taking a step–size $h = 0.4$, the third order RK formula (13.8) yields a numerical solution as depicted in Table 13.3. In this example, using the true solution the global error

Table 13.3: Solution of test problem with RK3Q

Stage	x	y	$f = -\tfrac{1}{2}y^3$
1	0.0	1.0	-0.5
2	0.2	0.9	-0.3645
3	0.4	0.9084	-0.374802
$x_1 = 0.4, \; y(x_1) \simeq y_1 = 0.844480$			

can be determined as

$$\varepsilon_1 = y_1 - y(x_1) = -0.000674.$$

To form an estimate of the global error using classical extrapolation (13.1), two halved steps of size 0.2 must be computed. Details of these steps are given in Table 13.4. From equation (13.1), the global error estimate is

$$\varepsilon_1 \simeq \varepsilon(0.4; 0.4) = \frac{2^3(y_1 - z_1)}{2^3 - 1} = -0.000718,$$

which is accurate to one significant figure, a fairly good result considering that the error is quite large with this step–size.

The extrapolation technique also gives an error estimate for the more accurate solution $(h = 0.2)$ as

$$\varepsilon(0.4; 0.2) = -0.000090.$$

Compared with the true value (-0.000046), this is not as impressive as the earlier result. Generally, the step halving procedure yields much better values than does the related step doubling extrapolation scheme.

Table 13.4: Two steps of size $h = 0.2$ with RK3Q

Step	Stage	x	y	$f = -\frac{1}{2}y^3$
	1	0.0	1.0	-0.5
1	2	0.1	0.95	-0.428688
	3	0.2	0.928525	-0.400268
$x_{\frac{1}{2}} = 0.2,\ y(x_{\frac{1}{2}}) \simeq z_{\frac{1}{2}} = 0.912833$				
	1	0.2	0.912833	-0.380315
2	2	0.3	0.874802	-0.334733
	3	0.4	0.855003	-0.312516
$x_1 = 0.4,\ y(x_1) \simeq z_1 = 0.845108$				

13.5 The correction technique

Before detailing the numerical aspects of this calculation, consider the
RK estimator formula which is valid in the RK3Q case. It is assumed
that the interpolant P is identical to the relevant continuous extension of
order 3 in this example. Since (13.8) is 4th order for the correction, the
estimator can be the same formula as used for the integrator. This is not
typical, but it provides an example with relatively simple parameters to
introduce the technique. The correction, to be solved with the estimator
formula, is

$$\varepsilon' = P' - f(x, P - \varepsilon) = \bar{f}(x, \varepsilon), \quad \varepsilon(0) = 0,$$

where

$$P(x) \;=\; y_{n+\sigma} = y_n + \sigma h \sum_{i=1}^{s} b_i^* f_i, \qquad \sigma = (x - x_n)/h,$$

$$P'(x) \;=\; \sum_{i=1}^{s} \left(b_i^* + \sigma \frac{db_i^*}{d\sigma} \right) f_i \;\;=\; \sum_{i=1}^{s} d_i^* f_i. \qquad (13.10)$$

Adding an FSAL stage to the RK3Q formula (13.8) allows the derivation
of a 3rd order C^1 continuous extension. Table 13.5 shows the complete
formula including the coefficients of the derivative P'. Since $\bar{f}$ contains
the P and P' functions, it is necessary for these to be evaluated three
times per step. Thus, for the RK3Q formula which has $c_2 = \frac{1}{2}$ and $c_3 = 1$,

Table 13.5: RK3Q with continuous extension and derivative

c_i	a_{ij}			b_i	b_i^*	d_i^*
0				$\frac{1}{6}$	$1 - \frac{3}{2}\sigma + \frac{2}{3}\sigma^2$	$1 - 3\sigma + 2\sigma^2$
$\frac{1}{2}$	$\frac{1}{2}$			$\frac{2}{3}$	$\sigma(2 - \frac{4}{3}\sigma)$	$4\sigma(1 - \sigma)$
1	-1	2		$\frac{1}{6}$	$\frac{1}{6}\sigma(3 - 2\sigma)$	$\sigma(1 - \sigma)$
1	$\frac{1}{6}$	$\frac{2}{3}$	$\frac{1}{6}$	0	$\sigma(\sigma - 1)$	$\sigma(3\sigma - 2)$

the quantities

$$\left.\begin{array}{l} P(x_n), P(x_n + \frac{1}{2}h), P(x_n + h), \\ P'(x_n), P'(x_n + \frac{1}{2}h), P'(x_n + h), \end{array}\right\} n = 0, 1, \ldots$$

must be computed. Fortunately the three values of σ are the same at each step and so the relevant dense output weights may be computed once and for all before the estimation process begins. For the specified process, Table 13.5 gives the relevant polynomials and the required values of b^* and d^* to be substituted in formulae (13.10) are shown in Table 13.6.

Table 13.6: Coefficients for evaluating P and P' with the three stage estimator RK3Q

σ	$\sigma b^*(\sigma)$	$d^*(\sigma)$
0	$(1, 0, 0, 0)$	$(1, 0, 0, 0)$
$\frac{1}{2}$	$\left(\frac{5}{24}, \frac{1}{3}, \frac{1}{12}, -\frac{1}{8}\right)$	$\left(0, 1, \frac{1}{4}, -\frac{1}{4}\right)$
1	$\left(\frac{1}{6}, \frac{2}{3}, \frac{1}{6}, 0\right)$	$(0, 0, 0, 1)$

All the formulae required for the global error estimation are now in place. As in the case of classical extrapolation, the test problem is equation (13.9) with step-size $h = 0.4$. The function values f_1, f_2, f_3 already have been computed in the integrator step shown in Table 13.3 and these are required for the evaluation of P and P', contained in the correction $\varepsilon' = \bar{f}(x, \varepsilon)$. In addition the FSAL stage f_4 is involved and must be evaluated before the correction. Actually this would be the first stage of the next step which will demand only a further three evaluations f_2, f_3

and f_4. Noting that the FSAL stage gives

$$f_4 = f(0.4, 0.844480) = -0.301119,$$

the implementation of the estimator is now shown in Table 13.7. This

Table 13.7: The estimator step with RK3Q

Stage	x	ε	$\bar{f} = P' + \frac{1}{2}(P - \varepsilon)^3$
1	0.0	0.0	0.0
2	0.2	0.0	-0.003275
3	0.4	-0.002620	0.002811
$x_1 = 0.4,\ \varepsilon_1 \simeq \varepsilon_{h1} = -0.000686$			

result is slightly better than the one obtained earlier with classical extrapolation, and it has been achieved by computing only a single step rather than two steps.

Typically, the quality of estimation from $O(h^q)$ correction is similar to that of Richardson extrapolation. For high quality global error estimation with RK3Q, the four stage estimator formula RK3E5 given in Table 13.2 should be selected. This requires new sets of b^* and d^*, for $\sigma = 0, (5 \pm \sqrt{5})/10, 1$, to replace those values in Table 13.6. Application of the formula to the same problem as above leads to the global error estimate

$$\varepsilon_{h1} = -0.000672,$$

a considerable improvement over the previous one-term estimates, at an extra compuational cost of one function evaluation. This impressive result is typical of *two-term estimation* as confirmed by extensive testing of Dormand and Prince (1992). However, the direct correction scheme outlined here is not the most convenient from a computational viewpoint and so further evidence of the high quality and reliability of the correction process is deferred at present.

13.6 Global embedding

It will be recalled that the estimation of local error in Runge-Kutta and in multistep formulae was usually followed by local extrapolation. This procedure, although popular with practitioners, does not find favour with all numerical analysts. It may be felt unsatisfactory to use, say, a fourth order error to control the step-size of a fifth order process. However, the most telling argument in favour of local extrapolation is that it delivers

superior results for a given computational cost. If the higher order formula of the RK embedded pair does not yield better accuracy than its
lower order partner, then the error estimate based on their difference is
not much use in step-size prediction or rejection, whatever the mode of
operation. Thus if one is prepared to estimate the error using the pair,
one is also forced to recognise the superiority of the high order formula
and, as a consequence, local extrapolation.

A similar argument may be applied in the global error context. If a
reliable global error estimate is available, why not use it for extrapolation?
Assuming that

$$E_n = \varepsilon_{hn} - \varepsilon_n \ll \varepsilon_n,$$

it is natural to form

$$\widetilde{y}_n = y_n - \varepsilon_{hn}.$$

Then the extrapolated value satisfies

$$\widetilde{y}_n - y(x_n) = -E_n$$

and, if the global error estimate is valid, $\widetilde{y}_n$ has a higher order of accuracy
than y_n. The correction process of the last section can be used to find
$\widetilde{y}_n$ at each step, thus providing solutions of different orders of accuracy,
as could be achieved by carrying out parallel integrations using entirely
different formulae. Since the error estimation process being used in the
present scheme is reminiscent of conventional RK embedding, the term
global embedding is an appropriate one.

A much more convenient computational algorithm can be developed
by combining the integrator–continuous extension–estimator processes.
The extrapolated solution can be expressed in increment function fashion involving the correction stages, which can be reduced to normal function evaluations. The global embedding process can be stated as follows
(Dormand, Gilmore and Prince, 1994).

Combining the RK3Q integrator and estimator with $\bar{s}$ stages (for
RK3E5, $\bar{s} = 4$), the solutions of differing orders, 3 and r, are given by

$$y_{n+1} \;=\; y_n + h \sum_{i=1}^{4} b_i f_i \qquad (13.11)$$

$$\widetilde{y}_{n+1} \;=\; \widetilde{y}_n + h \sum_{i=5}^{4+\bar{s}} b_i f_i, \qquad (13.12)$$

where

$$f_i = \begin{cases} f(x_n + c_i h, y_n + h \sum_{j=1}^{i-1} a_{ij} f_j, & i \le 4 \\[2mm] f(x_n + c_i h, \widetilde{y}_n + h \sum_{j=1}^{i-1} a_{ij} f_j, & 5 \le i \le 4 + \bar{s}. \end{cases}$$

Assuming that the 'bar' in $\bar{b}$ indicates a parameter from the estimator formula, then

$$b_{i+4} = \bar{b}_i, \quad c_{i+4} = \bar{c}_i, \quad i = 1, \ldots, \bar{s}.$$

This formula is only a little different from the 'standard' Runge-Kutta model. The new coefficients appearing in the function evaluations 5 onwards depend on the b^* and d^* values discussed earlier. It may be shown that

$$a_{4+k,j} = \begin{cases} \bar{c}_k b_j^*(\bar{c}_k) - \displaystyle\sum_{m=1}^{k-1} \bar{a}_{k,m} d_j^*(\bar{c}_m), & j = 1, 2, \ldots, 4 \\ \bar{a}_{k,j-4}, & j = 5, \ldots, 3+k \end{cases}$$

$$k = 1, 2, \ldots, \bar{s}. \tag{13.13}$$

The complete set of coefficients for an eight stage formula, designated GEM53, is shown in Table 13.8. Since stage 4 is the FSAL stage, to provide the continuous extension for the RK3 integrator, the effective stage count is seven per step, compared with six per step for a conventional RK5. Viewed as a fifth order formula the GEM53 does not seem particularly expensive. However it has the advantage of providing an estimate of the global error of a third order integrator.

Table 13.8: GEM53 in 8 stages

c_i	a_{ij}							b_i
0								$\frac{1}{6}$
$\frac{1}{2}$	$\frac{1}{2}$							$\frac{2}{3}$
1	-1	2						$\frac{1}{6}$
1	$\frac{1}{6}$	$\frac{2}{3}$	$\frac{1}{6}$					0
0	0	0	0	0				$\frac{1}{12}$
$\frac{5-\sqrt{5}}{10}$	$\frac{29\sqrt{5}-95}{300}$	$\frac{25-7\sqrt{5}}{75}$	$\frac{25-7\sqrt{5}}{300}$	$\frac{\sqrt{5}-5}{50}$	$\frac{5-\sqrt{5}}{10}$			$\frac{5}{12}$
$\frac{5+\sqrt{5}}{10}$	$\frac{25-17\sqrt{5}}{300}$	$\frac{\sqrt{5}-5}{75}$	$\frac{\sqrt{5}-5}{300}$	$\frac{\sqrt{5}}{25}$	0	$\frac{5+\sqrt{5}}{10}$		$\frac{5}{12}$
1	$\frac{35-6\sqrt{5}}{30}$	$\frac{6\sqrt{5}-20}{15}$	$\frac{3\sqrt{5}-10}{30}$	$\frac{5-3\sqrt{5}}{10}$	$\frac{\sqrt{5}-3}{2}$	0	$\frac{5-\sqrt{5}}{2}$	$\frac{1}{12}$

An important objective of local error estimation is the prediction of steplength, and so the question of step-size control based on global error estimates should be considered. The principle of step rejection, extended to global embedding, implies the rejection of all previous steps should the tolerated bound on the global error be exceeded. However, it is best to use a local error estimate to predict step-size. Since this is a local phenomenon there is no good reason to base the steplength on the global error.

Satisfactory results may be obtained by retaining the conventional embedded pair for the integrator to control step-size. A disadvantage of this is the implication that the local error of a formula of order $q - 1$ is used to control the globally extrapolated solution of order $q + 2$. With such a large difference in orders the usual assumption of tolerance proportionality is unjustified. An alternative scheme would be based on an estimator embedding. This implies extra stages in order to construct a pair of distinct estimators.

13.7   A global embedding program

Only a few modifications are required to incorporate global embedding in the RKpair program in Appendix D. First, it is necessary to modify the code used to read in the coefficients, and then the extra function evaluations are made following acceptance of the unextrapolated step. Figure 13.2 shows some of the extra code and the complete program (Gem) is listed in Appendix D.

Some results from the Gem program using GEM53 are illustrated in Figure 13.3. The test problem is the gravitational two-body simulation with eccentricity $e = 0.1$ integrated for $x \in [0, 20]$. The step-size control depends on an embedded 2nd order RK paired with the RK3. The error norms are taken over all steps and components, and the almost linear data provide a good check of the asymptotic properties of the 3rd order and extrapolated 5th order solutions. A surprising feature of the results, from the point of view of error estimation, is the high quality of the solution at very lax tolerances. The first point on the curves in Figure 13.3 is obtained from a local error tolerance $T = 0.01$, and the 3rd order solution has a maximum global error 2.09, a value larger than the maximum of any true solution component. The corresponding extrapolated solution has maximum error 0.0973, and so one could claim to have predicted successfully two significant figures in the global error of a very bad solution!

The situation is illustrated in Figure 13.4 which shows a plot of the 3rd order and extrapolated solutions for the first component (^1y) of y. The 3rd order curve shows a damped oscillation with decreasing period. After three periods ($x = 6\pi$) the bad solution is approximately π out of phase. This type of behaviour is not untypical in the numerical solution of oscillatory problems. In the n–body systems encountered in celestial mechanics, it is common practice to 'check' the numerical solution by testing the energy and angular momentum integrals, whose true values are available in non-dissipative cases. While an accurate solution would

```
. . . . . . . . .
    IF(delta < tol) THEN! Accepts step on local error basis
!   compute extra function values
      DO i = s+1, sg
        IF (i > star) THEN
!           Extra stage for estimator
            w = yt + h* MATMUL(a(i, 1: i-1), f(1: i-1, :))
        ELSE
!           Extra stage for continuous extension
            w = y +  h* MATMUL(a(i, 1: i-1), f(1: i-1, :))
        END IF
        f(i, :) = Fcn(x + c(i)*h, w)
      END DO
!     Extrapolated value
      yt = yt + h*MATMUL(b(star+1:), f(star+1:,))
!     Integrator value
      y = y + h*MATMUL(b(1:s), f(1:s,:))
      x = x + h
      gerr = y - yt                    ! Global error estimate
. . . . . . .
```

Figure 13.2: Extra code for global embedding

display good conservative properties, it should be remembered that conservation in itself does not imply small errors in the system components.
In the 2-body case, every point on the orbit will possess the same energy
and angular momentum. Therefore the numerical solution could be π
out of phase with the true value and still give 'acceptable' indications of
the two standard integrals. For the subject of Figure 13.4, the maximum
errors in angular momentum and energy are 0.10 and 0.11 respectively,
rather smaller than the actual solution error.

Figure 13.5 shows a comparison between two of the error measures.
The angular momentum error, being about 10% in a relative sense, and
the energy error (not shown) increase monotonically in this instance.
Bearing in mind the chaotic nature of celestial systems, one should not
place too much credence on numerical solutions whose quality is measured
purely in terms of conservation. The correction process outlined here is
much more valuable than the conservation law in determining the quality
of a numerical solution.

For the 2-body system considered above, the largest error is actually
associated with the component ^4y but the others are of the same magni-

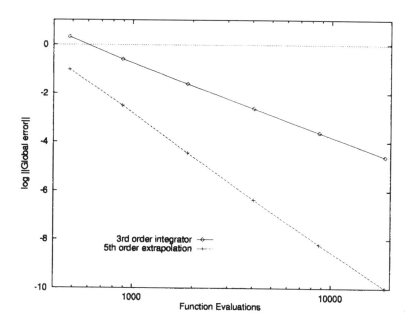

Figure 13.3: Errors of integrator and extrapolated solution for GEM53

tude. Even better global error estimation is obtained at more stringent tolerances, but the remarkable ability to cope successfully with poor solutions underlines the robustness of the correction technique.

The global embedding process may be applied to higher order formulae and Dormand, Gilmore and Prince have constructed a five stage estimator to give two-term estimation, or extrapolate to 7th order, when combined with the DOPRI5 integrator and a 5th order continuous extension. Since DOPRI5 is of FSAL type the effective number of function evaluations per step is 13. This GEM75 formula is not tabulated here but it is included in the program GEM90 given in Appendix D. Unlike most of the programs included in this text GEM90 comprises a package which, in a manner similar to STEP90 from Appendix B, is designed to be linked as an external unit to a user program. All parameters required by the GEM75 are included and therefore the package requires no external files. As well as extrapolated (7th order) solutions, the package can provide solutions at specified points using the 5th order continuous extension or by modifying the step-size to 'hit' the desired abscissa. In addition, Newton iteration on the dense output interpolant is used to locate a particular solution.

The step-size control in GEM90 is more sophisticated than that of RKpair. The algorithm is one used by Brankin et al. (1989), and avoids many of the step rejections suffered by some schemes.

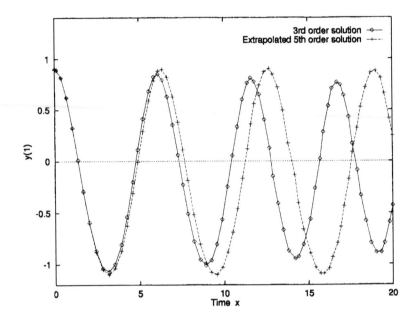

Figure 13.4: RK3 and extrapolated solution (first component) for GEM53

The GEM90 package is recommended as suitable for most non–stiff applications, either in 5th order form with dense output and global error estimates, or considered as a 7th order integrator. In the latter case the global embedding still provides a valuable error diagnostic.

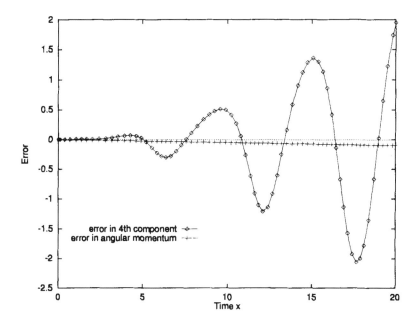

Figure 13.5: Error in angular momentum and a solution component with RK3

13.8   Problems

1. Given the equation $y' = x + y$, $y(0) = 1$, compute two steps of size $h = 0.1$ using Euler's method

$$y_{n+1} = y_n + hf(x_n, y_n).$$

Form the Lagrangian interpolant p of degree 2 for the three points (x_i, y_i), $i = 1, 2, 3$. Hence construct the differential equation

$$\varepsilon' = p' - f(x, p - \varepsilon)$$

satisfied by the global error of the solution of the main problem. By solving this equation with Euler's method, estimate the global error of the $\{y_i\}$ values. Also use the classical extrapolation method to approximate the same global errors, and compare with the exact values obtained from the true solution $y(x) = 2e^x - x - 1$. Finally, use an RK2 formula to solve the global error equation and compare with previous results.

2. The table below shows a 3–stage FSAL RK2 with a C^1 interpolant.

Find the derivative interpolant d^*.

c_i	a_{ij}		b_i	b_i^*
0			$\frac{1}{4}$	$1 - \frac{5}{4}\sigma + \frac{1}{2}\sigma^2$
$\frac{2}{3}$	$\frac{2}{3}$		$\frac{3}{4}$	$\frac{3}{4}\sigma(3 - 2\sigma)$
1	$\frac{1}{4}$	$\frac{3}{4}$	0	$\sigma(\sigma - 1)$

Will this formula be suitable as a global error estimator with itself as the integrator? Apply the formula to the differential equation from the question above, computing solutions and global error estimates for two steps of size 0.1.

3. Derive an estimator formula which will give valid global error estimation when applied to the RK4(3)5M formula given in Table 6.4. Assume that the fourth order dense output formula will be used to form P and P'. Test your formula on the problem (13.9) using the step-size $h = 0.4$.

4. Construct the GEM43 formula arising from the RK3 (13.8) being used as integrator and estimator. Compute two steps of size 0.4 with the GEM43 for the problem (13.9). Check your results against those obtained in §13.5.

5. Produce an appropriate data file for the GEM43 process and apply the Gem program from Appendix D to the two-body orbit problem with a range of tolerances $T \in [10^{-5}, 10^{-2}]$. Compare the quality of the fourth order solution with the fifth order from the GEM53 in Table 13.8.

6. Given the system $y' = f(x, y)$, y_0 given, is solved by a standard RK integrator of s stages, and the error equation $\varepsilon' = \bar{f}(x, \varepsilon)$ formed according to (13.3) and (13.10), an estimator formula

$$\varepsilon_{hn+1} = \varepsilon_{hn} + h \sum_{i=1}^{\bar{s}} \bar{b}_i \bar{f}_i, \quad \text{where}$$

$$\bar{f}_i = \bar{f}(x_n + \bar{c}_i h, \varepsilon_{hn} + h \sum_{j=1}^{i-1} \bar{a}_{i,j} \bar{f}_j), \quad i = 1, 2, \ldots, \bar{s},$$

is applied. Assuming standard notation, the continuous extension is contained within the s stages, and that $\tilde{y}_n = y_n - \varepsilon_{hn}$, show that

$$\bar{f}_1 = f_1 - f(x_n, \tilde{y}_n) = f_1 - f_{s+1},$$

$$\bar{f}_k = \sum_{i=1}^{s} d_i^*(\bar{c}_k) f_i - f_{s+k}, \quad k = 1, 2, \ldots, \bar{s},$$

where

$$f_{s+k} = f(x_n + \bar{c}_k h, \tilde{y}_n + h \sum_{j=1}^{s+k-1} a_{s+k,j} f_j), \quad k = 1, 2, \ldots, \bar{s}.$$

Hence establish the relations (13.12, 13.13) which characterise the GEM formula.

7. A parachutist jumps from an aeroplane travelling with speed v m/s in level flight at an altitude of y m. After a period of free-fall, the parachute is opened at a height y_p. The equations of motion of the skydiver are

$$
\begin{aligned}
\dot{x} &= v\cos\theta, \\
\dot{y} &= v\sin\theta, \\
\dot{v} &= -D/M - g\sin\theta, \quad D = \frac{1}{2}\rho C_D A v^2, \\
\dot{\theta} &= -g\cos\theta/v,
\end{aligned}
$$

where x is the horizontal coordinate, θ is the angle of descent, D is the drag force, and A is the reference area for the drag force, given by

$$A = \begin{cases} \sigma, & \text{when } y \geq y_p \\ S, & \text{when } y < y_p \end{cases}.$$

The constants are

$$
\begin{aligned}
g &= 9.81 \text{ m/s}, \quad M = 80 \text{ kg}, \quad \rho = 1.2 \text{ kg/m}^3, \\
\sigma &= 0.5 \text{ m}^2, \quad\quad S = 30 \text{ m}^2, \quad C_D = 1.
\end{aligned}
$$

Use GEM90 to simulate the descent of the parachutist. Use the interpolation facilities to determine the critical height y_p and the time of impact. Also estimate the minimum velocity of the descent prior to parachute deployment.

Chapter 14

Second order equations

14.1 Introduction

The numerical methods so far encountered have been designed for first order differential systems of the form given in equation (2.1)

$$y' = f(x, y), \quad y(x_0) = y_0, \quad y \in \mathbb{R}^k,$$

but many practical problems give rise to second order equations. In particular, dynamical systems are based on forces which cause acceleration, the second order derivative of position. The gravitational two-body problem, used as a test case in earlier chapters, falls into this category. It is a simple matter to convert second order problems into systems of first order equations amenable to conventional numerical integrators but, since they are relatively common, it is natural to seek new methods which can be applied directly with greater convenience. In this chapter we consider the derivation of numerical processes for direct application to the initial value problem

$$y''(x) = g(x, y(x), y'(x)), \quad y(x_0) = y_0, \ y'(x_0) = y_0', \ y \in \mathbb{R}^M. \quad (14.1)$$

Since equation (14.1) is easily converted to the form of (2.1) with $k = 2M$, it is clear that a valid approach to this problem is via a transformation of a standard first order integrator. A more general approach assumes the second order form (14.1) from the start. Using the Runge-Kutta basis, both approaches are discussed below.

Consideration is given also to the special second order problem

$$y''(x) = g(x, y(x)), \quad y(x_0) = y_0, \ y'(x_0) = y_0', \ y \in \mathbb{R}^M, \quad (14.2)$$

which is not explicitly dependent on the first derivative of the solution. Equations of this type are frequently encountered in celestial mechanics and many special Runge-Kutta methods have been constructed for such non-dissipative systems. The absence from the function g of the first derivative, y', yields a simpler numerical method and great savings in computational cost are possible as a result.

Runge-Kutta methods designed for second order differential equations are usually termed Runge-Kutta-Nyström formulae (RKN) since their introduction in 1925 by E.J.Nyström.

Special multistep formulae for second order problems also may be derived. These have become particularly popular in dynamical astronomy applications, although the use of RKN methods has become more popular in recent years.

14.2   Transformation of the RK process

Consider first the 'general' second order equation (14.1). Splitting it into first order equations, the system may be expressed as

$$\frac{d}{dx}\begin{pmatrix} y \\ y' \end{pmatrix} = \begin{pmatrix} f(x,y,y') \\ g(x,y,y') \end{pmatrix}$$

where $f = y'$. Application of a standard s-stage RK formula, as modelled by Table 3.3, to this system yields

$$y_{n+1} \;=\; y_n + h\sum_{i=1}^{s} b_i f_i, \tag{14.3}$$

$$y'_{n+1} \;=\; y'_n + h\sum_{i=1}^{s} b_i g_i, \tag{14.4}$$

where

$$g_i \;=\; g(x_n + c_i h, y_n + h\sum_{j=1}^{i-1} a_{ij} f_j, y'_n + h\sum_{j=1}^{i-1} a_{ij} g_j), \tag{14.5}$$

$$f_i \;=\; y'_n + h\sum_{j=1}^{i-1} a_{ij} g_j, \tag{14.6}$$

$$i \;=\; 1, 2, \ldots, s.$$

Substituting equation (14.6) in the increment formula (14.3), and using the result $\sum_{i=1}^{s} b_i = 1$, necessary for a consistent RK formula, one obtains

$$y_{n+1} \;=\; y_n + h y'_n + h^2 \sum_{i=1}^{s} b_i \sum_{j=1}^{i-1} a_{ij} g_j$$

$$= y_n + hy'_n + h^2 \sum_{i=1}^{s} \bar{b}_i g_i, \tag{14.7}$$

where

$$\bar{b}_i = \sum_{j=1}^{s} b_j a_{ji}, \quad i = 1, 2, \ldots, s. \tag{14.8}$$

Note that $\bar{b}_s = 0$ when the formula(14.7) is based on any conventional explicit RK process.

In a similar fashion, the y component of g_i in (14.5) can be transformed. Since $\sum_{j=1}^{i-1} a_{ij} = c_i$, equation (14.5) gives

$$g_i = g(x_n + c_i h, \quad y_n + c_i hy'_n + h^2 \sum_{j=1}^{i-1} \bar{a}_{ij} g_j, \quad y'_n + h \sum_{j=1}^{i-1} a_{ij} g_j), \tag{14.9}$$

where

$$\bar{a}_{ij} = \sum_{k=j+1}^{i-1} a_{ik} a_{kj}, \quad j = 1, \ldots, i-1; \quad i = 2, \ldots, s \tag{14.10}$$

and $\bar{a}_{i,i-1} = 0$, $i = 2, \ldots, s$, in this case.

When the transformation is applied to the three-stage RK3M given in (4.7), one obtains the RKNG formula of Table 14.1. The designa-

Table 14.1: A 'transformed' RKNG3 formula

c_i	a_{ij}		b_i	$\bar{a}_{ij}$		$\bar{b}_i$
0			$\frac{2}{9}$			$\frac{1}{6}$
$\frac{1}{2}$	$\frac{1}{2}$		$\frac{1}{3}$	0		$\frac{1}{3}$
$\frac{3}{4}$	0	$\frac{3}{4}$	$\frac{4}{9}$	$\frac{3}{8}$	0	0

tion 'RKNG' is used because the shorter 'RKN' abbreviation is usually reserved for formulae applicable to the special equation (14.2).

The RKNG3 of Table 14.1 is mathematically equivalent to the RK3M formula being applied to the corresponding first order system of differential equations. However, the method of transformation forces $\bar{b}_s = 0$, $\bar{a}_{i,i-1} = 0$, conditions which need not be imposed on the formulae (14.7) and (14.9). It is possible to show that the formulae for the $\bar{b}_i$ and

$\bar{a}_{ij}$ may be replaced by the simpler relations

$$
\begin{aligned}
\bar{b}_i &= b_i(1 - c_i), \quad i = 1, \ldots, s, \\
\bar{a}_{ij} &= a_{ij}(c_i - c_j), \quad i = 3, \ldots, s,
\end{aligned}
\tag{14.11}
$$

when the Runge-Kutta basis formula satisfies the Butcher simplifying relations (4.13, 4.14), and the Nyström row sum condition

$$
\sum_{j=1}^{i-1} \bar{a}_{ij} = \tfrac{1}{2}c_i^2, \quad i = 2, \ldots, s,
\tag{14.12}
$$

is satisfied.

14.3 A direct approach to RKNG processes

As in Chapter 3, where the Runge-Kutta process was expanded, general order conditions for the RKNG method can be found from the direct expansion of the local truncation error. The RKNG process may be expressed as

$$
\begin{aligned}
y_{n+1} &= y_n + h\Phi(x_n, y_n, y_n', h), \\
y_{n+1}' &= y_n' + h\Phi'(x_n, y_n, y_n', h),
\end{aligned}
$$

where the increment functions are

$$
\Phi = y' + h\sum_{i=1}^{s} \bar{b}_i g_i, \quad \Phi' = \sum_{i=1}^{s} b_i g_i,
$$

and g_i is defined in formula (14.9). If Δ is the Taylor series increment function the local truncation errors of the solution and derivative may be obtained by substituting the true solution $y(x)$ of the equation (14.1) into the RKNG increment function. This gives

$$
t_{n+1} = h[\Phi - \Delta], \quad t_{n+1}' = h[\Phi' - \Delta'].
$$

As before, these expressions are best given in terms of elementary differentials, and the Taylor series increment may be written

$$
\Delta = y' + \tfrac{1}{2}hF_1^{(2)} + \tfrac{1}{6}h^2 F_1^{(3)} + \tfrac{1}{24}h^3(F_1^{(4)} + F_2^{(4)} + F_3^{(4)}) + O(h^4),
$$

where, for the scalar case, the first few elementary differentials are

$$
F_1^{(2)} = g
$$

$$
\begin{aligned}
F_1^{(3)} &= g_x + g_y y' + g_{y'} g \\
F_1^{(4)} &= g_{xx} + 2g_{xy} y' + 2g_{xy'} g + 2g_{yy'} y' g + g_{yy}(y')^2 + g_{y'y'} g^2 \\
F_2^{(4)} &= g_{y'}(g_x + g_y y' + g_{y'} g) \\
F_3^{(4)} &= g g_y.
\end{aligned}
$$

Using these terms the increment function Φ' for a three stage formula becomes

$$
\begin{aligned}
\sum_{i=1}^{3} b_i g_i &= \sum_{i=1}^{3} b_i F_1^{(2)} + h \sum_{i=2}^{3} b_i c_i F_1^{(3)} \\
&+ h^2 \left(\frac{1}{2} \sum_{i=2}^{3} b_i c_i^2 F_1^{(4)} + b_3 a_{32} c_2 F_2^{(4)} + \sum_{i=2}^{3} \sum_{j=1}^{i-1} b_i \bar{a}_{ij} F_3^{(4)} \right) \\
&+ O(h^3).
\end{aligned}
\tag{14.13}
$$

The expressions for the local truncation errors in the y solution and the derivative are

$$
t_{n+1} = h^2 \left[\sum_{i=1}^{3} \bar{b}_i g_i - \left(\frac{1}{2} F_1^{(2)} + \frac{1}{6} h F_1^{(3)} + \cdots \right) \right],
\tag{14.14}
$$

$$
t'_{n+1} = h \left[\sum_{i=1}^{3} b_i g_i \right.
\tag{14.15}
$$

$$
\left. - \left(F_1^{(2)} + \frac{1}{2} h F_1^{(3)} + \frac{1}{6} h^2 (F_1^{(4)} + F_2^{(4)} + F_3^{(4)}) + \cdots \right) \right],
$$

and so, using equation (14.13), the order conditions for a three-stage, third order RKNG process are

$$
\sum_{i=1}^{3} \bar{b}_i = \frac{1}{2}
$$

$$
\sum_{i=2}^{3} \bar{b}_i c_i = \frac{1}{6}
$$

for the y component, and

$$
\sum_{i=1}^{3} b_i = 1
$$

$$
\sum_{i=2}^{3} b_i c_i = \frac{1}{2}
$$

$$\frac{1}{2}\sum_{i=2}^{3} b_i c_i^2 = \frac{1}{6}$$

$$b_3 a_{32} c_2 = \frac{1}{6}$$

$$\sum_{i=2}^{3}\sum_{j=1}^{i-1} b_i \bar{a}_{ij} = \frac{1}{6}$$

for the y' component. To simplify these equations it is usual, although this is not essential, to impose the condition (14.12), making the last equation of the set identical to an earlier one. This simplification also renders the set for y' identical to that for a standard RK3. Assuming that the y' parameters have been obtained, the remaining four parameters applicable to the y solution must satisfy only two conditions. A simple solution, shown in Table 14.2, is to choose the $\bar{b}_i$ according to equation (14.11) and to set $\bar{a}_{31} = 0$.

Table 14.2: A 'direct' RKNG3 formula

c_i	a_{ij}		b_i	$\bar{a}_{ij}$		$\bar{b}_i$
0			$\frac{2}{9}$			$\frac{2}{9}$
$\frac{1}{2}$	$\frac{1}{2}$		$\frac{1}{3}$	$\frac{1}{8}$		$\frac{1}{6}$
$\frac{3}{4}$	0	$\frac{3}{4}$	$\frac{4}{9}$	0	$\frac{9}{32}$	$\frac{1}{9}$

Since the direct approach offers free parameters, they can be chosen to optimise the RKNG formula in the same way as the RK methods of Chapter 4. The possibility exists of increasing the order of the y solution above that of the y' formula.

The general formulae for the components of the local truncation error for the RKNG processes are very similar to those of the standard RK as detailed in Chapter 3. In this case

$$t_{n+1} = \sum_{i=2}^{\infty} h^i \sum_{j=1}^{n_i} \tau_j^{(i)} F_j^{(i)}, \tag{14.16}$$

$$t'_{n+1} = \sum_{i=2}^{\infty} h^{i-1} \sum_{j=1}^{n_i} \tau_j^{(i-1)'} F_j^{(i)}. \tag{14.17}$$

The close relationship between the τ and τ' coefficients will be apparent from the third order example above. Consequently it is unnecessary to tabulate both sets of error coefficients. Table 14.3 shows the y error coefficients up to order 5 for the RKNG process.

To generate the y' coefficients up to order 4 from the table, the rule is as follows:

If the y error coefficient of order i is

$$\tau_j^{(i)} = \psi(a, \bar{a}, \bar{b}, c) - \frac{1}{\zeta}, \qquad (14.18)$$

then the corresponding y' coefficient of order $i - 1$ can be expressed as

$$\tau_j^{(i-1)'} = \psi(a, \bar{a}, b, c) - \frac{i}{\zeta}. \qquad (14.19)$$

Table 14.3: RKNG independent error coefficients to 5th order for y

1.	$\tau_1^{(2)} = \sum_i \bar{b}_i - \frac{1}{2}$
2.	$\tau_1^{(3)} = \sum_i \bar{b}_i c_i - \frac{1}{6}$
3.	$\tau_1^{(4)} = \frac{1}{2} \sum_i \bar{b}_i c_i^2 - \frac{1}{24}$
4.	$\tau_2^{(4)} = \sum_{ij} \bar{b}_i a_{ij} c_j - \frac{1}{24}$
5.	$\tau_1^{(5)} = \frac{1}{6} \sum_i \bar{b}_i c_i^3 - \frac{1}{120}$
6.	$\tau_2^{(5)} = \sum_{ij} \bar{b}_i c_i a_{ij} c_j - \frac{1}{40}$
7	$\tau_3^{(5)} = \sum_{ij} \bar{b}_i \bar{a}_{ij} c_j - \frac{1}{120}$
8.	$\tau_4^{(5)} = \frac{1}{2} \sum_{ij} \bar{b}_i a_{ij} c_j^2 - \frac{1}{120}$
9.	$\tau_5^{(5)} = \sum_{ijk} \bar{b}_i a_{ij} a_{jk} c_k - \frac{1}{120}$

A number of RKNG formulae have been constructed by Fehlberg (1974), Fine(1987), and Sharp and Fine (1992a, b). In particular the RKNG5(4) embedded pair of Fine is based on a model which has much in common with the DOPRI5 formula of Dormand and Prince (1980). Indeed the parameters a_{ij} and b_i, applicable to the y' component, satisfy identical relations to those of the DOPRI5 process, and so it is possible to derive a Nyström extension to DOPRI5 according to Fine's specification. Such an extension, which can be appended to Table 5.4, is shown in Table 14.4. Note that $\widehat{\bar{b}}_i$ and $\bar{b}_i$ refer to the 5th and 4th order members of the pair and

Table 14.4: The RK5(4)7FM (DOPRI5)Nyström extension

$\bar{a}_{ij}$						$\widehat{b}_i$	$\bar{b}_i$
						$\frac{35}{384}$	$\frac{5179}{57600}$
$\frac{1}{50}$						0	0
$\frac{9}{400}$	$\frac{9}{400}$					$\frac{50}{159}$	$\frac{7571}{238500}$
$\frac{8}{225}$	0	$\frac{64}{225}$				$\frac{25}{192}$	$\frac{393}{3200}$
$\frac{1276}{59049}$	$\frac{16}{81}$	$\frac{37312}{295245}$	$\frac{4876}{98415}$			$-\frac{243}{6784}$	$-\frac{10233}{339200}$
$\frac{19}{132}$	$\frac{9}{22}$	$-\frac{14}{55}$	$\frac{133}{660}$	0		0	0
$\frac{35}{384}$	0	$\frac{50}{159}$	$\frac{25}{192}$	$-\frac{243}{6784}$	0	0	0

that, as a consequence, error estimates may be computed for both components of the solution. This process is not mathematically identical to the DOPRI5 applied to the equivalent system of first order equations but its performance is quite similar to the conventional formula. Fine demonstrates some improvement in efficiency with his choice of parameters and also presents continuous extensions to the y and the y' solutions.

A program RKNG implementing the RKNG method is given in Appendix E. The problem selected is that of the restricted three-body motion which was the subject of Problem 11 from Chapter 9. Although this is a pure gravitational model with no dissipative effects, the choice of a rotating coordinate system introduces first derivative terms in the equations of motion.

14.4 The special second order problem

Many second order systems do not contain explicitly any first derivative terms. As an example consider the restricted three-body gravitational problem which was the subject of an exercise from Chapter 9 and the cited program RKNG. In an inertial frame of reference this problem can be written as

$$y'' = -\frac{E(y - y_E)}{|y - y_E|^3} - \frac{M(y - y_M)}{|y - y_M|^3}, \tag{14.20}$$

where

$$y = \begin{pmatrix} {}^1y \\ {}^2y \\ {}^3y \end{pmatrix}, \quad y_E = \begin{pmatrix} M\cos(x+\pi) \\ M\sin(x+\pi) \\ 0 \end{pmatrix}, \quad y_M = \begin{pmatrix} E\cos x \\ E\sin x \\ 0 \end{pmatrix}, \quad E = 1-M.$$

For this, and any other system of the form $y'' = g(x, y)$, the a_{ij} parameters in the RKNG process would be entirely redundant. In fact it is possible to derive special formulae which have less parameters but also less order conditions than would arise in the general case. This offers the prospect of fewer stages for a given algebraic order with concomitant saving in computational cost. From formulae (14.4, 14.7, 14.9) the special process becomes

$$
\begin{aligned}
y_{n+1} &= y_n + hy_n' + h^2\sum_{i=1}^{s} \bar{b}_i g_i, \\
y_{n+1}' &= y_n' + h\sum_{i=1}^{s} b_i g_i,
\end{aligned}
\tag{14.21}
$$

where

$$g_i = g\Big(x_n + c_i h, \quad y_n + c_i h y_n' + h^2\sum_{j=1}^{i-1} \bar{a}_{ij} g_j\Big).$$

An examination of Table 14.3 shows that the error coefficients

$$\tau_2^{(4)}, \ \tau_2^{(5)}, \ \tau_4^{(5)}, \ \tau_5^{(5)},$$

which depend on a_{ij}, must be absent for the special (RKN) formula. The fourth order RKN parameters need to satisfy just 3 conditions for the y component and 5 for the y' solution. Since a three stage formula involves 9 independent parameters when the row sum condition (14.12) is applied, it can attain 4th order accuracy. This is one stage less than the similar RK or RKNG methods. The error coefficients up to order 7 for y, and implicitly to order 6 for y', are given in Table 14.5. This table assumes the row sum condition holds and includes dependent coefficients and so it may be used to assess the error norms as well as provide relevant equations of condition.

Let us consider a fifth order RKN occupying four stages. Assuming the usual relation (14.11) between solution and derivative weights,

$$\bar{b}_i = b_i(1 - c_i), \quad i = 1, 2, 3, 4,$$

the problem reduces to one of solving 8 equations with 10 parameters available. These are

$$\sum_{i=1}^{4} b_i c_i^k = \frac{1}{k+1}, \quad k = 0, 1, 2, 3, 4,
\tag{14.22}$$

Table 14.5: RKN error coefficients to 7th order for y

1.	$\tau_1^{(2)} = \sum_i \bar{b}_i - \frac{1}{2}$
2.	$\tau_1^{(3)} = \sum_i \bar{b}_i c_i - \frac{1}{6}$
3.	$\tau_1^{(4)} = \frac{1}{2}\sum_i \bar{b}_i c_i^2 - \frac{1}{24}$
4.	$\tau_2^{(4)} = \tau_1^{(4)}$
5.	$\tau_1^{(5)} = \frac{1}{6}\sum_i \bar{b}_i c_i^3 - \frac{1}{120}$
6.	$\tau_2^{(5)} = 3\tau_1^{(5)}$
7.	$\tau_3^{(5)} = \sum_{ij} \bar{b}_i \bar{a}_{ij} c_j - \frac{1}{120}$
8.	$\tau_1^{(6)} = \frac{1}{24}\sum_i \bar{b}_i c_i^4 - \frac{1}{720}$
9.	$\tau_2^{(6)} = 6\tau_1^{(6)}$
10.	$\tau_3^{(6)} = 3\tau_1^{(6)}$
11.	$\tau_4^{(6)} = \sum_{ij} \bar{b}_i c_i a_{ij} c_j - \frac{1}{180}$
12.	$\tau_5^{(6)} = \frac{1}{2}\sum_{ij} \bar{b}_i a_{ij} c_j^2 - \frac{1}{720}$

13.	$\tau_6^{(6)} = \tau_5^{(6)}$
14.	$\tau_1^{(7)} = \frac{1}{120}\sum_i \bar{b}_i c_i^5 - \frac{1}{5040}$
15.	$\tau_2^{(7)} = 10\tau_1^{(7)}$
16.	$\tau_3^{(7)} = 15\tau_1^{(7)}$
17.	$\tau_4^{(7)} = \frac{1}{2}\sum_{ij} \bar{b}_i c_i^2 \bar{a}_{ij} c_j - \frac{1}{504}$
18.	$\tau_5^{(7)} = \tau_4^{(7)}$
19.	$\tau_6^{(7)} = \frac{1}{2}\sum_{ij} \bar{b}_i c_i \bar{a}_{ij} c_j^2 - \frac{1}{1008}$
20.	$\tau_7^{(7)} = \frac{1}{6}\sum_{ij} \bar{b}_i \bar{a}_{ij} c_j^3 - \frac{1}{5040}$
21.	$\tau_8^{(7)} = \tau_6^{(7)}$
22.	$\tau_9^{(7)} = 3\tau_7^{(7)}$
23.	$\tau_{10}^{(7)} = \sum_{ijk} \bar{b}_i \bar{a}_{ij} \bar{a}_{jk} c_k - \frac{1}{5040}$

and

$$b_3\bar{a}_{32}c_2 + b_4(\bar{a}_{42}c_2 + \bar{a}_{43}c_3) = \frac{1}{24} \qquad (14.23)$$

$$b_3 c_3 \bar{a}_{32}c_2 + b_4 c_4(\bar{a}_{42}c_2 + \bar{a}_{43}c_3) = \frac{1}{30} \qquad (14.24)$$

$$b_3\bar{a}_{32}c_2^2 + b_4(\bar{a}_{42}c_2^2 + \bar{a}_{43}c_3^2) = \frac{1}{60}. \qquad (14.25)$$

Selecting c_2, c_3 as the two free parameters, the quadrature equations (14.22) yield

$$c_4 = \frac{12 - 15(c_2 + c_3) + 20c_3 c_2}{15 - 20(c_2 + c_3) + 30c_3 c_2}, \qquad (14.26)$$

and the four b_i weights. The equations (14.23, 14.24, 14.25) are linear in the interior weights $\bar{a}_{32}, \bar{a}_{42}$ and $\bar{a}_{43}$ and so these are also obtained in terms of c_2 and c_3. The choice of parameters is determined by considering the principal error norm as outlined in Chapter 4 with reference to the standard Runge-Kutta process. A near optimal choice is

$$(c_2, c_3) = (\tfrac{1}{5}, \tfrac{3}{5}),$$

and the coefficients are displayed in the Table 14.6. The fifth order

Table 14.6: RKN5(4) coefficients

c_i	$\bar{a}_{ij}$			$\widehat{\widehat{b}}_i$	$\widehat{b}_i$	$\bar{b}_i$
0				$\dfrac{23}{432}$	$\dfrac{23}{432}$	$\dfrac{3}{1352}$
$\dfrac{1}{5}$	$\dfrac{1}{50}$			$\dfrac{25}{94}$	$\dfrac{125}{376}$	$\dfrac{1475}{4056}$
$\dfrac{3}{5}$	$-\dfrac{3}{250}$	$\dfrac{24}{125}$		$\dfrac{125}{756}$	$\dfrac{625}{1512}$	$\dfrac{125}{1352}$
$\dfrac{12}{13}$	$\dfrac{3708}{28561}$	$\dfrac{3525}{28561}$	$\dfrac{4935}{28561}$	$\dfrac{2197}{142128}$	$\dfrac{28561}{142128}$	$\dfrac{169}{4056}$

weights for y and y' are contained in the columns headed $\widehat{\widehat{b}}_i$ and $\widehat{b}_i$, respectively. Since local error estimates will be used for practical step-size control, a fourth order solution for y is provided under $\bar{b}_i$.

It is interesting to compare the efficiency of the RKN5(4) above with that of the standard RK5(4). The two-body gravitational problem is suitable and this must be split into a system of four first order equations as shown in Chapter 5 for the classical Runge-Kutta process. With eccentricity $e = 0.5$ and $x \in [0, 20]$, the problem has been integrated with DOPRI5 and the above RKN5(4) for tolerances $T \in [10^{-10}, 10^{-3}]$ to give the efficiency results of Figure 14.1. The maximum global error plotted is taken over all steps and variables. Both efficiency curves indicate fifth order formulae but the Nyström formula is cheaper in every instance. Since the coding for the Nyström formula is slightly simpler than that for standard RK processes the efficiency gain is marginally better than indicated. This is a typical result and problems of the special second order form are best tackled with RKN formulae.

Although the fifth order formula given here yields satisfactory results for many problems, it has one deficiency — there is no embedding for the y' solution. Thus the step-sizes will depend only on the local error estimates for the y component. A more robust control strategy should involve y' error estimates but, with only four stages, a distinct fourth order solution for y' cannot be achieved. Dormand, El-Mikkawy and Prince (1987a, 1987b) have presented a number of RKN processes, up to order 12, with $[y, y']$ embedding. The formula pairs of orders above 6

Table 14.7: RKN4(3)4FM coefficients

c_i	$\bar{a}_{ij}$			$\widehat{\widehat{b}}_i$	$\widehat{b}_i$	$\bar{b}_i$	b_i
0				$\frac{1}{14}$	$\frac{1}{14}$	$-\frac{7}{150}$	$\frac{13}{21}$
$\frac{1}{4}$	$\frac{1}{32}$			$\frac{8}{27}$	$\frac{32}{82}$	$\frac{67}{150}$	$-\frac{20}{27}$
$\frac{7}{10}$	$\frac{7}{1000}$	$\frac{119}{500}$		$\frac{25}{189}$	$\frac{250}{567}$	$\frac{3}{20}$	$\frac{275}{189}$
1	$\frac{1}{14}$	$\frac{8}{27}$	$\frac{25}{189}$	0	$\frac{5}{54}$	$-\frac{1}{20}$	$-\frac{1}{3}$

are unusual in that the lower order members have orders $p = q - 2$. All these pairs have been optimised according to the truncation error criteria introduced in Chapter 4 for Runge-Kutta methods. Table 14.7 contains the coefficients for the four stage FSAL RKN4(3)4FM formula.

Two of the formulae, of orders 6 and 12, were the basis of the RKN package described by Brankin et al. (1989) which is now included in the NAG library of FORTRAN77 subroutines. The 17 stage RKN12(10)17M is particularly suitable for accurate work.

14.5   Dense output for RKN methods

The construction of continuous extensions to RKN formulae is closely related to the same procedure for RK methods as seen in Chapter 6. Consider the four stage RKN5(4) derived earlier. Since a fifth order process for y has 5 independent error coefficients given in Table 14.5 an extra stage is essential. Fortunately an FSAL stage with

$$c_5 = 1, \quad \bar{a}_{5i} = \widehat{b}_i, \quad i = 1, \dots, 4$$

is able to provide the fifth parameter. The y interpolant formula is expressed as

$$y(x_n + \sigma h) \simeq y_{n+\sigma} = \widehat{y}_n + \sigma h \widehat{y}_n' + \sigma^2 h^2 \sum_{i=1}^{5} \bar{b}_i^* g_i, \qquad (14.27)$$

where $\bar{b}_i^* = \bar{b}_i^*(\sigma)$. To achieve common function evaluations for the interpolant RKN one must substitute c_i/σ and $\bar{a}_{ij}/\sigma^2$ for c_i and $\bar{a}_{ij}$ in the error coefficients in Table 14.5. In this case a unique solution for $\bar{b}_i^*(\sigma)$ given in Table 14.8 is obtained. To obtain a dense output formula for y'

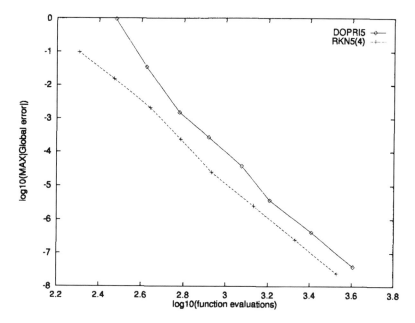

Figure 14.1: Efficiency curves for DOPRI5 and RKN5(4) applied to the orbit problem with eccentricity $e = 0.5$

it is sufficient to differentiate formula (14.27) yielding

$$y'_{n+\sigma} = \widehat{y}'_n + \sigma h \sum_{i=1}^{5} b_i^* g_i,$$

although the result will have local error only of order four. In this instance, the dense output formula has global continuity up to its second

Table 14.8: Dense output coefficients for RKN5(4)

$\bar{b}_i^*(\sigma)$
$-\dfrac{147\sigma^3 - 464\sigma^2 + 510\sigma - 216}{432}$
$\dfrac{25\sigma(9\sigma^2 - 25\sigma + 20)}{376}$
$-\dfrac{125\sigma^2(3\sigma - 5)}{1512}$
$-\dfrac{2197\sigma(33\sigma^2 - 76\sigma + 42)}{142128}$
$\dfrac{\sigma(\sigma - 1)^2}{2}$

derivative (C^2). For a detailed discussion of this topic the reader should consult the paper on RKN Triples by Dormand and Prince (1987).

14.6   Multistep methods

The integration procedure which was used to construct the Adams formulae in Chapter 9 can be extended to yield multistep methods for second order systems of equations. A single integration of the equation

$$y'' = g(x, y, y')$$

gives the conventional Adams explicit and implicit formulae which, in the present application, gives the first derivative solution. In backward difference form, these formulae are

$$\nabla y'_{n+1} = h \sum_{k=0}^{p} \gamma_k \nabla^k g_n, \tag{14.28}$$

$$\nabla y'_{n+1} = h \sum_{k=0}^{p} \gamma_k^* \nabla^k g_{n+1}, \tag{14.29}$$

where $g_j = g(x_j, y_j, y'_j)$, and γ_k, γ_k^* are given by

$$\gamma_0 = 1, \quad \gamma_k = \int_0^1 \binom{s+k-1}{k} ds, \quad k = 1, \ldots, p,$$

and

$$\gamma_0^* = 1, \quad \gamma_k^* = \gamma_k - \gamma_{k-1}, \quad k = 1, \ldots, p,$$

as detailed in Chapter 9. The formulae above are both of order $p + 1$.

Carrying out a double integration one obtains Störmer's explicit and implicit formulae which can be expressed as

$$\nabla^2 y_{n+1} = h^2 \left[g_n + \sum_{k=2}^{p-1} \chi_k \nabla^k g_n \right], \tag{14.30}$$

$$\nabla^2 y_{n+1} = h^2 \left[g_n + \sum_{k=2}^{p-1} \chi_k^* \nabla^k g_{n+1} \right], \tag{14.31}$$

where the coefficients χ, χ^* follow from

$$\left. \begin{array}{l} \chi_k = G(k, 1) + G(k, -1), \\ \chi_k^* = G(k, -2) - 2G(k, -1), \end{array} \right\} k = 2, \ldots, p, \tag{14.32}$$

and

$$G(k, s) = \int \int \binom{s+k-1}{k} ds.$$

The formulae (14.30, 14.31) have order $p + 1$. Given the starting values $y_n, y_{n-1}, \ldots, y_{n-p}$ and $y'_n, y'_{n-1}, \ldots, y'_{n-p}$, the above formulae can be applied in Predictor-Corrector mode. For the special second order problem, in which there is no explicit dependence on the first derivative, there is no need to compute y' at each step but, if this is required, there is no need to evaluate the Predictor for y'. Some Störmer coefficients are given in Table 14.9.

Table 14.9: Coefficients for Störmer's explicit and implicit formulae

k	χ_k	χ_k^*
2	$\frac{1}{12}$	$\frac{1}{12}$
3	$\frac{1}{12}$	0
4	$\frac{19}{240}$	$-\frac{1}{240}$
5	$\frac{3}{40}$	$-\frac{1}{240}$
6	$\frac{863}{12096}$	$-\frac{221}{60480}$
7	$\frac{275}{4032}$	$-\frac{19}{6048}$
8	$\frac{33953}{518400}$	$-\frac{9829}{3628800}$

The Störmer process is quite popular in astronomical applications which usually contain second order systems. With starting positional data based on accurate observations without a direct knowledge of velocity components the above ∇^2 procedure is particularly attractive. A common variant of the process is based on first and second sums, ∇^{-1} and ∇^{-2}, and is called the Gauss-Jackson method. Operating on equations (14.28, 14.29) with ∇^{-1} yields

$$y'_{n+1} = h \sum_{k=0}^{p} \gamma_k \nabla^{k-1} g_n,$$

$$y'_{n+1} = h \sum_{k=0}^{p} \gamma_k^* \nabla^{k-1} g_{n+1}.$$

Similarly one obtains

$$y_{n+1} = h^2 \left[\nabla^{-2} g_n + \sum_{k=2}^{p-1} \chi_k \nabla^{k-2} g_n \right],$$

$$y_{n+1} = h^2 \left[\nabla^{-2} g_n + \sum_{k=2}^{p-1} \chi_k^* \nabla^{k-2} g_{n+1} \right].$$

This method was described by Jackson (1924) and more recently by Herrick(1972). It has the advantage of requiring less differences than the mathematically equivalent Störmer process but the initialisation of the first and second sums, which satisfy

$$\nabla^{-1} g_j = \nabla^{-2} g_j - \nabla^{-2} g_{j-1}, \quad g_j = \nabla^{-1} g_j - \nabla^{-1} g_{j-1},$$

is an extra complication. For this purpose, Merson (1974) derives the formulae based on finite difference operations

$$\nabla^{-1} g_{j-1} = h^{-1} y_j' + \tfrac{1}{2} g_j + \tfrac{1}{12} \nabla g_{j+1} - \tfrac{1}{24} \nabla^2 g_{j+1} - \tfrac{11}{720} \nabla^3 g_{j+2}$$
$$+ \tfrac{11}{1440} \nabla^4 g_{j+2} + \cdots$$

$$\nabla^{-2} g_j = h^{-2} y_j - \tfrac{1}{12} g_j + \tfrac{1}{240} \nabla^2 g_{j+1} - \tfrac{31}{60480} \nabla^4 g_{j+2} + \cdots .$$

The value of j is in the centre of the difference table which is then extended to $\nabla^{-1} g_n$ and $\nabla^{-2} g_n$. An alternative scheme is based on a rearrangement of the Corrector formula to give

$$\nabla^{-2} g_{n-1} = h^{-2} y_n - \sum_{k=2}^{p-1} \chi_k^* \nabla^{k-2} g_n$$

with a similar expression for $\nabla^{-2} g_{n-2}$ to allow the initialisation of $\nabla^{-1} g_{n-1}$. This method requires an extra starter value so that both sums have the same order of accuracy but it does avoid the need for special formulae.

It must be said that the Gauss-Jackson procedure was developed specially for the convenience of manual operation and it offers little or no advantage over the Störmer process using modern computing methods.

14.7 Problems

1. Use the relations (14.8, 14.10) to construct the RKNG formula based on the fourth order RK process

c_i		a_{ij}		b_i
0				$\frac{1}{8}$
$\frac{1}{3}$	$\frac{1}{3}$			$\frac{3}{8}$
$\frac{2}{3}$	$-\frac{1}{3}$	1		$\frac{3}{8}$
1	1	-1	1	$\frac{1}{8}$

Note that this formula does not satisfy the Butcher relation (4.14). Use your RKNG4 formula to compute one step of size $h = 0.3$ of the solution of the Lane-Emden equation

$$y'' = \begin{cases} -\dfrac{2}{x}y' + R, & x > 0 \\[2mm] \frac{1}{3}R, & x = 0 \end{cases}$$

where $R = -y^5$ and the initial condition $y(0) = 1$, $y'(0) = 0$. Compare your result with the true solution

$$y(x) = \frac{1}{\sqrt{1 + \frac{1}{3}x^2}}.$$

This equation arises in the study of stellar structure.

2. Embed a third order formula for the y solution in the RKNG4 from the previous problem. Use the RKNG4(3) in the program RKNG from Appendix E to solve the Lane-Emden equation for $x \in [0, 6]$. Compare the results with the true solution when tolerances in the range 10^{-5} to 10^{-3} are applied.

3. Show that the relation (14.8) reduces to

$$\bar{b}_i = b_i(1 - c_i)$$

when the RK formula satisfies the Butcher condition (4.13).

4. Using the error coefficients of Table 14.3 and the conditions

$$\sum_j a_{ij}c_j^k = \frac{1}{k+1}c_i^{k+1}, \quad k = 0, 1, 2$$

$$\sum_j \bar{a}_{ij}c_j^k = \frac{1}{(k+1)(k+2)}c_i^{k+2}, \quad k = 0, 1,$$

$$i = 3, 4, 5,$$

where $b_2 = 0$, show that the equations of condition for the y' RKNG solution of order 4 in 5 stages reduce to

$$\sum_{i=3}^{5} b_i c_i^k = \frac{1}{k+1}, \quad k = 0, 1, 2, 3$$

$$\sum_{i=3}^{5} b_i a_{i2} = 0.$$

Outline a method of solution in terms of the free parameters (c_3, c_4, c_5) and hence obtain the complete RKNG4 formula when $(c_3, c_4, c_5) = (\frac{1}{3}, \frac{2}{3}, 1)$.
This formula is related to those of Fine (1985).

5. Find an embedded 3rd order formula to pair with the 4th order formula from the previous problem. Also construct a locally 4th order interpolant for the y solution. Apply the formula in the program RKNG from Appendix E to the Lane-Emden equation from problem 1 above in which $R = -y$. Use the dense output formula to estimate x such that $y(x) = 0$.

6. Apply the RKNG program to solve the van der Pol equation

$$y'' = \mu(1 - y^2)y' - y, \quad y(0) = 0.1, y'(0) = 0,$$

for $x \in [0, 50]$ with (a) $\mu = 2$, (b) $\mu = 5$. Draw phase plane diagrams to show the behaviour of the two solution components.

7. Show that the number of parameters available for the s stage RKN formula for the special equation $y'' = g(x, y)$ is $s(s + 3)/2$.

8. Consider the construction of a 3-stage RKN of order 4 for the special problem $y'' = g(x, y)$, using the error coefficients given in Table 14.5. Show that the quadrature conditions yield the relation

$$c_3 = \frac{4c_2 - 3}{2(3c_2 - 3)}$$

and, when the free parameter $c_2 = \frac{1}{2}$, a fourth order RKN is

c_i	a_{ij}		b_i	d_i
0			$\frac{1}{6}$	$\frac{1}{6}$
$\frac{1}{2}$	$\frac{1}{8}$		$\frac{1}{3}$	$\frac{2}{3}$
1	0	$\frac{1}{2}$	0	$\frac{1}{6}$

9. Investigate the provision of a dense output formula of local order 3 or 4 for the RKN4 in the previous problem. Also consider embedding a third order formula for step–size control purposes.

10. Compute two steps of size $h = 0.2$ of the solution of

$$y'' = -1/y^2, \quad y(0) = 1, y'(0) = 0$$

using the RKN4. Reconsider your calculation tolerance $T = 10^{-4}$ when an error estimate from an embedded third order formula is available.

11. Halley's comet passed through perihelion (closest point to the sun) on February 9th 1986. Its heliocentric position and velocity at this time were

$$
\begin{aligned}
r &= (0.325514, -0.459460, 0.166220) \text{ AU} \\
\dot{r} &= (-9.096111, -6.916686, -1.305721) \text{ AU/year}
\end{aligned}
$$

where the astronomical unit (AU) is the Earth's mean distance from the sun. The equation of motion of the comet is

$$\ddot{r} = -4\pi^2 r / |r|^3.$$

Use the RKN6(4)FM formula in Table 14.10 from Dormand, El-Mikkawy and Prince (1987a) in RKNG to solve the equation to determine the maximum value of r and the orbital period of the comet. A y interpolant will be most useful for these tasks.

12. Repeat problem 12.7 using the RKN6(4)6FM.

13. Apply a fourth order Störmer process to the Lane-Emden equation with $R = -y^5$ with $h = 0.25$. Determine appropriate starter values from the true solution and then compute two steps with the multistep formula. Write a program to implement the Störmer method with any order. The Adams program from Appendix B should serve as a model for your code.

Table 14.10: RKN6(4)6FM coefficients

c_i	$\bar{a}_{ij}$				
0					
$\frac{1}{10}$	$\frac{1}{200}$				
$\frac{3}{10}$	$-\frac{1}{2200}$	$\frac{1}{22}$			
$\frac{7}{10}$	$\frac{637}{6600}$	$-\frac{7}{110}$	$\frac{7}{33}$		
$\frac{17}{25}$	$\frac{225437}{1968750}$	$-\frac{30073}{281250}$	$\frac{65569}{281250}$	$-\frac{9367}{984375}$	
1	$\frac{151}{2142}$	$\frac{5}{116}$	$\frac{385}{1368}$	$\frac{58}{168}$	$-\frac{6250}{28101}$

$\bar{\bar{b}}_i$	$\hat{\bar{b}}_i$	$\bar{b}_i$	b_i
$\frac{151}{2142}$	$\frac{151}{2142}$	$\frac{1349}{157500}$	$\frac{1349}{157500}$
$\frac{5}{116}$	$\frac{25}{522}$	$\frac{7873}{50000}$	$\frac{7873}{45000}$
$\frac{385}{1368}$	$\frac{275}{684}$	$\frac{192199}{900000}$	$\frac{27457}{90000}$
$\frac{58}{168}$	$\frac{275}{252}$	$\frac{521683}{2100000}$	$\frac{521683}{630000}$
$-\frac{6250}{28101}$	$-\frac{78125}{112404}$	$-\frac{16}{125}$	$-\frac{2}{5}$
0	$\frac{1}{12}$	0	$\frac{1}{12}$

Chapter 15

Partial differential equations

15.1 Introduction

Many physical processes important in science and engineering are' modelled by partial differential equations. The numerical analysis of these is usually presented quite separately (and differently) from that devoted to ordinary differential equations. Finite element techniques have dominated the recent research effort in problems involving several independent variables but finite difference methods are still popular in practical situations. The importance of numerical stability is somewhat greater in this field than in the ODE case and, for this reason, relatively low orders of consistency are the norm. The problem of non-linearity is more severe than in the one variable case and this can make the application of implicit finite difference schemes fairly daunting. Consequently, it is common for parabolic equations to be solved numerically with explicit finite difference methods, closely related to Euler's method for ordinary differential equations, which might seem excessively crude to users of high order Runge-Kutta and multistep processes developed in earlier chapters.

The finite difference method has been covered extensively in a number of specialized texts (e.g. Mitchell and Griffiths, 1980; Smith, 1985; Morton and Mayers, 1994) and so we do not propose to describe this conventional technique in detail here. Rather we consider the application of methods described in earlier chapters. This can be achieved very easily following the process known as semi-discretisation in which some, but not all, of the partial derivatives are replaced by differences to create systems of ordinary differential equations. In this chapter we will consider only those cases yielding initial value problems but, alternatively, boundary value ODEs could be formed. Any process depending on semi-discretisation processes is usually classified as a *Method of Lines*.

15.2   Finite differences

Consider the parabolic partial differential equation which models transient heat conduction in one space dimension (X). The temperature $u = u(X, T)$, where T is the time, satisfies the equation

$$c\rho \frac{\partial u}{\partial T} = \frac{\partial}{\partial X} \left(\kappa \frac{\partial u}{\partial X} \right), \quad X \in [0, L], \quad T \geq 0, \qquad (15.1)$$

with typical boundary and initial conditions

$$u(0, T) = b_l(T), \ u(L, T) = b_r(T); \quad u(X, 0) = f(X), \ X \in [0, L].$$

The physical properties c, ρ, and κ, representing specific heat, density, and thermal conductivity, generally will be variable and dependent on the

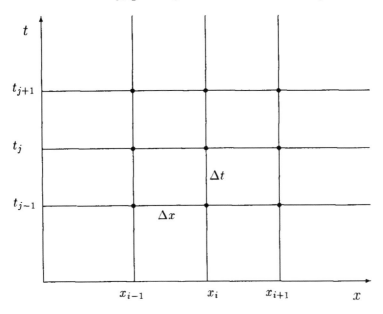

Figure 15.1: A finite difference mesh

temperature u. The simplest case occurs when the physical parameters are constant. In this case the equation reduces to

$$\frac{\partial u}{\partial T} = \alpha \frac{\partial^2 u}{\partial X^2}, \quad \alpha = \frac{\kappa}{c\rho},$$

and, writing $t = \alpha T/L^2$, it further simplifies to give the equation

$$\frac{\partial u}{\partial t} = \frac{\partial^2 u}{\partial x^2}, \quad x \in [0, 1], \quad t \geq 0. \tag{15.2}$$

The quantity u and its derivatives occur only in the first degree in this equation which is thus said to be linear. To form a finite difference approximation to the partial differential equation (15.2), we construct a mesh in the $x - t$ plane as shown in Figure 15.1. Writing

$$\begin{aligned}
x_i &= i\Delta x, \quad i = 0, 1, \ldots, m \\
t_j &= j\Delta t, \quad j = 0, 1, 2, \ldots \\
u_{i,j} &\simeq u(x_i, t_j)
\end{aligned}$$

then the derivatives may be approximated at the mesh point (x_i, t_j) by

$$\begin{aligned}
\frac{\partial u}{\partial t}(x_i, t_j) &\simeq \frac{u_{i,j+1} - u_{i,j}}{\Delta t} \\
\frac{\partial^2 u}{\partial x^2}(x_i, t_j) &\simeq \frac{u_{i-1,j} - 2u_{i,j} + u_{i+1,j}}{(\Delta x)^2}.
\end{aligned}$$

The consistency of these expressions is easily determined by Taylor expansions. Replacing the derivatives in equation (15.2) with the above difference formulae gives the explicit formula

$$u_{i,j+1} = u_{i,j} + \frac{\Delta t}{(\Delta x)^2}(u_{i-1,j} - 2u_{i,j} + u_{i+1,j}) \tag{15.3}$$

for an approximation to the solution at a single mesh point at time $t = t_j + k$. A single time step is completed when a complete row of the mesh has been updated. This serves as the initial condition for the next step.

The local truncation error of this solution can be expressed as

$$t_{ij} \leq A(\Delta t)^2 + B\Delta t(\Delta x)^2,$$

where A and B are independent of Δt and Δx. The explicit nature of the finite difference equation (15.3) is very attractive since it is easily applied even when the variable physical properties render the differential equation non-linear.

Improved orders of consistency and unconditional stability are a feature of implicit finite difference schemes but these can be difficult to apply in the non-linear situation. The alternative formulation as a method of lines, which will be described below, permits the application of explicit techniques for ordinary differential equations.

15.3   Semi-discretisation of the heat equation

In this application, one of the independent variables (x) from the equation (15.2) is discretised. This gives

$$\frac{\partial u}{\partial t}(x_i, t) \simeq \frac{1}{(\Delta x)^2}\{u(x_{i-1}, t) - 2u(x_i, t) + u(x_{i+1}, t)\},$$
$$i = 1, 2, \ldots, m - 1,$$

but the time derivative should now be considered of the ordinary type. Thus a system of ordinary differential equations

$$\frac{du_i}{dt} = \frac{1}{(\Delta x)^2}(u_{i-1} - 2u_i + u_{i+1}), \quad i = 1, 2, \ldots, m - 1, \tag{15.4}$$

with solution $u_i = u_i(t)$, will be used to approximate $u(x_i, t)$ from the partial differential equation (15.2). Any of the methods described in earlier chapters could be applied to this system. With this approach, the explicit process (15.3) is equivalent to the application of Euler's method.

Intuition tells us that the replacement of (15.2) by (15.4) is a reasonable approach. However, it is straightforward to compare the analytical solutions of the PDE and of the ODE system in a simple case. Consider (15.2) with the boundary and initial conditions

$$u(0, t) = u(1, t) = 0; \quad u(x, 0) = \sin \pi x, \ x \in [0, 1]. \tag{15.5}$$

The method of separation of variables yields the solution

$$u(x, t) = \exp(-\pi^2 t) \sin \pi x$$

in this case. The linear system (15.4) with the corresponding initial condition
$$u_i(0) = \sin(i\pi/m), \ i = 1, 2, \ldots, m - 1,$$

and $u_0(t) = u_m(t) = 0$, has the solution

$$u_i(t) = \exp\{-4m^2 \sin^2(\pi/2m)t\} \sin(i\pi/m). \tag{15.6}$$

It is clear that (15.6) will converge to the true solution of (15.2) as $\Delta x = 1/m \to 0$.

The system (15.4) becomes Stiffer as the number of intervals m in x increases, suggesting that a conventional explicit ordinary differential equation solver may be unsuitable for problems of this type. In matrix form equation (15.4) may be written as

$$\frac{du}{dt} = Bu, \quad u = (u_1, u_2, \ldots, u_{m-1})^T,$$

where

$$B = \frac{1}{(\Delta x)^2} \begin{bmatrix} -2 & 1 & 0 & \cdots & 0 \\ 1 & -2 & 1 & \cdots & 0 \\ 0 & 1 & -2 & \cdots & 0 \\ \vdots & \vdots & \vdots & \ddots & \vdots \\ 0 & 0 & \cdots & 1 & -2 \end{bmatrix}, \quad \Delta x = 1/m.$$

It may be shown (Smith, 1985) that B has eigenvalues

$$\lambda_k = -\frac{4}{(\Delta x)^2} \sin^2 \left(\frac{k\pi}{2m} \right), \quad k = 1, 2, \ldots, m - 1,$$

and so

$$-\frac{4}{(\Delta x)^2} < \lambda_k < 0. \tag{15.7}$$

Consider now the application of an ordinary differential equation solver with a real negative absolute stability interval $[-l, 0]$. Thus for the ODE steplength Δt, the inequality

$$-l \le \Delta t \lambda_k \le 0$$

must be satisfied. From the result (15.7), and assuming a positive time step, this gives

$$\Delta t \le \frac{l}{4} (\Delta x)^2$$

and, in the terminology familiar to finite difference practitioners, where the *mesh ratio* is defined to be $R = \Delta t / (\Delta x)^2$,

$$0 < R \le \tfrac{1}{4} l.$$

For Euler's method the absolute stability limit is $l = 2$ and so the limit on the mesh ratio is $R = \frac{1}{2}$. This case is equivalent to the finite difference formula (15.3). Since the limit applies to the mesh ratio, a reduction in the spacing Δx must be accompanied by an appropriate reduction in the time step-size and so an improvement in the spatial approximation implies more integration steps.

Reference to Chapter 7 confirms that a two-stage RK2 would be subject to the same limiting step-size, but DOPRI5 from Table 5.4 would permit $R \le 0.82$, a very modest improvement in view of the extra computational cost of a step with the higher order process. Of course, the solution accuracy would be greatly improved with the fifth order method. In order to solve efficiently a system such as (15.4), in which stability is limiting the step-size, we must seek formulae with extended stability properties. Conventional explicit RK and multistep processes are obviously not appropriate.

15.4   Highly stable explicit schemes

Although implicit Runge-Kutta and multistep processes can be unconditionally stable, their computational cost per step is rather higher explicit
methods. In Chapter 7, explicit RK methods with extended stability
were constructed. These are particularly suitable for application to the
method of lines. The 3-stage first order formula given by Table 7.2 may
be expressed in the form

$$
\begin{aligned}
y_{n+1} &= y_n + hf_3, \\
f_1 &= f(x_n, y_n), \\
f_2 &= f(x_n + \tfrac{1}{27}h, y_n + \tfrac{1}{27}hf_1), \\
f_3 &= f(x_n + \tfrac{4}{27}h, y_n + \tfrac{4}{27}hf_2).
\end{aligned}
$$

The diagonal feature of this and other highly stable formulae makes

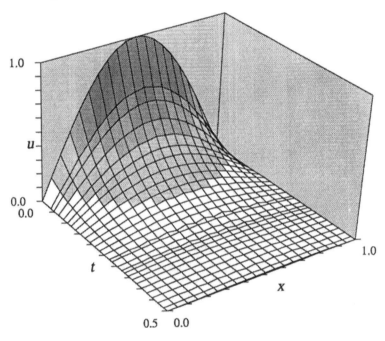

Figure 15.2: Solution of the heat equation (15.2) with the highly stable
RK2(1)3S in Table 7.4

their application to an autonomous system such as (15.4) particularly
straightforward. Writing the mesh ratio

$$
R = \Delta t / (\Delta x)^2,
$$

a two stage RK1 process with maximal stability ($l = 8$) applied to the system (15.2), expressed in the style of (15.3), is

$$u_{i,j+1} = u_{i,j} + R(u^{(1)}_{i-1} - 2u^{(1)}_i + u^{(1)}_{i+1}), \qquad (15.8)$$

$$u^{(1)}_i = u_{i,j} + \tfrac{1}{8}R(u_{i-1,j} - 2u_{i,j} + u_{i+1,j}).$$

Similarly the three stage process above takes the form

$$u_{i,j+1} = u_{i,j} + R(u^{(2)}_{i-1} - 2u^{(2)}_i + u^{(2)}_{i+1}), \qquad (15.9)$$

$$u^{(2)}_i = u_{i,j} + \tfrac{4}{27}R(u^{(1)}_{i-1} - 2u^{(1)}_i + u^{(1)}_{i+1}),$$

$$u^{(1)}_i = u_{i,j} + \tfrac{1}{27}R(u_{i-1,j} - 2u_{i,j} + u_{i+1,j}).$$

Figure 15.2 shows the solution of (15.2) subject to the boundary and initial conditions (15.5) using the high stability three stage RK2(1)3S given in Table 7.4. The mesh ratio with this pair has an upper limit of $R = 1.5$. This is three times the critical mesh ratio applicable with the conventional explicit scheme and the order of accuracy is one higher for the time integration.

If the optimal two stage first order RK (15.8) is applied to the same problem the limit on the mesh ratio is $R = 2$. The computational cost per step is double that of formula (15.3), but the overall cost for a fixed time interval is halved. With the three stage formula ($l = 18$) the saving is much greater, provided the step-size is limited by stability.

For non-linear equations the stability requirement is variable and so some step-size control is desirable. Embedded RK pairs with extended stability can be applied. An example of these is the RK2(1)3S shown in Table 7.4, but more highly stable pairs employing more stages are relatively easy to construct. However it is necessary to ensure that the members of any such pair have matched stability characteristics because the step-size will be limited by the least stable formula. When a step is rejected by breaching the stability boundary the embedding error estimate can be very large and so the optimal reduction formula discussed in Chapter 5 is not very reliable.

15.5 Equations with two space dimensions

Many practical problems are based on 2 or 3 space variables rather than the one dimensional case considered above. The extension of the semi-discretisation process is straightforward. Thus with $u = u(x, y, t)$ satisfying the equation

$$\frac{\partial u}{\partial t} = \nabla^2 u$$

in a rectangular domain $0 \le x \le 1$, $0 \le y \le 1$, and mesh points

$$
\begin{aligned}
x_i &= i\Delta x, \quad i = 0, 1, 2, \ldots, m, \\
y_j &= j\Delta y, \quad j = 0, 1, 2, \ldots, n,
\end{aligned}
$$

and $u_{i,j}(t) \simeq u(x_i, y_j, t)$, semi-discretisation yields the system

$$
\begin{aligned}
\frac{du_{i,j}}{dt} &= \frac{1}{(\Delta x)^2}(u_{i-1,j} - 2u_{i,j} + u_{i+1,j}) \\
&+ \frac{1}{(\Delta y)^2}(u_{i,j-1} - 2u_{i,j} + u_{i,j+1}), \quad (15.10) \\
&\quad i = 0, 1, 2, \ldots, m, \quad j = 0, 1, 2, \ldots, n.
\end{aligned}
$$

In the case $\Delta y = \Delta x$, stability limits the mesh ratio to a half that permissible in the one dimensional problem. Greater accuracy and stability may be achieved by using a nine-point approximation for the space derivatives.

15.6   Non-linear equations

If the equation (15.1) has the form

$$
\frac{\partial u}{\partial t} = \frac{\partial}{\partial x}\left\{ K(u)\frac{\partial u}{\partial x} \right\},
$$

then the semi-discretisation technique still may be applied directly. In this case we can form a system

$$
\begin{aligned}
\frac{du_i}{dt} &= \frac{1}{4h^2}(K(u_{i+1}) - K(u_{i-1}))(u_{i+1} - u_{i-1}) \\
&+ \frac{K(u_i)}{h^2}(u_{i+1} - 2u_i + u_{i-1}), \\
&\quad i = 0, 1, 2, \ldots, m-1.
\end{aligned}
$$

Since the method of lines is explicit, this system presents no more computational difficulty than the linear case. A suitable RK formula pair will enable reliable step-size control.

For some problems it is convenient, if not essential, to vary the space interval. This technique can be useful whether or not the method of lines is employed. The solution of Burgers' equation,

$$
\varepsilon\frac{\partial^2 u}{\partial x^2} = \frac{\partial u}{\partial t} + u\frac{\partial u}{\partial x}, \quad x \in [0,1], \ t \in [0,T], \quad (15.11)
$$

$$
u(0,t) = u(1,t) = 0, \quad u(x,0) = \left\{ \begin{array}{ll} 2x, & 0 \le x \le \frac{1}{2} \\ 2(1-x), & \frac{1}{2} < x \le 1 \end{array} \right. ,
$$

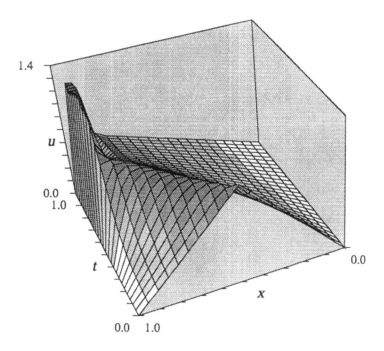

Figure 15.3: A poor solution of Burgers' equation with constant Δx

presents a potentially severe problem when ε is small ($\ll 1$). As t_j increases the maximum $(u_x(x,t_j) = 0)$ moves to the right with a steepening front. Very small mesh intervals in x are essential to cope with the boundary layer near $x = 1$ where $|u_x| \sim \varepsilon^{-1}$. A characteristic of inadequate numerical procedures is the development of *ripples* with maximum amplitude as close to the boundary as the grid allows. Shown in Figure 15.3 is a graphical surface display of a poor numerical solution when $\varepsilon = 0.01$, $\Delta x = 0.05$. The surface is a projection of $u_{i,j} = u_{i,j}(x,t)$. The large *maximum* near $x = 1$ is an artifact, but not due to instability, of the numerical method. In this case the method (15.8) has been applied with mesh ratio $R = 0.4$.

The ripples would decay after further time steps but they are certainly not a feature of the true solution to the problem. To compute an accurate numerical solution, a small interval size near $x = 1$ is essential and, since there is no need of small intervals at the other extreme, this is best arranged by allowing the Δx spacing to vary. A convenient scheme uses a geometric spacing in which the mesh points may be defined as

$$x_i - x_{i-1} = d_i, \quad i = 1, 2, \ldots, m, \quad d_i = \beta d_{i-1}.$$

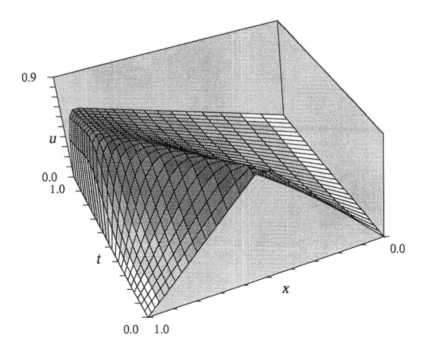

Figure 15.4: A good solution of Burgers' equation with variable spacing

Based on this scheme, Taylor series expansions yield

$$\frac{\partial^2 u}{\partial x^2}(x_i, t) \simeq \frac{u_{i+1}(t) + \beta u_{i-1}(t) - (1 + \beta)u_i(t)}{\frac{1}{2}\beta(\beta + 1)d_i^2}, \tag{15.12}$$

with a corresponding expression for the first derivative. The order of accuracy of these expressions is lower than those based on conventional finite differences but, in practice, values of β close to unity are capable of yielding acceptable accuracy with a small number (m) of intervals. With 20 intervals, as in the earlier case, $\beta = 0.9$ gives $d_1 = 0.1138, \ldots, d_{20} = 0.01538$. With the same Δt the variable interval scheme with formula (15.8) produces the solution illustrated in Figure 15.4.

As in the numerical solution of ordinary differential equations the importance of variable step-sizes must be emphasised. With functions of two or more independent variables it is sometimes advantageous to vary the space mesh with time. However the fixed geometric mesh is particularly effective in the above case. It is easy to modify the program RKpair from Appendix A to solve the problems in this section.

15.7 Hyperbolic equations

The success of the semi-discretisation process for parabolic equations suggests that it might be applied to hyperbolic equations which arise frequently in fluid flow modelling. The simplest case is the linear advection equation

$$\frac{\partial u}{\partial t} + c\frac{\partial u}{\partial x} = 0, \quad x > 0, t > 0 \tag{15.13}$$

with boundary and initial conditions $u(t,0) = 0, u(x,0) = f(x)$. This equation can be reduced to ordinary differential equations without any discretisation. Suppose $t = t(x)$ is a curve in the $x - t$ plane. On such a curve the solution to the equation (15.13) may be expressed as $u = u(x, t(x))$, and so the total derivative

$$\frac{du}{dx} = \frac{\partial u}{\partial x} + \frac{\partial u}{\partial t}\frac{dt}{dx}$$

can be defined. Comparing this with equation (15.13), it is seen that for a curve satisfying $\dfrac{dt}{dx} = \dfrac{1}{c}$, the PDE reduces to the trivial equation $\dfrac{du}{dx} = 0$. Thus the solution u is constant on such a curve, which is called a *characteristic* of the PDE (15.13). A family of characteristic curves is formed by varying the constant of integration. If c is constant the characteristics are a set of parallel lines

$$x - ct = x_i, \quad \text{a constant,}$$

as shown in Figure 15.5. The true solution at a point on a characteristic cutting the x-axis at $x = x_i$ is simply $u = f(x_i)$, leading to the result

$$u(x, t) = f(x - ct).$$

For non-homogeneous equations with $c = c(x, t, u)$, the characteristics may be non-linear with variable solutions, but the ordinary differential equations are amenable to solution by methods already discussed.

An alternative to the characteristic technique is based on semi-discretisation. The advection equation (15.13) can be replaced by the system of ordinary differential equations

$$\frac{du_i}{dt} = -c(u_x(t))_i, \quad i = 1, 2, \ldots, m,$$

with $x_i = i\Delta x$ and $u_i(0) = f(x_i)$. The choice of finite difference approximation for $(u_x(t))_i$ is of some importance. A simple choice is to select the backward difference formula which is based on

$$(u_x(t))_i = \frac{u_i - u_{i-1}}{\Delta x} + O(\Delta x).$$

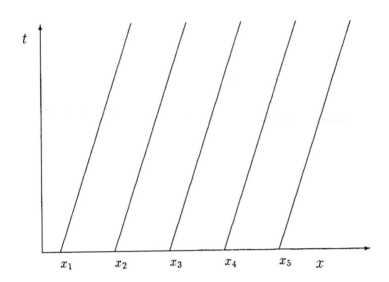

Figure 15.5: Characteristics for advection equation with c constant

Assuming $x \in [0,1]$ and $\Delta x = 1/m$, the system of equations can be expressed as

$$\frac{d\boldsymbol{u}}{dt} = B\boldsymbol{u}, \quad \boldsymbol{u} = (u_1, u_2, \ldots, u_{m-1})^T,$$

where

$$B = \frac{c}{\Delta x} \begin{bmatrix} -1 & 0 & 0 & \cdots & 0 \\ 1 & -1 & 0 & \cdots & 0 \\ 0 & 1 & -1 & \cdots & 0 \\ \vdots & \vdots & \vdots & \ddots & \vdots \\ 0 & \cdots & \cdots & 1 & -1 \end{bmatrix}.$$

Since B has only one eigenvalue $(-c/\Delta x)$, this system can be solved with a conventional method such as the extended stability RK formulae applied to parabolic equations. The finite difference approximation for u_x is only first order and so the solution will not be particularly accurate unless a very small interval Δx is chosen. Choice of the backward difference approximation for u_x, often called an *upwind* procedure, introduces a dissipative effect which reduces the amplitude of $f(x - ct)$ as t increases. Figure 15.6 illustrates this effect when the initial condition is

$$u(x,0) = f(x) = \begin{cases} (15x - 2)(2 - 5x)^3, & 0 \le x \le \frac{2}{5}, \\ 0, & x > \frac{2}{5}, \end{cases}$$

and $m = 40$.

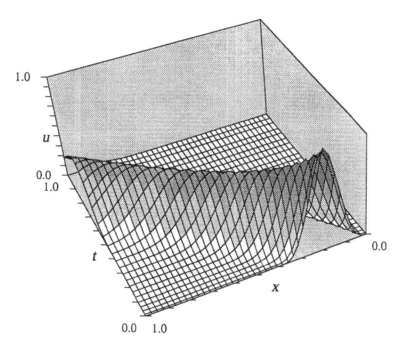

Figure 15.6: Solution of advection equation using backward difference approximation

A better approximation, in the algebraic sense, to the x-derivative is the central difference formula

$$(u_x(t))_i = \frac{u_{i+1} - u_{i-1}}{2\Delta x} + O((\Delta x)^2).$$

With this formula the eigenvalues of the resulting system are predominantly imaginary, and a suitable integrator must be absolutely stable on or near the imaginary axis of the complex plane. This is not a typical property of Runge-Kutta processes and even Backward Differentiation formulae are not very suitable.

Fortunately, one can construct appropriate RK formulae from a consideration of the test equation

$$\frac{dy}{dx} = \sqrt{-1}\omega y \quad y(x_0) = y_0. \tag{15.14}$$

Van der Houwen and Sommeijer (1987) have derived some RK formulae for solving oscillatory problems to give reduced phase errors or dispersion and improved dissipation qualities. These methods are particularly valuable for hyperbolic equations.

An s stage RK formula applied with constant step h to equation (15.14) yields the solution

$$y_n = [P(r)]^n y_0, \quad r = wh,$$

where the polynomial P, split into real and imaginary parts, can be expressed as

$$P(r) = F(r^2) + \sqrt{-1}\, r G(r^2),$$

where F and G are polynomials. The true solution to the differential equation is

$$y(x_n) = y_0 \exp(\sqrt{-1}\, nr).$$

Using these terms it is convenient to define the *dispersion* or *phase error*, and the *amplification* error of the RK process, as

$$\phi(r) = r - \arg P(r), \qquad \alpha(r) = 1 - |P(r)|.$$

A method with $\phi(r) = O(r^{q+1})$, and $\alpha(r) = O(r^{w+1})$, is said to be dispersive of order q and dissipative of order w. These values are readily determined from the definitions since

$$\phi(r) = r - \arctan\left(r \frac{G_s(r^2)}{F_s(r^2)}\right),$$

and

$$\alpha(r) = 1 - \sqrt{F_s^2 + r^2 G_s^2}.$$

Table 15.1 contains the parameters of a five stage RK3(2) pair, given by van der Houwen and Sommeijer (1987), which has $(q, w) = (6, 3)$ for both formulae, and an imaginary stability interval $(0, 2.66)$. Applying the central difference formula to the advection equation yields the system

$$u' = \begin{bmatrix} 0 & -1 & 0 & \cdots & 0 \\ 1 & 0 & -1 & \cdots & 0 \\ 0 & 1 & 0 & \cdots & 0 \\ \vdots & \vdots & \vdots & \ddots & \vdots \\ 0 & \cdots & -1 & 4 & -3 \end{bmatrix} u,$$

where the last component of u_x is approximated by a second order backward difference to avoid reference to a point outside the solution domain. The solution of this system with the formula (15.1) is pictured in Figure 15.7. The dissipative feature of the earlier solution based on upwinding with the same space interval is absent from the new solution.

Table 15.1: Coefficients of an embedded RK3(2) with dispersive and dissipative orders 6 and 3 respectively

c_i	a_{ij}				$\hat{b}_i$	b_i
0					$-\frac{5}{56}$	0
$\frac{1}{5}$	$\frac{1}{5}$				$\frac{25}{112}$	0
$\frac{1}{3}$	0	$\frac{1}{3}$			$\frac{75}{112}$	0
$\frac{1}{2}$	0	0	$\frac{1}{2}$		$-\frac{1}{14}$	1
1	0	0	0	1	$\frac{15}{56}$	0

15.8 Problems

1. Establish the local truncation error of the explicit method (15.3) for the solution of the partial differential equation (15.2).

2. Use the 4th degree Chebyshev polynomial to construct a 4-stage first order formula with real negative stability interval $[-32, 0]$.

3. Write a program to solve the equation (15.2) using a multistage first order RK with optimal stability and a constant step-size in t. Compare the efficiency of methods with 2, 3, and 4 stages by comparing solutions with the true solution.

4. Write a module for the computer program RKPair suitable for the solution of the transient heat conduction equation (15.2). Apply the optimally stable 3 stage RK2(1)3S from Table 7.4 to solve this equation with boundary conditions (15.5) for $t \in [0, 1]$ with 20 space intervals and a tolerance $T = 0.01$. Compare this solution with one based on the 5th order DOPRI5 with the same tolerance.

   Suggest and test a modification of the step-size control process to improve the efficiency of RKPair when it is applied to this type of system.

5. The function $u(x, t)$ satisfies a non-linear equation

$$\frac{\partial u}{\partial t} = \frac{\partial^2 u}{\partial x^2} + \left(\frac{\partial u}{\partial x}\right)^2, \quad 0 \le x \le 1,$$

which has boundary and initial conditions

$$\frac{\partial u}{\partial x}(0, t) = 1, \quad u(1, t) = 0, t > 0, \quad u(x, 0) = x(1 - x).$$

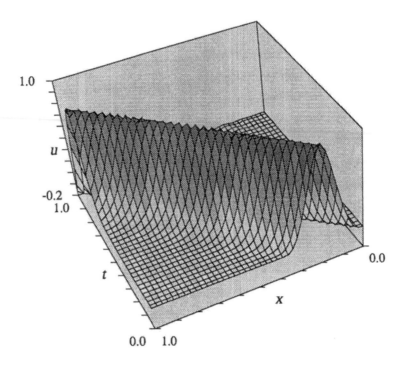

Figure 15.7: Solution of advection equation using central difference approximation

Construct a system of ordinary differential equations to approximate this PDE, using a central difference approximation for the first derivative boundary condition, which should be used to simplify the equation for u at $x = 0$. Use an appropriate program to solve the system for $0 < t \leq 1$.

6. Verify the formula (15.12) and find a corresponding formula for the first derivative. Hence write down the system of equations arising from the semi-discretisation of Burgers' equation based on a variable spacing.

7. The equation of heat conduction within a sphere may be written

$$\frac{\partial u}{\partial t} = \frac{\partial^2 u}{\partial r^2} + \frac{2}{r}\frac{\partial u}{\partial r} + H(r,t) , \quad r \in [0,1] , \quad t > 0 ,$$

with boundary conditions

$$\frac{\partial u}{\partial r}(0,t) = 0, \quad u(1,t) = U_s, \quad t > 0,$$

where $u(r, t)$ is the temperature at radius r and time t, and $H(r, t)$ represents heat sources within the sphere. Show that the transformation $w = ru$ yields the simpler equation

$$\frac{\partial w}{\partial t} = \frac{\partial^2 w}{\partial r^2} + rH(r, t).$$

Obtain an explicit 3-stage algorithm with extended stability for the transformed differential equation based on N unequal space intervals h_i satisfying

$$h_0 = h , \ r_{i+1} = r_i + h_i , \ i = 0, 1, \ldots, N - 1,$$

where $h_{i+1} = \theta h_i$, $0 < \theta < 1$. If $N = 10$ and $\theta = 0.8$ find h and h_{10}. What criterion would you apply for choosing the time step?

Your difference approximation from above will not yield solutions for $u(0, t)$. Describe a method for approximating the central temperature.

8. Find the imaginary stability boundaries of: (i) a three stage RK3 and (ii) a four stage RK4. Find the order of dissipation of the RK3.

9. Use the method of characteristics to show that the true solution of the inviscid Burgers' equation

$$\frac{\partial u}{\partial t} + u\frac{\partial u}{\partial x} = 0, \quad x \geq 0, \quad t \geq 0,$$

with $u(0, t) = 0, t > 0$ and the initial condition $u(x, 0) = x$, is

$$u(x, t) = \frac{x}{1 + t}.$$

Compute a solution for $0 < x \leq 1, \quad 0 < t \leq 1$, by the method of lines using the program RKPair from Appendix A. Compare your solution with the true solution.

10. Apply semi-discretisation to compute a numerical solution to the wave equation

$$\frac{\partial^2 u}{\partial t^2} = \frac{\partial^2 u}{\partial x^2}, \quad 0 \leq x \leq 1,$$

for $0 < t \leq 1$, given the boundary and initial conditions

$$u(0, t) = u(1, t) = 0, \ t > 0, \quad u(x, 0) = 16x^2(1 - x)^2, \ \frac{\partial u}{\partial t}(x, 0) = 0.$$

Since this is a second order partial differential equation, it must be split into a system of first order ODEs before the RKPair program can be applied.

$$\boxed{\sigma\upsilon\nu \ \ \tau\psi \ \ \Theta\epsilon\psi \ \ \tau\epsilon\lambda o\varsigma}$$

Appendix A

Programs for single step methods

A.1 A variable step Taylor method

The program below was used to provide data described in §5.4. A scalar equation is solved with and without local extrapolation.

```
PROGRAM StepCntrl  ! Computes solution to   y'=y^3-y,
                   ! y(0) = 1/SQRT(1+3EXP(-10)), using
                   ! 2(3) or 3(2) Taylor series procedure
                   ! and standard step control prediction
                   ! formula.
!------------------------------------J R Dormand, 1/1996
LOGICAL :: locext
INTEGER :: reject
REAL :: x, y, yd(3), w, h, h1, delta, tol, xend
 PRINT*, 'Enter tolerance, trial step, and xend '
 READ*, tol, h, xend
 PRINT*, ' Local extrapolation?(t/f) '; READ*, locext
 x = 0;  y = 1.0/SQRT(1.0 + 3.0*EXP(-10.0))
 reject = 0
 DO WHILE (x < xend)
   CALL Derivs(3, x, y, yd)       ! Evaluates 3 derivatives
   w = y + h*(yd(1) + 0.5*h*yd(2))          ! 2nd order step
   delta = h**3*yd(3)/6.0                  ! 3rd order term
   h1 = 0.9*h*(tol/ABS(delta))**0.3333333       ! predict h
   IF(ABS(delta) < tol) THEN      ! update if |error| < tol
     x = x + h
```

```
      IF(locext) THEN                    ! Local extrapolation
         y = w + delta
      ELSE
         y = w                           ! No extrapolation
      END IF
      PRINT '(1x, 4f10.6)', x, h, y, y - True(x)
   ELSE                                  ! reject step
      reject = reject + 1
      PRINT*,'Step rejected '
   END IF
   h = h1                                ! new step value
END DO
PRINT*, 'No of rejects is ',reject
END PROGRAM StepCntrl
!
SUBROUTINE Derivs(n, x, y, yd)          ! Compute derivatives
INTEGER :: n
REAL :: x, y, yd(n), f
   yd(1) = y*(y*y - 1)                   ! First derivative
   f = 3.0*y*y - 1.0
   yd(2) = f*yd(1)                       ! Second derivative
   yd(3) = 6.0*y*yd(1)**2 + f*yd(2)      ! Third derivative
END SUBROUTINE Derivs
!
FUNCTION True(x)                         ! True solution
REAL :: x, True
   True = 1.0/SQRT(1.0 + 3.0*EXP(2.0*x-10.0))
END FUNCTION True !-------------------------------------
```

A.2   An embedded Runge-Kutta program

The program below was used to generate the test data for the solution
of the elliptic orbit problem described in Chapter 5. Any Runge-Kutta
pair can be specified and its coefficients must be placed in an appropriate
data file, an example of which follows the code.

In principle, the program can deal with any system of equations by
making changes to Noeqns, Initial, Fcn, and True. Of course the true
solution is included for the test problem here but the numerical process
does not use this. For a general problem whose analytical solution is
unknown, the True function would be omitted.

```
PROGRAM RKmbed    ! Solves a system of ODEs with variable
                  ! step using an embedded pair whose
                  ! coefficients are input from a data file.
                  ! FSAL is enabled with a suitable pair.
                  ! Prints solution or error statistics from
                  ! varying tolerance
!-------------------------------------J R Dormand,  1/1996
   IMPLICIT NONE
   INTERFACE                     ! Explicit interface required
     FUNCTION Fcn(x, y)
        REAL (KIND = 2) :: x, y(:), Fcn(SIZE(y))
     END FUNCTION Fcn
     FUNCTION True(neq, e, x)
        INTEGER :: neq
        REAL (KIND = 2) :: e, x, True(neq)
     END FUNCTION True
   END INTERFACE
   LOGICAL :: done, fsal, solout
   INTEGER :: s, i, j, p, q, mcase, neq, k, nfunc, nrej, &
                                     Noeqns
   REAL (KIND = KIND(1.0D0)), PARAMETER :: z = 0.0D0, &
                         one = 1.0D0, sf = 0.6D0
   REAL (KIND = 2) :: x, h, xend, tol, ttol, delta, alpha, &
                                     oop, hs, e, errmax
!--------------------------- Dynamic arrays------------
   REAL (KIND = 2), ALLOCATABLE :: y(:), w(:), err(:), f(:,:)
   REAL (KIND = 2), ALLOCATABLE :: a(:, :), b(:), bl(:), &
                                     c(:), to(:), bo(:)
   CHARACTER(LEN = 30) :: rkname, outfil
!
! Read Runge-Kutta coefficients from a data file
!
   PRINT*, 'Enter file containing RK coefficients'
   READ '(a)', rkname
   OPEN(13, FILE = rkname)
   READ(13, *) s, q, p, fsal
   ALLOCATE(a(2:s, 1:s-1), b(s), bl(s), c(2:s), to(s), bo(s))
   c = z
   DO i = 2, s
      READ(13, *) (to(j), bo(j), j = 1, i - 1)
      a(i, 1:i-1) = to/bo
      c(i) = SUM(a(i, 1:i-1))
   END DO
```

```
READ(13, *) (to(i), bo(i), i = 1, s)
b = to/bo
READ(13, *) (to(i), bo(i), i = 1, s)
bl = to/bo
oop = one/(1+p)
CLOSE(13)
PRINT*, 'Output file '; READ '(a)', outfil
OPEN(14, FILE = outfil)
PRINT*, 'Print error/solution (t) or error statistics(f)'
READ*, solout
neq = Noeqns()
ALLOCATE( y(neq), w(neq), err(neq), f(s, neq))
PRINT*, 'Enter orbital eccentricity '; READ*, e
PRINT*, 'Enter xend, trial step, tolerance, No. of cases'
READ*, xend, hs, tol, mcase
IF(.NOT.solout) THEN
   PRINT '(a)', '  logtol  nfunc nrej   end error'
END IF
DO k = 1, mcase                    ! Loop over tolerances
   ttol = tol*sf
   errmax = 0.0D0
   CALL Initial(neq, e, x, y)
   h = hs
   f(1, :) = Fcn(x, y)             ! First function evaluation
   nfunc = 1; nrej = 0
   done = .FALSE.
   IF (solout) WRITE(14, 9999) x, h, z, z, z, z, y
   DO                              ! Loop over steps
     DO i = 2, s
       w = y + h*MATMUL(a(i, 1: i-1), f(1: i-1, :))
       f(i, :) = Fcn(x + c(i)*h, w)
     END DO
     nfunc = nfunc + s - 1
     delta = h*MAXVAL(ABS(MATMUL(bl - b, f)))! Local error
     alpha = (delta/ttol)**oop                ! Step ratio
     IF(delta < tol) THEN                     ! Accepts step
       y = y + h*MATMUL(b, f)
       x = x + h
       IF(done) EXIT
       IF(fsal) THEN                          ! FSAL true
         f(1, :) = f(s, :)      ! Set 1st f for next step
       ELSE                                   ! Not FSAL
         f(1, :) = Fcn(x, y); nfunc = nfunc + 1
       END IF
```

```
            h = h/MAX(alpha, 0.1D0)              ! Update step
            IF(x + h > xend) THEN                ! Hit end point
               h = xend - x
               done = .TRUE.
            ENDIF
          ELSE                                   ! Reject step
            nrej = nrej + 1
            h = h/MIN(alpha, 10.0D0)             ! Reduced step
            IF(done) done = .FALSE.
          END IF
          err = y - True(neq, e, x)              ! Global error
          errmax = MAX(errmax, MAXVAL(ABS(err)))
          IF (solout) WRITE(14, 9999) x, h, err, y
        END DO
        err = y - True(neq, e, x)
        IF(solout) THEN
           WRITE(14, 9999) x, h, err, y
        ELSE
           PRINT 9998, LOG10(tol), nfunc, nrej, LOG10(errmax)
           WRITE(14, 9998) LOG10(tol), nfunc, nrej, &
                                    LOG10(errmax)
        END IF
        tol = tol/10
     END DO
9999 FORMAT(1X, F6.3, 1P9E10.2)
9998 FORMAT(1X, F6.1, 2I6, F10.4)
     END PROGRAM RKmbed
!
     FUNCTION Noeqns()
        INTEGER :: Noeqns
          Noeqns = 4
     END FUNCTION Noeqns
!
     SUBROUTINE Initial(neq, e, x, y)
        INTEGER :: neq
        REAL (KIND = 2) :: e, x, y(neq)

          x = 0.0D0
          y(1) = 1.0D0 - e; y(2) = 0.0D0
          y(3) = 0.0D0; y(4) = SQRT((1.0D0 + e)/(y(1)))
     END SUBROUTINE Initial
!
     FUNCTION Fcn(x, y)         ! Derivatives for 2-body problem
        REAL (KIND = 2) :: x, y(:), Fcn(SIZE(y)), r3
```

```
      r3 = SQRT(y(1)**2 + y(2)**2)**3
      Fcn = (/ y(3), y(4), -y(1)/r3, -y(2)/r3 /)
   END FUNCTION Fcn
!
   FUNCTION True(neq, e, x) ! True solution by Kepler's equn
      INTEGER :: neq, n
      REAL (KIND = 2) :: e, x, True(neq)
      REAL (KIND = 2) :: ecca, c, s, den, b
      REAL (KIND = KIND(1.0D0)), PARAMETER :: &
                              tole = 1.0D-10, one = 1.0D0

      n = 0
      ecca = x
      dele = one
      DO WHILE (ABS(dele) > tole)              ! Newton iteration
        n = n + 1
        dele = (ecca - e*sin(ecca) - x)/(one - e*cos(ecca))
        ecca = ecca - dele
        IF(n > 25) THEN
           PRINT*, 'True solution not converged '
           EXIT
        END IF
      END DO
      c = COS(ecca); s = SIN(ecca)
      den = one/(one - e*c); b = SQRT(one - e*e)
      True = (/ c - e, b*s, -s*den, b*c*den /)
   END FUNCTION True !-------------------------------------
```

A.3   A sample RK data file

The data below defines the RK5(4)7FM (DOPRI5) from Chapter 5 and
can be used with the program coded above.

```
7 5 4 t          ! s, q, p, FSAL
1,5              ! a(2,1)
3,40 9,40                  ! a(3,1), a(3,2)
44,45 -56,15 32,9
19372,6561 -25360,2187 64448,6561 -212,729
9017,3168 -355,33  46732,5247 49,176 -5103,18656
35,384 0,1 500,1113 125,192 -2187,6784 11,84  ! a(7,j)
35,384 0,1 500,1113 125,192 -2187,6784 11,84 0,1  ! bhat(i)
5179,57600 0,1 7571,16695 393,640 -92097,339200 187,2100 1,40
-------------rk5(4)7fm for RKmbed.f90------------------
```

A.4   An alternative Runge–Kutta scheme

The program in this section is very similar to RKmbed, but a different structure is used. In this case the differential system is defined in a MODULE which is then USEd by the RKpair program. This is applicable to different systems without the necessity of changes. There are some other minor differences between RKpair and RKmbed. The former uses a specific Runge-Kutta pair and solves the system for a single tolerance. Also the data format is slightly different.

```
MODULE twobody          ! Defines the 2-body gravitational
                        ! problem with eccentricity e
                        ! True solution by Kepler's equation
!--------------------------------J R Dormand,    1/1996
  LOGICAL :: printsol
  INTEGER, PARAMETER :: neq = 4
  INTEGER :: dev, nfunc
  REAL (KIND=2) :: err(neq), e, bb, gerrmax
  CHARACTER(33) :: fmt, outfil
CONTAINS
  !
  SUBROUTINE Initial(x, y)      ! Sets initial values and
    REAL (KIND = 2) :: x, y(:) ! output format
      WRITE (fmt, '(a,2(i1,a))' ) '(1x, f10.4, 2x,', &
                    neq,'f10.4,',neq,'e12.4)'
      gerrmax = SPACING(1.0d0)
      PRINT*, 'Print solution(t/f) ' ; READ*, printsol
      IF(printsol) THEN
      PRINT*, 'Output filename ("vdu" for screen) '
      READ '(a)', outfil
      IF(outfil == 'vdu') THEN
         dev = 6
      ELSE
         dev = 14
         OPEN(dev, FILE = outfil)
      ENDIF
      ENDIF
      PRINT*, 'Enter eccentricity'; READ*, e
      x = 0.0d0
      bb = SQRT(1.0d0 - e*e)
      y(1) = 1.0d0 - e; y(2) = 0.0d0
```

```fortran
      y(3) = 0.0d0; y(4) = SQRT((1.0d0 + e)/(y(1)))
      nfunc = 0
  END SUBROUTINE Initial
!
  SUBROUTINE Output(x, y)         ! Outputs solution with
    REAL (KIND = 2) :: x, y(:) !   error using True
        IF(printsol) THEN
          err = y - True(x)
          WRITE(dev, fmt) x, y, err
          gerrmax = MAX(gerrmax, MAXVAL(ABS(err)))
        END IF
  END SUBROUTINE Output
!
  FUNCTION Fcn(x, y)              ! Defines derivative
    REAL (KIND = 2) :: x, y(:), Fcn(SIZE(y)), r3
      nfunc = nfunc + 1
      r3 = SQRT(y(1)**2 + y(2)**2)**3
      Fcn = (/ y(3), y(4), -y(1)/r3, -y(2)/r3 /)
  END FUNCTION Fcn
!
  FUNCTION True(x)                ! True solution y(x)
    INTEGER ::  n
    REAL (KIND = 2) :: x, True(neq)
    REAL (KIND = 2) :: ecca, c, s, den
    REAL (KIND = KIND(1.0D0)), PARAMETER :: &
                        tole = 1.0D-10, one = 1.0D0
    n = 0
    ecca = x
    dele = one
    DO WHILE (ABS(dele) > tole) ! Newton iteration loop
      n = n + 1
      dele = (ecca - e*SIN(ecca) - x)/(one - e*COS(ecca))
      ecca = ecca - dele
      IF(n > 25) THEN
         PRINT*, 'True solution not converged '
         EXIT
      END IF
    END DO
    c = COS(ecca); s = SIN(ecca)
    den = one/(one - e*c)
    True = (/ c - e, bb*s, -s*den, bb*c*den /)
  END FUNCTION True
!
END MODULE twobody
```

```
!
!-----------------------------------------------------------
PROGRAM RKpair    ! Solves a system of ODEs defined in a
                  ! MODULE using variable step RK547FM.
                  ! Absolute error criterion
!------------------------------- J R Dormand,    1/1996
USE twobody       ! Module defining system of ODEs
  IMPLICIT NONE
  LOGICAL :: done, fsal
  INTEGER :: s, i, p, q, neqs, ns, nrej
  REAL (KIND = Kind(1.0d0)), PARAMETER :: z = 0.0D0, &
                          one = 1.0D0, sf = 0.6D0
  REAL (KIND = 2) :: x, h, xend, tol, ttol, delta, alpha, &
                  oop, hs
!-----------------Dynamic arrays-------------------------
  REAL (KIND = 2), ALLOCATABLE :: y(:), w(:), f(:,:)
  REAL (KIND = 2), ALLOCATABLE :: a(:, :), bcap(:), b(:), &
                                  c(:)
  neqs = neq                      ! ----- System  Dimension
  CALL RKcoeff                    ! ----- Read RK data file
  ALLOCATE( y(neqs), w(neqs), f(s, neqs))
  PRINT*, 'Enter xend, starting step, tolerance'
  READ*, xend, hs, tol
  ttol = tol*sf
  CALL Initial(x, y)
  h = hs; ns = 0
  f(1, :) = Fcn(x, y);  nrej = 0  ! ---- First f evaluation
  done = .FALSE.
  CALL Output(x, y)
  DO                    ! ----------------- Loop over steps
    DO i = 2, s
      w = y + h*MATMUL(a(i, 1: i-1), f(1: i-1, :))
      f(i, :) = Fcn(x + c(i)*h, w)
    END DO
    delta = h*MAXVAL(ABS(MATMUL(b - bcap, f))) !Local error
    alpha = (delta/ttol)**oop        ! ------ Step ratio
    IF(delta < tol) THEN             ! ---- Accepts step
      y = y + h*MATMUL(bcap, f)      ! -------- Solution
      x = x + h
      ns = ns + 1
      CALL Output(x, y)                      ! ------ Output
      IF(done) EXIT
      IF(fsal) THEN                          ! Test for FSAL
        f(1, :) = f(s, :)                    !- Re-use stage
```

```
         ELSE
            f(1, :) = Fcn(x, y)                    ! New 1st stage
         END IF
         h = h/MAX(alpha, 0.1D0)                   !- Predict step
         IF(x + h > xend) THEN                     ! Hit end point
            h = xend - x
            done = .TRUE.
         ENDIF
       ELSE                                        ! - Reject step
         nrej = nrej + 1
         h = h/MIN(alpha, 10.0D0)                  ! - Reduce step
         IF(done) done = .FALSE.
      END IF
    END DO
    PRINT 109, ns, nfunc, nrej, LOG10(gerrmax)    ! Statistics
109 FORMAT(1x, i5, 2i6, f10.4)
CONTAINS
SUBROUTINE RKcoeff ! Read RK coefficients from a data file
 CHARACTER(30) :: rkfile
 REAL (KIND = 2) :: cden
 REAL (KIND = 2), ALLOCATABLE :: to(:)
 PRINT*, 'Enter name of file containing coefficients '
 READ '(a)', rkfile
 OPEN(13, FILE = rkfile, STATUS = 'OLD')
 fsal = .FALSE.
 READ(13, *) s, q, p
 ALLOCATE(a(2: s, 1: s-1), bcap(s), b(s), c(2:s), to(s))
 c = z
 DO i = 2, s
    READ(13, *)  cden, to(1: i-1)
    a(i, 1:i-1) = to/cden
    c(i) = SUM(a(i, 1: i-1))
 END DO
 READ(13, *)  cden, to
 bcap = to/cden
 READ(13, *)  cden, to
 b = to/cden
 oop = one/(1 + p)
 fsal = SUM(ABS(a(s, 1:s-1) - bcap(1:s-1))) < 1.0d-10
 CLOSE(13)
 PRINT 100, s, q, p, fsal
100 FORMAT(1x, i2,' stage, RK',i1,'(',i1,'), FSAL = ',l1)
 END SUBROUTINE RKcoeff
END PROGRAM RKpair ! -----------------------------------
```

A.5   Runge–Kutta with dense output

The program below is closely related to RKpair and provides dense output
at equal intervals within a specified interval. The problem solved is the
gravitational two-body problem and the same MODULE as defined in the
previous section is used. An important difference between this and the
earlier program is in the entry of the RK data which now includes the
dense output polynomial coefficients. An example data file is given below;
the format is nearly identical to one suitable for RKpair.

```
PROGRAM RKden       ! Solves a system of ODEs with a variable
                    ! step RK pair and dense output provision
                    !
!--------------------------------- J R Dormand,     1/1996
USE twobody
  IMPLICIT NONE
  LOGICAL :: done, fsal, extra
  INTEGER :: s, star, i, j, p, q, neqs, nrej, deg, ns, &
                                    nden, dencount
  REAL (KIND = KIND(1.0d0)), PARAMETER :: z = 0.0D0, &
                    one = 1.0D0, sf = 0.6D0
  REAL (KIND = 2) :: x, h, xend, tol, ttol, delta, alpha, &
                    oop, xst, hst, xa, xb, sh, sigma
!-------------------Dynamic arrays ----------------------
  REAL (KIND = 2), ALLOCATABLE :: y(:), w(:), f(:,:)
  REAL (KIND = 2), ALLOCATABLE :: a(:, :), bcap(:), b(:), &
                    c(:), dd(:, :), yst(:), bstar(:)
  neqs = neq
  CALL RKcoeff                    ! ----- Read RK data file
  ALLOCATE( y(neqs), yst(neqs), w(neqs), f(star, neqs))
  PRINT*, 'Enter xend, starting step, tolerance'
  READ*, xend, h, tol
  PRINT*, 'Dense output interval, no of sub-intervals? '
  READ*, xa, xb, nden
  hst = (xb - xa)/nden
  xst = xa          ! --------- First dense output point
  dencount = -1
  extra = .TRUE.
  ttol = tol*sf
  CALL Initial(x, y)   ! ------------------ Initial values
  f(1, :) = Fcn(x, y);  nrej = 0; ns = 0
```

```
done = .FALSE.
CALL Output(x, y)
DO                          ! ------------------ Loop over steps
  DO i = 2, s               ! ----- Loop over discrete RK stages
    w = y + h*MATMUL(a(i, 1:i - 1), f(1:i - 1, :))
    f(i, :) = Fcn(x + c(i)*h, w)
  END DO
  delta = h*MAXVAL(ABS(MATMUL(b(1:s) - bcap(1:s), &
                     f(1:s, :)))) ! --------- Error estimate
  alpha = (delta/ttol)**oop         ! ------ Step ratio
  IF(delta < tol) THEN              ! ---- Accepts step
!-----------------------------------------------------------
!---- Loop over all dense output points within current step
  DO WHILE (x + h > xst .AND. dencount < nden)
      dencount = dencount + 1
      IF (extra) THEN               ! Compute extra f's once
        DO i = s + 1, star
          w = y + h*MATMUL(a(i, 1:i - 1), f(1:i - 1, :))
          f(i, :) = Fcn(x + c(i)*h, w)
        END DO
      END IF
      sh = xst - x
      sigma = sh/h                  ! Interpolation parameter
      bstar = dd(:, deg)
      DO j = deg - 1, 0, -1  ! Compute b*(1:star) vector
         bstar = bstar * sigma + dd(:, j)
      END DO
      yst = y + sh*MATMUL(bstar, f)  ! - Solution at xst
      CALL Output(xst, yst)
      extra = .FALSE.
      xst = xst + hst               ! Next output point
   END DO
!-----------------------------------------------------------
   extra = .TRUE.
   y = y + h*MATMUL(bcap(1:s), f(1:s, :)) ! Discrete sol
   x = x + h
   ns = ns + 1
   CALL Output(x, y)
   IF(done) EXIT
   IF(fsal) THEN
     f(1, :) = f(s, :)
   ELSE
     f(1, :) = Fcn(x, y)
   END IF
```

```
      h = h/MAX(alpha, 0.1D0)
      IF(x + h > xend) THEN
         h = xend - x
         done = .true.
      ENDIF
   ELSE
      nrej = nrej + 1
      h = h/MIN(alpha, 10.0D0)
      IF(done) done = .false.
   END IF
 END DO
 PRINT 109,  ns, nfunc, nrej, LOG10(gerrmax)  ! Statistics
109 FORMAT(1x, i5, 2i6, f10.4)
 CONTAINS
 SUBROUTINE RKcoeff
  CHARACTER(30) :: rkfile
!
! Read Runge-Kutta  coefficients from a data file
!
  INTEGER :: deg1
  REAL (KIND = 2) :: cden
  REAL (KIND = 2), ALLOCATABLE :: to(:)
  PRINT*, 'Enter name of file containing coefficients '
  READ '(a)', rkfile
  OPEN(13, FILE = rkfile, STATUS = 'OLD')
  fsal = .FALSE.
  READ(13, *) star, q, p, s
  ALLOCATE(a(2: star, 1: star-1), bcap(star), b(star), &
                     c(2:star), to(star),  bstar(star))
  c = z
  DO i = 2, star
     READ(13, *)  cden, to(1: i-1)
     a(i, 1:i-1) = to/cden
     c(i) = SUM(a(i, 1: i-1))
  END DO
  READ(13, *)  cden, to
  bcap = to/cden
  READ(13, *)  cden, to
  b = to/cden
  oop = one/(1 + p)
  fsal = SUM(ABS(a(s, 1:s-1) - bcap(1:s-1))) < 1.0d-10
  READ(13, *) deg1; deg = deg1 - 1  ! ------- degree of b*
  ALLOCATE(dd(1:star, 0:deg))
  DO i = 1, star          ! Enter b* polynomial coefficients
```

```
      READ(13, *) cden, to(1: deg1)
      dd(i, 0: deg) = to(1: deg1)/cden
   END DO
   CLOSE(13)
   PRINT 100, star, q, p, fsal
   100 FORMAT(1x, i2,' stage dense RK',i1,'(',i1,'),   &
                              FSAL = ',l1)
   END SUBROUTINE RKcoeff
   END PROGRAM RKden   ! ---------------------------------
```

A.6   A sample continuous RK data file

The data below defines the RK5(4)7FM (DOPRI5) from Chapter 5, complete with coefficients for the continuous extension in Table 6.5, and can be used with the RKden program coded above.

```
9 5 4 7
5  1
40   3  9
45   44   -168   160
6561  19372  -76080  64448  -1908
167904  477901   -1806240  1495424  46746  -45927
142464  12985  0  64000  92750  -45927  18656
2544000 275971 0 294400 -209350 159651 -93280 81408
325632 32489 0 128000 -13250 24057 -18656 10176 0
142464  12985  0  64000  92750  -45927  18656  0 0 0
21369600  1921409  0  9690880  13122270  -5802111  1902912
534240 0 0
5
384 384 -1710 3104 -2439 696
1 0 0 0 0 0
1113 0 3000 -16000 25500 -12000
192 0 750 -4000 6375 -3000
6784 0 -13122 69984 -111537 52488
84 0 66 -352 561 -264
8 0 -7 38 -63 32
24 0 125 -250 125 0
3 0 -16 80 -112 48
-------rk5(4)7fm  + 5th order dense for RKden.f90----------
```

Appendix B

Multistep programs

B.1   A constant steplength program

The following program applies a Predictor–Corrector algorithm to the logistic equation (8.12). The program is valid for systems of equation and not just the scalar example provided here. To define the problem to be solved it is necessary only to rewrite the MODULE logistic and reference the new version with a new USE command. Separate compilation of the PROGRAM and MODULE might be an advantage. The MODULE used here is compatible with that shown in Appendix A for the RKpair program.

```
MODULE logistic ! Defines the logistic equation
                ! with initial value y(0) specified
                ! True solution defined
                !                       J R Dormand,   1/1996
!---------------------------------------------------------------
  LOGICAL :: printsol
  INTEGER, PARAMETER :: neq = 1
  INTEGER :: dev, nfunc
  REAL (KIND=2) :: const
  CHARACTER(33) :: fmt, outfil
CONTAINS
!
  FUNCTION Fcn(x, y)
!
! Logistic equation derivative
!
    REAL (KIND = 2) :: x, y(:), Fcn(SIZE(y))
      Fcn = (/ y(1)*(2.0d0 - y(1)) /)
      nfunc = nfunc + 1
```

```
  END FUNCTION Fcn
!
  FUNCTION True(x)
    REAL (KIND = 2) :: x, True(neq)
    True = (/ 1.0d0/(0.5d0 + const*EXP(-2.0d0*x)) /)
  END FUNCTION True
   !
  SUBROUTINE Initial(x, y)
    REAL (KIND = 2) :: x, y(:)
      WRITE (fmt, '(a,2(i1,a))' ) '(1x, f10.4, 2x,', &
                    neq,'f10.4,',neq,'e12.4)'
      PRINT*, 'Print solution(t/f) ' ; READ*, printsol
      IF(printsol) THEN
      PRINT*, 'Output filename ("vdu" for screen) '
      READ '(a)', outfil
      IF(outfil == 'vdu') THEN
         dev = 6
      ELSE
         dev = 14
         OPEN(dev, FILE = outfil)
      ENDIF
      ENDIF
      x = 0.0d0
      PRINT*, 'Enter initial y '; READ*, y(1)
      const = 1.0d0/y(1) - 0.5d0 ! const of integration
      nfunc = 0
  END SUBROUTINE Initial
!
  SUBROUTINE Output(x, y)
    REAL (KIND = 2) :: x, y(:)
      IF(printsol) THEN
        WRITE(dev, fmt) x, y, y - True(x)
      END IF
  END SUBROUTINE Output
!
END MODULE logistic
!
!-------------------------------------------------------------
PROGRAM PmeCme  ! Solves a system of ODEs with constant step
                ! using a Predictor-Corrector scheme and
                ! optional Modifiers
                ! RK4 starter scheme provided
!-------------------------------- J R Dormand,    1/1996
!
```

```
  USE logistic                  ! Defines system of ODEs
  IMPLICIT NONE
  LOGICAL :: modp, modc
  INTEGER :: i, j, p, ns, neqs
  REAL (KIND = KIND(1.0d0)), PARAMETER :: z = 0.0D0, &
                                          one = 1.0D0
  REAL (KIND = 2) :: x, h, xend, den, pmod, cmod
!------------------Dynamic arrays------------------------
  REAL (KIND = 2), ALLOCATABLE :: y(:, :), v(:), w(:), &
                                  pmc(:), f(:, :), ff(:, :)
  REAL (KIND = 2), ALLOCATABLE :: a(:), b(:), bc(:), ac(:)
  CHARACTER(LEN = 30) :: pcname
!
! Read multistep coefficients from a data file
!
  PRINT*, 'Enter file containing PC coefficients'
  READ '(a)', pcname
  PRINT*, 'Modify predictor(t/f), corrector(t/f)?'
  READ*, modp, modc
  OPEN(13, FILE = pcname)
  READ(13, *)  p
  ALLOCATE(a(0: p), b(0: p), ac(0: p), bc(0: p + 1))
  READ(13, *) a, den;   a = a/den
  READ(13, *) b, den;   b = b/den
  READ(13, *) ac, den; ac = ac/den
  READ(13, *) bc, den; bc = bc/den
  READ(13, *) pmod, cmod
  CLOSE(13)
  IF(.NOT.modp) pmod = z
  IF(.NOT.modc) cmod = z
  neqs = neq
  ALLOCATE( y(0: p, 1: neqs), v(neqs), w(neqs), pmc(neqs),&
                 f(0: p+1,1: neqs),  ff(5, neqs))
  CALL Initial(x, y( 0, :))           ! Initial value
  f(0, :) = Fcn(x, y(0, :))
  pmc = z
  CALL Output(x, y(0, :))
  PRINT*, 'Enter stepsize, xend '
  READ*, h, xend
  DO i = 1, p    ! Starter values using RK4 solution
     y(i, :) = Rk4(y(i-1, :))
     f(i, :) = Fcn(x + h, y(i, :))
     CALL Output(x + h, y(i, :))
     x = x + h
```

```
      END DO
      ns = (xend - x)/h + 0.5d0; x = p*h
!
      DO i = p, ns                                    ! Steps Loop
        w = MATMUL(a, y) + h*MATMUL(b, f(0: p, :))    ! Predictor
        f(p + 1, :) = Fcn(x + h, w + pmod*pmc)        ! Evaluate
        v = MATMUL(ac, y) + h*MATMUL(bc, f)           ! Correct
        pmc = w - v                                   ! Modifier
        w = v + cmod*pmc
        x = x + h
        CALL Output(x, w)
        f(0: p - 1, :) = f(1: p, :)                   ! Update f's
        y(0: p - 1, :) = y(1: p, :)                   ! New starters
        y(p, :) = w                                   ! New step y
        f(p, :) = Fcn(x, w)                           ! New y'
      END DO
      CONTAINS
!
      FUNCTION Rk4(y)
!
! Compute a single RK step size h with 4th order formula
!
      REAL (KIND = 2) :: y(:), Rk4(SIZE(y))
      REAL (KIND = KIND(1.0d0)), DIMENSION(2: 5, 4), &
         PARAMETER :: a =  RESHAPE((/0.2d0, z, 1.2d0, &
               -2.125d0, z, 0.4d0, -2.4d0, 5.0d0, z,&
         z, 2.0d0, -2.5d0, z, z, z, 0.625d0/), (/4, 4/))
      REAL (KIND = KIND(1.0d0)), DIMENSION(5), PARAMETER :: &
        b = (/0.135416666666667d0,z,0.520833333333333d0, &
            0.260416666666667d0, 0.0833333333333333d0/), &
              c = (/z, 0.2d0, 0.4d0, 0.8d0, one/)
        ff(1, :) = Fcn(x, y)
        DO j = 2, 5
          w = y + h*MATMUL(a(j, 1: j-1), ff(1: j-1, :))
          ff(j, :) = Fcn(x + c(j)*h, w)
        END DO
        Rk4 = y + h*MATMUL(b, ff)
      END FUNCTION Rk4
!
      END PROGRAM PmeCme !-------------------------------------
```

A data file, suitable for PmeCme, defining the third order Predictor and Corrector formulae derived in Chapter 8, is shown below.

```
1                    p
5 -4 1               a
2  4 1               b
0  1 1               ac
-1 8 5 12            bc
-0.8d0 0.2d0         pmod, cmod
```

B.2   A variable step Adams PC scheme

The program in this section applies a variable step Adams procedure to solve the same system of equations defined in the MODULE twobody which is given in Appendix A. The algorithm is essentially the same as that in PmeCme from the previous section, with the added possibility of Corrector iterations. The Krogh method described in §9.6 is used to halve and double the step-size as required, and the multistep coefficients are generated for the specified order, rather than being read-in from a file.

```
PROGRAM Adams  ! Solves a system of ODEs with variable step
               ! using Adams-Bashforth-Moulton Predictor-
               ! Corrector scheme and optional local
               ! extrapolation. Variable step by doubling
               ! and halving according to Krogh's method.
               ! RK4 starter scheme provided, PE(CE)^k
!--------------------------------------J R Dormand,     1/1996
!
USE twobody     ! See Appendix A for this module
  IMPLICIT NONE
  LOGICAL :: modp, modc, noiter
  INTEGER :: i, j, k, p, neqs, iter, nhalf, ndouble, ns
  REAL (KIND = KIND(1.0d0)), PARAMETER :: z = 0.0D0, &
                one = 1.0D0, two = 2.0d0, half = 0.5d0

  REAL (KIND = 2) :: x, h, xend,   pmod, cmod, tol, letol,&
                let2, errest
!--------------Dynamic arrays----------------------------------
  REAL (KIND = 2), ALLOCATABLE :: y(:, :), v(:), w(:), &
                     pmc(:), f(:, :), ff(:, :)
  REAL (KIND = 2), ALLOCATABLE :: b(:), bc(:)
```

```
!---------------------------------------------------------
!
! Specify mode of operation
!
  PRINT*, 'Modify predictor(t/f)', ' corrector(t/f)?',&
        ' iterate(t/f)?'
  READ*, modp, modc, noiter
  IF(noiter) THEN
    PRINT*, 'Enter tolerance for corrector iteration'
    READ*, tol
  ENDIF
  noiter = .NOT.noiter
  PRINT*, 'Order of ABM formula, local error tolerance '
  READ*, k, letol
  let2 = letol/two**(k+1)      ! Tolerance for doublng step
  CALL Abm                     ! Generate Adams formulae
  IF(.NOT.modp) pmod = z ! Set modifier to zero if not used
  IF(.NOT.modc) cmod = z
  neqs = neq
  ALLOCATE( y(0: p, 1: neqs), v(neqs), w(neqs), pmc(neqs), &
            f(0: p + 1, 1: neqs), ff(5, neqs))
  CALL Initial(x, y( 0, :))                      ! Initial
  f(0, :) = Fcn(x, y(0, :))                      !   value
  pmc = z
  nhalf = 0; ndouble = 0; ns = 0
  PRINT*, 'Enter stepsize, xend '
  READ*, h, xend
  DO i = 1, p    ! Find starter values using  RK4
     y(i, :) = Rk4( y(i - 1, :))
     f(i, :) = Fcn(x + h, y(i, :))
     x = x + h
     CALL Output(x, y(i, :))
  END DO
  x = p*h

  DO WHILE(x < xend)                          ! --- Steps Loop
     w = y(p, :) + h*MATMUL(b, f(0: p, :)) ! ---- Predictor
     ff(1, :) = w + pmod*pmc                ! ------- Modify
     f(p + 1, :) = Fcn(x + h, ff(1, :))     ! ----- Evaluate
     iter = 0
     DO
       v = y(p, :) + h*MATMUL(bc, f(1: p + 1, :)) ! Correct
       iter = iter + 1
       IF(MAXVAL(ABS(ff(1, :) - v)) < tol .OR. noiter) THEN
```

```
      EXIT                                   ! Converged
    ELSE
      f(p + 1, :) = Fcn(x + h, v)            ! -Evaluate
      ff(1, :) = v
    END IF
  END DO
  pmc = w - v                                ! ------ Modifier
  errest = MAXVAL(ABS(pmc))                  ! ---- Error norm
  IF(errest < letol) THEN                    ! - Step accepted
    w = v + cmod*pmc                         ! -------- Update
    x = x + h; ns = ns + 1
    CALL Output(x, w)
    f(0: p - 1, :) = f(1: p, :)              ! ---- Update f's
    y(0: p - 1, :) = y(1: p, :)              ! Update starters
    y(p, :) = w                              ! ---- New step y
    f(p, :) = Fcn(x, w)                      ! -------- New y'
    IF(errest < let2) THEN                   ! --- double step
      CALL Vstep(.FALSE.)
      h = two*h
      ndouble = ndouble + 1
    ENDIF
  ELSE              ! Step rejected (halve step and repeat)
    CALL Vstep(.TRUE.)
    h = half*h
    nhalf = nhalf + 1
  END IF
END DO
print 101, ns, ndouble, nhalf
101 FORMAT(1x, 3i7)
CONTAINS

!------------------------------------------------------------
  SUBROUTINE Vstep(halve)  ! Halves or doubles steplength
                           ! using Krogh(1973) algorithm

!-------------------------------------J R Dormand, 1/1996
!
  LOGICAL :: halve
  INTEGER :: j, m
! Compute backward differences. These occupy back-value
! storage
  DO j = 1, p
    f(0: p - j, :) = f(1: p - j + 1, :) - f(0: p - j, :)
  END DO
  IF (halve) THEN    ! compute differences for half-step
```

```
   f(0, :) = half*f(0, :)
   DO j = p - 1, 1, -1
      f(0, :) = half*f(0, :)
      DO m = 1, p - j
         f(m, :) = half*(f(m, :) + f(m - 1, :))
      END DO
   END DO
 ELSE                    ! compute differences for double-step
   DO j = 1, p - 1
      DO m = p - j, 1, -1
         f(m, :) = two*f(m, :) - f(m - 1, :)
      END DO
      f(0, :) = two*f(0, :)
   END DO
   f(0, :) = two*f(0, :)
 END IF
! Convert back from differences to function values
 DO j = p - 1, 0, -1
    f(0: j, :) = f(1: j + 1, :) - f(0: j, :)
 END DO
 END SUBROUTINE Vstep
!
!------------------------------------------------------------
 SUBROUTINE Abm       ! Generate Adams coefficients
!--------------------------------------J R Dormand, 1/1996
!
 INTEGER ::    i, j, m
 REAL(KIND = 2), ALLOCATABLE :: g(:), gs(:), a(:)
 p = k - 1
 ALLOCATE (g(0: p + 1), gs(0: p + 1), a(0: p), b(0: p),&
          bc(0: p))
! First generate coefficients for backward difference form
 g = 1; gs(0) = 1
 DO i = 1, p + 1
   m = 1
   DO j = i - 1, 0, -1
     m = m + 1
     g(i) = g(i) - g(j)/m
   END DO
   gs(i) = g(i) - g(i - 1)
 END DO
! ---------------Compute modifier coefficients-------------
 pmod = -g(p + 1)/g(p); cmod = -gs(p + 1)/g(p)
! --Find coefficients for functional form of Adams formulae
```

```
a = 0; a(0) = 1; b = 0; bc = 0; b(p) = 1; bc(p) = 1
DO i = 1, p
  b(p)  = b(p) + g(i)
  bc(p) = bc(p) + gs(i)
  DO j = i, 1, -1
    a(j) = a(j) - a(j - 1)
    b(p - j)  = b(p - j)  +  a(j)*g(i)
    bc(p - j) = bc(p - j) + a(j)*gs(i)
  END DO
END DO
END SUBROUTINE Abm
!
!------------------------------------------------------------
FUNCTION Rk4(y)    ! Compute a single RK step of size h
                   ! using a RK4(3)5M formula. Tolerance
                   ! will determine steplength
!_____J R Dormand, 1/1996
!
REAL (KIND = 2) :: y(:), Rk4(SIZE(y))
REAL (KIND = KIND(1.0d0)), DIMENSION(2: 5, 4), &
PARAMETER :: a = RESHAPE((/0.2d0, z, 1.2d0, -2.125d0, z,&
     0.4d0, -2.4d0, 5.0d0, z, z, two, -2.5d0, z, z, z, &
     0.625d0/), (/4, 4/))
REAL (KIND = KIND(1.0d0)),DIMENSION(5), PARAMETER :: b ≈&
     (/ 13d0/96d0, z, 25d0/48d0, 25d0/96d0, 1d0/12d0 /),&
                  c = (/z, 0.2d0, 0.4d0, 0.8d0, one/),&
  d = (/ 3d0/192d0, z, -5d0/96d0, 15d0/192d0, -1d0/24d0 /)
  ff(1, :) = Fcn(x, y)
  DO                 ! ---- Loop until tolerance achieved
    DO j = 2, 5
      w = y + h*MATMUL(a(j, 1: j - 1), ff(1: j - 1, :))
      ff(j, :) = Fcn(x + c(j)*h, w)
    END DO
    IF(i == 1) THEN   ! -- Check local error at first step
      errest = h*MAXVAL(ABS(MATMUL(d, ff)))
      IF ( errest < letol) THEN  ! Accept if le < tol
        EXIT
      ELSE                          ! Reject
        h = 0.9d0*h*(letol/errest)**0.25  ! New step-size
        CYCLE
      ENDIF
    END IF
    EXIT
  END DO
```

```
      Rk4 = y + h*MATMUL(b, ff)
   END FUNCTION Rk4
!-----------------------------------------------------------
   END PROGRAM Adams !-------------------------------------------
```

B.3   A variable coefficient multistep package

The following code defines the STEP90 package modelled on the STEP
package of Shampine & Gordon (1975). The constants and most of the
variables are declared in a MODULE but the package interface is provided
by the SUBROUTINE De. The problem definition must be contained in a
MODULE named System which is essentially similar to those used in earlier
RK and multistep programs. An appropriate example is given below.

```
! STEP90 package  !
!----------------!
                   ! For complete documentation of the units
                   ! in this package the book by Shampine &
                   ! Gordon (1975) should be consulted. The
                   ! Fortran 90 units are essentially coded
                   ! from the STEP package.
!------------------------------------------------------------
MODULE DATA        ! Contains variables and constants for
                   ! STEP90 units.
                   !
!-------------------------------------- J R Dormand. 1/1996
!
  USE System    ! References MODULE containing current system
  IMPLICIT NONE
  LOGICAL :: start
  INTEGER ::  k, kold, ns, isn, isnold,  maxstep = 500
  REAL (KIND = 2), PARAMETER :: ze = 0.0d0, one = 1.0d0, &
                              half = 0.5d0
  REAL (KIND = 2) :: gstr(13), two(13), g(13), sig(13), &
            psi(12), twou, fouru, h, hold, told, x, delsgn
  REAL (KIND = 2), ALLOCATABLE :: yy(:), wt(:), p(:), &
                              yp(:), phi(:,:)
  LOGICAL :: phase1, nornd
  INTEGER :: fail, kp1, kp2, km1, km2, nsp1, im1,&
            limit1, limit2,  nsp2, nsm2, ip1, knew
```

```
      REAL (KIND = 2), DIMENSION(12) :: alpha, beta, w, v
      REAL (KIND = 2) :: erk, erkm1, erkm2, xold, erkp1, err,  &
                      p5eps, round, ssum, absh, hnew, realns
CONTAINS
!
  SUBROUTINE Setdat
    INTEGER :: i
    gstr = (/0.500, 0.0833, 0.0417, 0.0264, 0.0188, 0.0143,&
            0.0114, 0.00936, 0.00789, 0.00679, 0.00592, &
            0.00524, 0.00468/)
    ALLOCATE(yy(neq), wt(neq), p(neq), yp(neq), phi(neq,16))
    DO i = 1, 13;  two(i) = 2.0d0**i;   END DO
    g(1) = one;    g(2) = half;    sig(1) = one
    twou = 2.0d0*SPACING(one);    fouru = 2.0d0*twou
  END SUBROUTINE Setdat
END MODULE DATA
!
!-----------------------------------------------------------
SUBROUTINE De(neqs, y, ypout, t,tout, relerr, abserr, flag)
          ! This unit acts as user interface with package
!------------------------------INTEGRATES SYSTEM FROM t TO tout
!--------------------------- y(tout) = y, y'(tout) = ypout
! relerr, abserr ARE RELATIVE AND ABSOLUTE ERROR TOLERANCES
!    flag = +1 OR -1 TO INITIALISE , flag = -1 INDICATES NO
!               INTEGRATION BEYOND tout, flag = +1 NORMALLY
! ON RETURN flag = 2 normally (t = tout)
!                = 3 did not reach tout because
!                      abserr, relerr too small. Values
!                      increased for new entry
!                = 4 maxstep exceeded before tout
!                = 5 equations probably stiff
!                = 6 invalid parameters input
!-----------------------------------------------------------
USE DATA
  IMPLICIT NONE
  LOGICAL :: crash, stiff, fatal
  INTEGER :: flag, neqs
  INTEGER :: kle4, nostep
  REAL (KIND = 2), DIMENSION(neqs) :: y, ypout
  REAL (KIND = 2) :: tout, t, relerr, abserr
  REAL (KIND = 2) :: eps, del,tend, absdel, releps, abseps
  IF(ABS(flag) == 1) THEN
     CALL Setdat; delsgn = one
  END IF
```

```
eps = MAX(relerr, abserr)
isn = SIGN(1, flag);  flag = ABS(flag)
fatal = neqs < 1 .OR. t==tout .OR. relerr < ze &
         .OR.abserr < ze .OR. eps <= ze .OR. flag == 0
IF(flag > 1) fatal = fatal .OR. t /= told
IF(fatal) THEN
  PRINT*, neqs, t,tout, relerr, abserr, eps, flag
  flag = 6;   PRINT*, 'fatal error, invalid parameters'
  RETURN
END IF
del = tout - t;  absdel = ABS(del);  tend = t + 10.0d0*del
IF(isn < 0) tend = tout
nostep = 0;   kle4 = 0;   stiff = .FALSE.
releps = relerr/eps;  abseps = abserr/eps
IF(flag == 1 .OR. isnold < 0 .OR. delsgn*del <= ze) THEN
  start = .TRUE. ;   x = t;   yy = y
  delsgn = SIGN(one, del)
  h = SIGN(MAX(ABS(tout-x), fouru*ABS(x)), tout-x)
END IF
DO
IF(ABS(x-t) >= absdel) THEN
   CALL Intrp( yy, tout, y, ypout, neqs)
   flag = 2;  t = tout;  told = t;  isnold = isn
   RETURN
END IF
IF(isn <= 0 .AND. ABS(tout-x) < fouru*ABS(x)) THEN
   h = tout - x
   yp = Fcn(x, yy)
   y = yy + h*yp
   flag = 2;  t = tout;  told = t;  isnold = isn
   RETURN
ELSE IF(nostep >= maxstep) THEN
   flag = isn * 4;    PRINT*, 'Maximum steps exceeded'
   IF(stiff) THEN
     flag = isn*5
     PRINT*, 'System is probably stiff'
   END IF
   y = yy;  t = x;  told = t;  isnold = 1
   RETURN
 ELSE
   h = SIGN(MIN(ABS(h), ABS(tend-x)), h)
   wt = releps*ABS(yy) + abseps
   CALL Step(eps,  crash)          ! -  Compute a new step
   IF(crash) THEN
```

```
            flag = isn*3; relerr = eps*releps
            abserr = eps*abseps
            PRINT*, 'abserr & relerr too small'
            y = yy;  t = x;  told = t;  isnold = 1
            RETURN
        END IF
      END IF
    nostep = nostep + 1;  kle4 = kle4 + 1
    IF(kold > 4) kle4 = 0
    IF(kle4 >= 50) stiff = .TRUE.
    END DO
END SUBROUTINE De
!-----------------------------------------------------------

!-----------------------------------------------------------
SUBROUTINE Step(eps, crash)             ! Computes a new step
USE DATA
  IMPLICIT NONE
  LOGICAL ::  crash
  INTEGER :: i, j, iq
  REAL (KIND = 2) :: eps
  REAL (KIND = 2) ::  temp1, temp2, temp3, temp4, temp5,  &
                      temp6
  crash = .TRUE.
  IF(ABS(h) < fouru*ABS(x)) THEN
     h = SIGN(fouru*ABS(x), h)
     RETURN
  ENDIF
  p5eps = half*eps
  round = twou*SQRT(SUM((yy/wt)**2))
  IF(p5eps < round) THEN
    eps = 2.0d0*round*(one+fouru)
    RETURN
  END IF
  crash = .FALSE.
  IF(start) THEN
     yp = Fcn(x, yy)
     phi(:, 1) = yp; phi(:, 2) = ze
     ssum = SQRT(SUM((yp/wt)**2)); absh = ABS(h)
     IF(eps < 16*ssum*h*h) absh = 0.25d0*SQRT(eps/ssum)
     h = SIGN(MAX(absh, fouru*ABS(x)), h)
     hold = ze; k = 1; kold = 0; start = .FALSE.
     phase1 = .TRUE. ; nornd = .TRUE.
     IF(p5eps <= 100.0d0*round) THEN
       nornd = .FALSE. ; phi(:, 15) = ze
```

```
      END IF
   END IF
   fail = 0
!
!  Compute coefficients for current step
!
   Dostep:DO                              ! New or repeated step
      kp1 = k+1;  kp2 = k+2;  km1 = k-1;  km2 = k-2
      IF(h /= hold) ns= 0
      ns = MIN(ns+1, kold+1);  nsp1 = ns+1
      IF(k >= ns) THEN
         beta(ns) = one;  realns = ns
         alpha(ns) = one/realns; temp1 = h*realns
         sig(nsp1) = one
         DO i = nsp1, k
            im1 = i-1;  temp2 = psi(im1)
            psi(im1) = temp1
            beta(i) = beta(im1)*psi(im1)/temp2
            temp1 = temp2 + h;   alpha(i) = h/temp1
            sig(i+1) = i*alpha(i)*sig(i)
         END DO
         psi(k) = temp1
         IF(ns <= 1) THEN
            DO iq = 1, k
               temp3 = iq*(iq+1)
               v(iq) = one/temp3
               w(iq) = v(iq)
            END DO
         ELSE
            IF(k > kold) THEN
               temp4 = k*kp1;   v(k) = one/temp4
               nsm2 = ns-2
               DO j = 1, nsm2
                 i = k - j
                 v(i) = v(i) - alpha(j+1)*v(i+1)
               END DO
            END IF
            limit1 = kp1 - ns
            temp5 = alpha(ns)
            DO iq = 1, limit1
               v(iq) = v(iq) - temp5*v(iq+1)
               w(iq) = v(iq)
            END DO
            g(nsp1) = w(1)
```

```
      END IF
      nsp2 = ns + 2
      DO i = nsp2, kp1
         limit2 = kp2 - i
         temp6 = alpha(i-1)
         DO iq = 1, limit2
            w(iq) = w(iq) - temp6*w(iq+1)
         END DO
         g(i) = w(1)
      END DO
   END IF
!
!      Predict  solution, Estimate local error at orders
!      K, K-1, K-2
!
   DO i = nsp1, k
      temp1 = beta(i)
      phi(:, i) = temp1*phi(:, i)
   END DO
   phi(:, kp2) = phi(:, kp1);  phi(:, kp1) = ze;  p = ze
   DO j = 1, k
      i = kp1 - j
      ip1 = i + 1
      temp2 = g(i)
      p = p + temp2*phi(:, i)
      phi(:, i) = phi(:, i) + phi(:, ip1)
   END DO
   IF(nornd) THEN
      p = yy + h*p
   ELSE
      p = yy + h*p - phi(:, 15)
      phi(:, 16) = (one - h)*p - yy + phi(:, 15)
   END IF
   xold = x
   x = x + h;  absh = ABS(h)
   yp = Fcn(x, p)
   erkm2 = ze;    erkm1 = ze
   erk = SUM(((yp - phi(:, 1))/wt)**2)
   IF(km1 > 0) THEN
      erkm1 = SUM(((phi(:, k) + yp - phi(:, 1))/wt)**2)
      erkm1 = absh*sig(k)*gstr(km1)*SQRT(erkm1)
      IF(km2 > 0) THEN
       erkm2 = SUM(((phi(:,km1) + yp - phi(:, 1))/wt)**2)
       erkm2 = absh*sig(km1)*gstr(km2)*SQRT(erkm2)
```

```
      END IF
   END IF
   temp5 = absh*SQRT(erk)
   err = temp5*(g(k) - g(kp1))
   erk = temp5*sig(kp1)*gstr(k)
   knew = k
   IF(km2 > 0) THEN
      IF(MAX(erkm1, erkm2) <= erk) knew = km1
   ELSE IF(km2 == 0) THEN
      IF(erkm1 <= half*erk) knew = km1
   END IF
   IF(err <= eps) EXIT          ! Exit loop after good step
!
!
!          Restore variables following an unsuccessful step
!
   phase1 = .FALSE.
   x = xold
   DO i = 1, k
      temp1 = one/beta(i)
      ip1 = i + 1
      phi(:, i) = temp1*(phi(:, i) - phi(:, ip1))
   END DO
   DO i = 2, k ;  psi(i-1) = psi(i) - h;  END DO
   fail = fail + 1
   temp2 = half
   IF(fail >= 3) THEN
      IF(fail > 3) THEN
         IF(p5eps < 0.25d0*erk) temp2 = SQRT(p5eps/erk)
      END IF
      knew = 1
   END IF
   h = temp2 * h;  k = knew
   IF(ABS(h) < fouru*ABS(x)) THEN
      crash = .TRUE.
      h = SIGN(fouru*ABS(x), h)
      eps = 2*eps
      RETURN
   END IF
END DO Dostep
!
!
!          Correct after successful step, Update differences
!                 Determine order and step-size for next step
```

```
!
  kold = k;    hold = h
  temp1 = h*g(kp1)
  IF(nornd) THEN
     yy = p + temp1*(yp - phi(:, 1))
  ELSE
     yy = p + temp1*(yp - phi(:, 1)) - phi(:, 16)
     phi(:, 15) = (yy - p) - temp1*(yp - phi(:, 1)) &
                  + phi(:, 16)
  END IF
  yp = Fcn(x, yy)
  phi(:, kp1) = yp - phi(:, 1)
  phi(:, kp2) = phi(:, kp1) - phi(:, kp2)
  DO i = 1, k
     phi(:, i) = phi(:, i) + phi(:, kp1)
  END DO
  erkp1 = ze
  IF(knew == km1 .OR. k == 12) phase1 = .FALSE.
  IF(phase1) THEN
     CALL Newstep(kp1, erkp1); RETURN
  ELSE IF(knew == km1) THEN
     CALL Newstep(km1, erkm1); RETURN
  ELSE IF(kp1 > ns) THEN
      CALL Newstep(k, erk); RETURN
  ENDIF
  erkp1 = absh*gstr(kp1)*SQRT(SUM((phi(:, kp2)/wt)**2))
  IF(k <= 1) THEN
     IF(erkp1 >= half*erk) THEN
        CALL Newstep(k, erk)
     ELSE
        CALL Newstep(kp1, erkp1)
     END IF
   ELSE IF(erkm1 <= MIN(erk, erkp1)) THEN
     CALL Newstep(km1, erkm1)
   ELSE IF(erkp1 >= erk .OR. k == 12) THEN
     CALL Newstep(k, erk)
   ELSE
     CALL Newstep(kp1, erkp1)
   END IF
!                 END of SUBROUTINE Step
CONTAINS
!----------------------------------------------------------
  SUBROUTINE Newstep(knew, erknew)! Computes a new step-size
   IMPLICIT NONE
```

```
      INTEGER :: knew
      REAL(KIND = 2) :: erknew, r, temp
      k = knew; erk = erknew
      hnew = h + h
      IF(.NOT.phase1) THEN
         IF(p5eps < erk*two(k+1)) THEN
            hnew = h
            IF(p5eps < erk) THEN
               temp = k + 1
               r = (p5eps/erk)**(one/temp)
               hnew = absh*MAX(half, MIN(0.9d0, r))
               hnew = SIGN(MAX(hnew, fouru*ABS(x)), h)
            END IF
         END IF
      END IF
      h = hnew
   END SUBROUTINE Newstep
!
END SUBROUTINE Step

!-----------------------------------------------------------
SUBROUTINE Intrp( y, xout, yout, ypout, neqs)
! -------------------------Approximates solution at xout
USE DATA
   IMPLICIT NONE
   INTEGER :: neqs, ki, kip1, i,j, jm1
   REAL(KIND = 2), DIMENSION(neqs) :: y, yout, ypout
   REAL(KIND = 2) :: wint(13), rho(13), gi(13)
   REAL(KIND = 2) :: xout, gamma, term, eta,&
                     hi, psijm1, temp1, temp2, temp3
   rho(1) = one;  gi(1) = one
   hi = xout - x;  ki = kold + 1;   kip1 = ki + 1
   DO i = 1, ki
      temp1 = i;  wint(i) = one/temp1
   END DO
   term = ze
   DO j = 2, ki
      jm1 = j - 1;  psijm1 = psi(jm1)
      gamma = (hi + term)/psijm1
      eta = hi/psijm1;  limit1 = kip1 - j
      DO i = 1, limit1
         wint(i) = gamma*wint(i) - eta*wint(i + 1)
      END DO
      gi(j) = wint(1);  rho(j) = gamma*rho(jm1)
      term = psijm1
```

```
END DO
ypout = ze;    yout = ze
DO j = 1, ki
   i = kip1 - j
   temp2 = gi(i);    temp3 = rho(i)
   yout = yout + temp2*phi(:, i)
   ypout = ypout + temp3*phi(:, i)
END DO
yout = y + hi*yout
END SUBROUTINE Intrp ! -------------------------------------
```

The code below depicts a simple driver program for STEP90 applied to
the system

$$
\begin{aligned}
{}^1y' &= {}^2y\,{}^3y, & {}^1y(0) &= 0 \\
{}^2y' &= -{}^1y\,{}^3y, & {}^2y(0) &= 1 \\
{}^3y' &= -0.51\,{}^1y\,{}^2y, & {}^3y(0) &= 1.
\end{aligned}
$$

These are the Euler equations for an isolated rigid body and have the
Jacobi elliptic solutions

$$
{}^1y(x) = \mathrm{sn}(x|0.51), \quad {}^2y(x) = \mathrm{cn}(x|0.51), \quad {}^3y(x) = \mathrm{dn}(x|0.51).
$$

```
!------------------------------------------------------------
MODULE System   ! Euler equations of a rigid body without
                ! external forces. Quarter period is
                ! 1.862640802332738d0
!-----------------------------------------J R Dormand,    1/1996
   LOGICAL :: printsol
   INTEGER, PARAMETER :: neq = 3
   INTEGER :: dev, nfunc
   CHARACTER(33) :: fmt, outfil
CONTAINS
 !
   SUBROUTINE Initial(x, y)      ! Sets initial values and
     REAL (KIND = 2) :: x, y(:) ! output format
     WRITE (fmt, '(a,2(i1,a))' ) '(1x, f10.4, 2x,', &
                   neq,'f10.4,',neq,'e12.4)'
!--------------Open output file---------------------------
     PRINT*, 'Print solution(t/f) ' ; READ*, printsol
     IF(printsol) THEN
       PRINT*, 'Output filename ("vdu" for screen) '
```

```
      READ '(a)', outfil
      IF(outfil == 'vdu') THEN
         dev = 6
      ELSE
         dev = 14
         OPEN(dev, FILE = outfil)
      ENDIF
    ENDIF
!-------------Set initial values--------------------------
    x = 0.0d0
    y = (/0.0d0, 1.0d0, 1.0d0/)
    nfunc = 0
  END SUBROUTINE Initial
!
  SUBROUTINE Output(x, y)        ! Outputs solution with
    REAL (KIND = 2) :: x, y(:) ! error using True
    IF(printsol) THEN
       WRITE(dev, fmt) x, y
    END IF
  END SUBROUTINE Output
!
  FUNCTION Fcn(x, y)             ! Defines derivative
    REAL (KIND = 2) :: x, y(:), Fcn(SIZE(y))
    nfunc = nfunc + 1
    Fcn(1) = y(2)*y(3)
    Fcn(2) = -y(1)*y(3)
    Fcn(3) = -0.51d0*y(1)*y(2)
  END FUNCTION Fcn
!
END MODULE System
!------------------------------------------------------------
PROGRAM Testde    ! Simple example of main program to drive
                  ! the STEP90 package. The equations to be
                  ! solved are defined in the MODULE System
                  ! which is USEd also by the STEP90 DATA
                  ! MODULE. Based on the STEP package of
                  ! Shampine & Gordon (1975).
!---------------------------------------J.R.Dormand 1/1996
USE System
IMPLICIT NONE
  INTEGER :: flag, i, neqs, nout
  REAL (KIND = 2), ALLOCATABLE :: y(:), yd(:)
  REAL (KIND = 2):: t, relerr, abserr, tout, tinc
  neqs = neq
```

```
    ALLOCATE(y(neqs), yd(neqs))
    CALL Initial(t, y)
    flag = 1
    PRINT*, 'Enter relerr and abserr';  READ*, relerr, abserr
    tout = 0.0d0
    tinc = 1.862640802332738d0
    PRINT*, 'No of outs'; READ*,  nout
    DO i = 1, nout
       tout = tout + tinc
       DO
         CALL De(neqs, y, yd, t, tout, relerr, abserr, flag)
         IF(flag == 6) STOP
         IF(flag == 2) EXIT
       END DO
       CALL Output(t, y)
    END DO
END PROGRAM ! ---------------------------------------------------
```

Appendix C

Programs for Stiff systems

C.1 A BDF program

The program below applies a backward differentiation formula to a Stiff system defined in the MODULE Stiff11. A constant step-size is employed and the Corrector is solved with Newton's method. The MODULE is compatible with the non-stiff procedures listed in the previous appendices, although these will not call the Jacobian unit.

```
MODULE Stiff11 ! Defines a stiff system of 2 ODEs
               ! Stiffness increases with x. Jacobian has
               ! eigenvalues -x^4 and -2x^(-0.5)
               ! True solution given
!-----------------------------------J R Dormand,   1/1996
  LOGICAL :: printsol
  INTEGER, PARAMETER :: neq = 2
  INTEGER :: dev, nfunc, njac
  CHARACTER(33) :: fmt, outfil
CONTAINS
  !
  SUBROUTINE Initial(x, y)        ! Sets initial values and
    REAL (KIND = 2) :: x, y(:)    ! output format
      WRITE (fmt, '(a,2(i1,a))' ) '(1x, f10.4, 2x,', &
                  neq,'f10.4,',neq,'e12.4)'
!--------------Open output file--------------------------
      PRINT*, 'Print solution(t/f) ' ; READ*, printsol
      IF(printsol) THEN
```

```
      PRINT*, 'Output filename ("vdu" for screen) '
      READ '(a)', outfil
      IF(outfil == 'vdu') THEN
        dev = 6
      ELSE
        dev = 14
        OPEN(dev, FILE = outfil)
      ENDIF
      ENDIF
      x = 1.0d0
      y(1) = 1.0d0; y(2) = 1.0d0
      nfunc = 0; njac = 0
  END SUBROUTINE Initial
!
  SUBROUTINE Output(x, y)               ! Outputs solution with
    REAL (KIND = 2) :: x, y(:)          ! error using True
      IF(printsol) THEN
        WRITE(dev, fmt) x, y, y - True(x)
      END IF
  END SUBROUTINE Output
!
  FUNCTION Fcn(x, y)                 ! ----- Defines derivative
                              ! This stiffens as x increases
    REAL (KIND = 2) :: x, y(:), Fcn(SIZE(y))
      nfunc = nfunc + 1
      Fcn(1) = 1.0d0/y(1) - x**2 - 2.0d0/x**3
      Fcn(2) = y(1)/y(2)**2 - 1.0d0/x - 0.5d0/SQRT(x**3)
  END FUNCTION Fcn
!
  FUNCTION Jac(x, y)                 ! - Jacobian matrix of Fcn
    REAL (KIND = 2) :: x, y(:), Jac(neq, neq)
! A simple stiff example (JRD 1/95)
      njac = njac + 1
      Jac(1,1) = -1.0d0/y(1)**2
      Jac(1,2) =  0.0d0
      Jac(2,1) =  1.0d0/y(2)**2
      Jac(2,2) = -2.0d0*y(1)/y(2)**3
  END FUNCTION Jac
!
  FUNCTION True(x)                        ! True solution y(x)
    REAL (KIND = 2) :: x, True(neq)
    True = (/ 1.0d0/x**2,  1.0d0/SQRT(x) /)
  END FUNCTION True
!
```

```
END MODULE Stiff11

!------------------------------------------------------------
PROGRAM BDF
!                   ! Solves a system of ODEs using a backward
!                   ! differentiation formula with Newton
!                   ! iteration
!----------------------------------------- J R Dormand, 1/1996
  USE Stiff11
  IMPLICIT NONE
  INTEGER :: neqs, i, j, p, pp1, pm1, nstep, its, ldiff, &
             interv
  INTEGER, PARAMETER :: itmax = 10
  INTEGER, ALLOCATABLE :: order(:)
  REAL (KIND = 2), ALLOCATABLE :: y(:, :), amat(:, :),&
   ident(:,:), yp(:), bdel(:, :), a(:),&
   alpha(:), beta(:), c(:), dy(:)
  REAL (KIND = 2) :: x, h, hrk, b, rb, xend, errfun, tol, &
                     error
  PRINT*, 'Enter order of BDF (2-6)'; READ*, p
  pm1 = p - 1; pp1 = p + 1
  ALLOCATE(a(p), alpha(0:p), beta(p))
!------Generate BDF coefficients for required order--------
  alpha = 0.0d0; beta = 0.0d0; alpha(0) = 1; rb=0.0d0
  DO i = 1, p
     DO j = i, 1, -1
        alpha(j) = alpha(j) - alpha(j-1)
        beta(j)  = beta(j) + alpha(j)/i
     END DO
     rb = rb + 1.0d0/i; b = 1.0d0/rb
     a = -b*beta
  END DO
  neqs = neq
  tol = 1.0d-6               ! Tolerance for Newton iteration
  ALLOCATE( y(neqs, pp1), amat(neqs, neqs),  &
    yp(neqs), order(neqs),c(neqs), bdel(neqs, pp1),&
    dy(neqs), ident(neqs,neqs))
  ident = 0.0d0;  DO i = 1, neq; ident(i,i) = 1.0d0; END DO
  PRINT*, 'h, Xend? ';    Read*, h, xend
  CALL initial(x, y(:, 2))
  nstep = (xend - x)/h + 0.5
  PRINT*, 'Step interval/print ';   READ*, interv
!-------------------- Starter values --------------------
  DO i = 3, pp1
     hrk = h
```

```
      y(:, i) = RKstart(y(:, i-1), hrk)   ! Runge-Kutta
   END DO
   ldiff = 2   ! Extrapolation is lower order for 1st step
!--------------------- Start integration -----------------
   DO j = p, nstep                       ! Loop over steps
! --- Predictor is polynomial extrapolation formula
! ------------------------------- to avoid f evaluations
      bdel = y
      yp = y(:, pp1)
      DO i = 1, pp1 - ldiff
! -----------------------------------------Compute differences
         bdel(:, 1:pp1 - i) = bdel(:, 2: pp1-i+1)&
                                    - bdel(:, 1: pp1-i)
! -----------------Accumulate interpolant to form predictor
         yp = yp + bdel(:, pp1 - i)
      END DO
      ldiff = 1    ! Predictor order same as BDF after step 1
      c = 0
      DO i = 0, pm1;  c = c + a(i+1)*y(:, pp1 - i);  END DO
!----------------------------------------------------------
! -----Correct with BDF formula using Newton iteration ----
      its = 0
      x = x + h
      error = 1.0d0
      DO While(error > tol)        ! Commence Newton iteration
         amat = ident - h*b*Jac(x, yp)        ! Use Jacobian
         dy = -yp + h*b*Fcn(x, yp) + c        ! f evaluation
         errfun = MAXVAL(Abs(dy))
         CALL LU(amat, neq, order)        ! Solve Newton equns
         CALL FBSubst(amat, dy, neq, order)   ! by LU factoring
         error = MAXVAL(Abs(dy)) + errfun    ! error statistic
         yp = yp + dy;    its = its + 1;      ! Update solution
         IF(its > itmax) THEN
            PRINT*, 'Maximum iterations exceeded'
            STOP
         END IF
      END DO
      DO i = 1, p; y(:, i) = y(:, i+1); END DO ! Update
      y(:, pp1) = yp                       ! starter values
      IF(MOD(j, interv) == 0) THEN
         CALL Output(x, y(:, pp1))
      ENDIF
   END DO
   CONTAINS
```

```
!------------------------------------------------------------
    FUNCTION RKstart(y, h)     ! Carries out RK integration  to
                               ! hit x + h. Several steps are
                               ! computed if necessary according
                               ! to local tolerance 1.0e-6.
!-----------------------------------------J R Dormand, 1/1996
!
    IMPLICIT NONE
    INTEGER ::  ir, istep
    LOGICAL :: hit
    REAL (KIND = 2) :: y(:), h, Rkstart(SIZE(y))
    REAL (KIND = 2) :: letol, errest, xhit, beta
    REAL (KIND = KIND(1.0d0)), DIMENSION(2: 5, 4),&
        PARAMETER :: a =  RESHAPE((/0.2d0, 0.0d0, 1.2d0,&
            -2.125d0, 0.0d0, 0.4d0, -2.4d0, 5.0d0, 0.0d0,&
      0.0d0, 2.0d0, -2.5d0, 0.0d0, 0.0d0, 0.0d0, 0.625d0/),&
      (/4, 4/))
    REAL (KIND= KIND(1.0d0)), DIMENSION(5), PARAMETER :: b =&
     (/ 13d0/96d0, 0.0d0, 25d0/48d0, 25d0/96d0, 1d0/12d0 /),&
        c = (/0.0d0, 0.2d0, 0.4d0, 0.8d0, 1.0d0/), &
     d = (/ 3d0/192d0, 0.0d0, -5d0/96d0, 15d0/192d0,&
          -1d0/24d0 /)
    REAL (KIND=2), ALLOCATABLE :: ff(:,:), w(:), ys(:)
    ALLOCATE(ff(5, neqs), w(neqs), ys(neqs))
      ys = y
      xhit = x + h; hit = .TRUE.
      letol = 1.0d-6
      ir = 0; istep = 0
      ff(1, :) = Fcn(x, ys)
      DO
        DO j = 2, 5
          w = ys + h*MATMUL(a(j, 1: j - 1), ff(1: j - 1, :))
          ff(j, :) = Fcn(x + c(j)*h, w)
        END DO
        errest = h*MAXVAL(ABS(MATMUL(d, ff)))
        IF ( errest < letol) THEN
          ys = ys + h*MATMUL(b, ff)
          istep = istep + 1
          IF(hit) THEN
            x = xhit
            PRINT '(1x, a, i3)', 'Steps =', istep
            RKstart = ys
            RETURN
          ENDIF
```

```
         x = x + h
         beta = errest/letol
         h = MIN(0.9d0*h/beta**0.25, 2*h)
         IF(xhit < x + h) THEN
            h = xhit - x
            hit = .TRUE.
         END IF
         ff(1, :) = Fcn(x, ys)
      ELSE
         hit = .FALSE.
         beta = errest/letol
         h = MAX(0.9d0*h/beta**0.25, h/10)
         ir = ir + 1
         PRINT '(1x,a,i3)', 'Rejects in RKstart =', ir
      ENDIF
   END DO
  END FUNCTION Rkstart
 !
 END PROGRAM BDF
 !

!-------------------------------------------------------------
SUBROUTINE LU(a, n, order)
      ! Factorise a square matrix a = LU
      ! Elements of (a) are replaced by elements of factors
      ! Pivoting is effected by row swaps. The row order
      ! is an output array                   J.R.Dormand, 1/96
!-------------------------------------------------------------
INTEGER :: n, order(n), i, j, pivot, piv(1)
REAL (KIND=2) :: a(n, n), row(SIZE(order))
DO i = 1, n;  order(i) = i; END DO
DO i = 1, n-1
   piv = MAXLOC(ABS(a(i:n, i)))
   pivot = piv(1) + i - 1
   IF(pivot /= i) THEN
     row = a(i, :); a(i, :) = a(pivot, :); a(pivot, :) = row
     j = order(i); order(i) = order(pivot); order(pivot) = j
   END IF
   IF(ABS(a(i, i)) < 1.0D-10) THEN
      PRINT*, 'ZERO PIVOT'
      RETURN
   END IF
   DO j = i+1, n
      a(j, i) = a(j, i)/a(i, i)
      a(j, i+1:) = a(j, i+1:) - a(j, i)*a(i, i+1:)
```

```
      END DO
   END DO
   END SUBROUTINE LU
   !------------------------------------------------------------
   SUBROUTINE FBSubst(a, b, n, order)
    ! Carry out forward and backward substitution for LUx = b
    ! where (a) contains the elements of L and U. The pivoting
    ! order is supplied. The RHS vector (b) is replaced by the
    ! solution vector.                        J.R.Dormand, 1/96
   !------------------------------------------------------------
   INTEGER :: n, order(n), i
   REAL (KIND=2) :: a(n, n), b(n), x(SIZE(order))
   DO i = 1, n
      x(i) = b(order(i))
   END DO
   DO i = 2, n
      x(i:n) = x(i:n) - a(i:n, i-1)*x(i-1)
   END DO
   DO i = n, 1, -1
      b(i) = x(i)/a(i, i)
      x(1:i-1) = x(1:i-1) - a(1:i-1, i)*b(i)
   END DO
   END SUBROUTINE FBSubst
   !------------------------------------------------------------
```

C.2   A diagonally implicit RK program

The DIRK3 program below implements a third order diagonal implicit
RK method, as described in Chapter 11, with step-size variation based
on the convergence of a Newton iteration. The program is compatible
with the same MODULE format as used in the earlier multistep method,
but a different system of equations is illustrated here.

```
MODULE Stiffr   ! Defines a stiff system of 3 ODEs
                ! Robertson problem with fast transient
                ! near x = 0
                !
!--------------------------------------J R Dormand,   1/1996
   LOGICAL :: printsol
   INTEGER, PARAMETER :: neq = 3
   INTEGER :: dev, nfunc, njac
```

```
    REAL (KIND = KIND(1.0D0)), PARAMETER :: af =0.04D0, &
                              bf = 1.0D4, cf = 3.0D7
    CHARACTER(33) :: fmt, outfil
CONTAINS
!
  SUBROUTINE Initial(x, y)      ! Sets initial values and
    REAL (KIND = 2) :: x, y(:) ! output format
      WRITE (fmt, '(a,2(i1,a))' ) '(1x, e12.4, 2x,', &
                    neq,'e14.5,',neq,'e12.4)'
!--------------Open output file--------------------------
      PRINT*, 'Print solution(t/f) ' ; READ*, printsol
      IF(printsol) THEN
        PRINT*, 'Output filename ("vdu" for screen) '
        READ '(a)', outfil
        IF(outfil == 'vdu') THEN
          dev = 6
        ELSE
          dev = 14
          OPEN(dev, FILE = outfil)
        ENDIF
      ENDIF
      x = 0.0d0
      y(1) = 1.0d0; y(2) = 0.0d0; y(3) = 0.0d0
      nfunc = 0; njac = 0
  END SUBROUTINE Initial
!
  SUBROUTINE Output(x, y)      ! Outputs solution with
    REAL (KIND = 2) :: x, y(:) ! error using True
      IF(printsol) THEN
        WRITE(dev, fmt) x, y
      END IF
  END SUBROUTINE Output
!
  FUNCTION Fcn(x, y)    ! Defines derivative of stiff system
                        ! modeling chemical reactions
    REAL (KIND = 2) :: x, y(:), Fcn(SIZE(y))
      nfunc = nfunc + 1
      Fcn(1) = -af*y(1) + bf*y(2)*y(3)
      Fcn(2) = af*y(1) - bf*y(2)*y(3) - cf*y(2)**2
      Fcn(3) = cf*y(2)**2
  END FUNCTION Fcn
!
  FUNCTION Jac(x, y)   ! Jacobian matrix of Fcn
    REAL (KIND = 2) :: x, y(:), Jac(neq, neq)
```

```
        njac = njac + 1
        Jac(1,1) = -af
        Jac(1,2) = bf*y(3); Jac(1,3) = bf*y(2)
        Jac(2,1) = af; Jac(2,2) = -bf*y(3)-2.0d0*cf*y(2)
        Jac(2,3) = -bf*y(2)
        Jac(3,1) = 0.0d0; Jac(3,2) = 2.0d0*cf*y(2)
        Jac(3,3) = 0.0d0
     END FUNCTION Jac
!
END MODULE Stiffr
!-----------------------------------------------------------
PROGRAM Dirk3     ! Solves a non-linear system of ODEs using
                  ! diagonally-implicit 2-stage Runge-Kutta 3
                  ! formula. Jacobian formed at each step &
                  ! Newton iteration at each stage.
                  ! Steplength variation according to number
                  ! of iterations.
!-----------------------------------------J R Dormand, 1/96
   USE Stiffr
   IMPLICIT NONE
   INTEGER :: neqs, i,  s,  iter, itmax = 5
   INTEGER, ALLOCATABLE :: order(:)
   REAL (KIND = KIND(1.0D0)) :: half = 0.5D0, one = 1.0D0,&
                                zero = 0.0D0, tol =1.0d-6
!-------------Dynamic arrays--------------------------------
   REAL (KIND = 2), ALLOCATABLE :: y(:), amat(:, :), yy(:),&
     w(:), f(:, :), ident(:, :), a(:, :), b(:), c(:), dy(:)
   REAL (KIND = 2) :: x, z, h, xend, r3
   LOGICAL :: accept, hit = .FALSE. , second = .FALSE.
!------------- SET DIRK3  PARAMETERS -----------------------
   s = 2;        r3 = SQRT(3.0D0)
   ALLOCATE( a(s,s),  b(s), c(s))
   a = zero;   c = zero;    b = half
   a(1,1) = (3 + r3)/6;    a(2,2) = a(1,1);    a(2,1) = -r3/3
   c(1) = a(1,1);   c(2) = a(2,1) + a(2,2)
!--------------Allocate arrays------------------------------
   neqs = neq
   ALLOCATE(y(neqs), amat(neqs, neqs), yy(neqs), w(neqs),&
                dy(neqs), order(neqs), ident(neqs,neqs))
   ALLOCATE(f(s, neqs))
!------------- Set identity matrix -------------------------
   ident = zero;  DO i = 1, neqs;  ident(i,i) = one;  END DO
   PRINT*, 'h, Xend ? ';   READ*,  h, xend
   CALL Initial(x, y)                ! COMPUTE INITIAL VALUES
```

```fortran
! --------------START INTEGRATION ------------------------
DO WHILE(x < xend) !   LOOP OVER STEPS
  amat = ident - h*a(1, 1)*Jac(x, y) ! Jacobian once/step
  CALL LU(amat, neq, order)              ! FACTORISE MATRIX
  DO i = 1, s                            ! LOOP OVER STAGES
    z = x + c(i)*h
    w = y + h*MATMUL(a(i, 1:i-1), f(1:i-1, :))
    yy = y
    f(i, :) = Fcn(z, yy)
    accept = .FALSE.
    DO iter = 1, itmax                   ! NEWTON ITERATION
      dy = w + h*a(i, i)*f(i, :) - yy
      CALL FBSubst(amat, dy, neq, order)  ! SOLVE SYSTEM
      yy = yy + dy                       !  UPDATE SOLUTION
      f(i, :) = Fcn(z, yy)               ! RE-EVALUATE f
      IF(MAXVAL(ABS(dy)) < tol) THEN
        accept = .TRUE.        ! Exit loop if converged
        EXIT
      END IF
    END DO                     ! End Newton iteration
    IF(.NOT.accept) EXIT       ! if not converged
  END DO
  IF(accept) THEN
    y = y + h*MATMUL(b, f)               ! Updated SOLUTION
    x = x + h
    CALL Output(x, y)
    IF(hit) EXIT
    IF(iter < 3) THEN                    ! < 3 iterations
      IF(second) THEN                    ! Double step
        PRINT 100, x, h, ' too small, try doubling'
        h = 2.0d0*h
        second = .FALSE.
      ELSE
        second = .TRUE.
      END IF
    END IF
    IF(x + h - xend > zero) THEN
      hit = .TRUE.                       ! Hit endpoint
      h = xend - x
    END IF
  ELSE                                   ! Not converged
    PRINT 100, x, h, ' rejected, halve step'
    h = half*h                           ! Halve rejected step
  ENDIF
```

```
  END DO
  100 FORMAT(1x, 'At x =',f8.3, '  h =', f9.5, a)
  print*, 'nfunc=', nfunc,'  njac=',njac
END PROGRAM Dirk3!-----------------------------------------
```

Appendix D

Global embedding programs

D.1 The Gem global embedding code

The program here is an extension of the RKpair program from Appendix A to compute the extrapolated solution based on the correction process described in Chapter 13. Gem is compatible with the same MODULE defining the system to be solved encountered earlier. A small difference between this and RKpair is the data entry. Since a GEM is likely to involve 'less simple' coefficients the data may be input from a file containing floating point values. There is no need to specify denominators for all coefficients although a common denominator is present if the rational form is required. When data is given in floating point format this denominator, preceding the coefficients for each stage, should be set to unity.

```
PROGRAM Gem         ! Solves a system of ODEs with variable
                    ! step with local embedding for step-size
                    ! control and GLOBAL embedding to yield an
                    ! extrapolated solution
! ------------------------------------- J R Dormand, 1/1996
    USE System
    IMPLICIT NONE
    LOGICAL :: done
    INTEGER :: s, star, sg, i, j, p, q, neqs, nrej
    REAL (KIND = KIND(1.0d0)), PARAMETER :: z = 0.0D0, &
                                    one = 1.0D0, sf = 0.6D0
    REAL (KIND = 2) :: x, h, xend, tol, ttol, delta, alpha, &
```

```
            bot, oop, hs, maxerr
   REAL (KIND = 2), ALLOCATABLE :: y(:), yt(:), w(:), &
                                   ynew(:), err(:), f(:,:)
   REAL (KIND = 2), ALLOCATABLE :: a(:, :), b(:), bl(:), &
                                   c(:), bt(:), to(:)
   CHARACTER(LEN = 30) :: rkname
!
! ----------Read Runge-Kutta coefficients from a data file
! ----------NB slightly different from RKpair
   PRINT*, 'Enter file containing GEM coefficients'
   READ '(a)', rkname
   OPEN(13, FILE = rkname)
   READ(13, *) s, q, p, sg, star
   ALLOCATE(a(2:sg, 1:sg-1), b(s), bl(s), bt(sg), c(2:sg), &
                                   to(sg))
   c = z
   DO i = 2, sg
      READ(13, *) c(i), (to(j), j = 1, i - 1)
      a(i, 1:i-1) = to/c(i)     ! --- Common denominator form
      c(i) = SUM(a(i, 1:i-1))
   END DO
   READ(13, *) bot, to(1:s)
   b = to(1:s)/bot
   READ(13, *) bot, to(1:s)
   bl = to(1:s)/bot
   READ(13, *) bot, to; bt = to/bot
   oop = one/(1+p)
   CLOSE(13)
   neqs = neq
   ALLOCATE( y(neqs), yt(neqs), w(neqs), err(neqs), &
                                   ynew(neqs),f(sg, neqs))
   PRINT*, 'Enter xend, starting step, tolerance'
   READ*, xend, hs, tol
   ttol = tol*sf; maxerr = z
   CALL Initial(x, y); yt = y
   h = hs
   f(1, :) = Fcn(x, y); nrej = 0
   done = .false.
   DO                                     !-------- Loop over steps
     DO i = 2, s
        w = y + h*MATMUL(a(i, 1:i-1), f(1:i-1, :))
        f(i, :) = Fcn(x + c(i)*h, w)
     END DO
     ynew = w                             ! ------- NB Fsal formula
```

```fortran
    err = h*ABS(MATMUL(bl - b, f(1:s, :)))
    delta = MAXVAL(err/(one + ABS(w)))  ! ------ mixed error
    alpha = (delta/ttol)**oop
    IF(delta < tol) THEN                ! --------- Accepts step
!--------- compute extrapolated values --------------------
      DO i = s+1, sg
        IF (i > star) THEN
          w = yt + h* MATMUL(a(i, 1:i-1), f(1:i-1, :))
        ELSE
          w = y +  h* MATMUL(a(i, 1:i-1), f(1:i-1, :))
        END IF
        f(i, :) = Fcn(x + c(i)*h, w)
      END DO
! ----Extrapolated solution
      yt = yt + h*MATMUL(bt(star+1:), f(star+1:, :))
! ----Integrator solution
      y = ynew
      x = x + h
      err = y - yt                      ! ---- Error estimate
      maxerr = MAX(maxerr, MAXVAL(ABS(err)))
      CALL Output(x, y); CALL Output(x, yt)   ! --- Outputs
      IF(done) EXIT
      f(1, :) = f(s, :)
      h = h/MAX(alpha, 0.1D0)
      IF(x + h > xend) THEN
        h = xend - x
        done = .TRUE.
      ENDIF
    ELSE
      nrej = nrej + 1
      h = h/MIN(alpha, 10.0D0)
      IF(done) done = .FALSE.
    END IF
  END DO
  PRINT 9999, maxerr
9999 FORMAT('Maximum estimated global error =', 1pe12.3)
  END PROGRAM Gem
```

D.2   The GEM90 package with global embedding

The package below provides dense output and global extrapolation for a discrete solution. A range of options is described in the code listing. The problem to be solved must be defined in a MODULE essentially similar to the earlier examples. The unextrapolated solution and dense output is derived from the 5th order DOPRI5 pair and a solution component value to be 'hit' can be specified. The global error estimates, which are valid for the discrete solution only, have 7th order accuracy. An example driver program which can be linked with the GEM90 package is included below.

```
MODULE GEM57            ! Data and variables for GEM57 package
                        ! Uses DOPRI5 with dense output from
                        ! Table 6.5 and a global estimator by
                        ! Dormand, Gilmore & Prince, 1994.
                        ! 2-term estimation makes extrapolated
                        ! result O(h^7). A specific solution
                        ! can be obtained using Newton method
                        ! with dense output formula.
!  ------------------------------------- J R Dormand, 1/96
USE System
   IMPLICIT NONE
   INTEGER, PARAMETER :: sgem = 14, star = 9, s = 7, &
                    p = 4, q = 5, deg = 5
   INTEGER :: fsn, allstep = 0, nostep = 0, nrej = 0, &
              maxstep = 500, natt
   LOGICAL :: start, first, yhit = .FALSE., &
              dout = .FALSE. , toosmall = .FALSE., &
              newstart = .FALSE., optred
   REAL (KIND = KIND(1.0D0)), PARAMETER :: ze = 0.0d0, &
              one = 1.0d0, two = 2.0d0,  pnine = 0.9d0, &
              sfpred = 0.8d0, sffail=0.8d0
   REAL (KIND = 2) :: a(sgem, sgem-1), bc(s), b(s), &
              c(sgem), bt(sgem - star), ad(deg, star)
   REAL (KIND = 2) ::  yy, yold, w, yt, f, x, h, tol, epw,&
         told, small4, hold, xold, smallv, biggest, smallest
   ALLOCATABLE :: yy(:),  yold(:), w(:), yt(:), f(:, :)
CONTAINS
!-----------------------------------------------------------

   SUBROUTINE Init(neqs)              ! Initialise RK54 & Ext7
      INTEGER, INTENT(IN) :: neqs
```

```
      REAL (KIND = 2) :: s21
! -------Internal weights a(i,j) for DOPRI5 & cts extension
      a(2, 1)   = 0.2d0
      a(3, 1:2) = (/3d0/40d0, 9d0/40d0/)
      a(4, 1:3) = (/ 44d0/45d0, -56d0/15d0, 32d0/9d0/)
      a(5, 1:4) = (/ 19372d0/6561d0, -25360d0/2187d0, &
                     64448d0/6561d0, -212d0/729d0/)
      a(6, 1:5) = (/ 9017d0/3168d0, -355d0/33d0, &
                 46732d0/5247d0, 49d0/176d0, -5103d0/18656d0 /)
      a(7, 1:6) = (/ 35d0/384d0, 0d0, 500d0/1113d0, &
                     125d0/192d0, -2187d0/6784d0, 11d0/84d0/)
      a(8, 1:7) = (/ 5207d0/48000d0, ze, 92d0/795d0, &
                 -79d0/960d0, 53217d0/848000d0, -11d0/300d0, &
                 4d0/125d0 /)
      a(9, 1:8) = (/ 613d0/6144d0, ze, 125d0/318d0, &
                 -125d0/3072d0, 8019d0/108544d0, &
                             -11d0/192d0, 1d0/32d0, ze /)
!--------Coefficients of b(i)* polynomials----------------
      ad(:, 1) = (/one, -285d0/64d0, 1552d0/192d0, &
                             -813d0/128d0, 29d0/16d0 /)
      ad(:, 2) =  ze ! (/ ze, ze, ze, ze, ze /)
      ad(:, 3) = (/ ze, 1000d0/371d0, -16000d0/1113d0, &
                          8500d0/371d0, -4000d0/371d0 /)
      ad(:, 4) = (/ ze, 125d0/32d0, -125d0/6d0, &
                          2125d0/64d0, -125d0/8d0 /)
      ad(:, 5) = (/ ze, -6561d0/3392d0, 2187d0/212d0, &
                          -111537d0/6784d0,  6561d0/848d0 /)
      ad(:, 6) = (/ ze, 11d0/14d0, -88d0/21d0, 187d0/28d0,&
                                        -22d0/7d0 /)
      ad(:, 7) = (/ze, -7d0/8d0, 19d0/4d0, -63d0/8d0, 4d0 /)
      ad(:, 8) = (/ ze, 125d0/24d0, -125d0/12d0, &
                                125d0/24d0, ze /)
      ad(:, 9) = (/ ze, -16d0/3d0, 80d0/3d0, -112d0/3d0, &
                                16d0 /)
!--------------------5th order weights--------------------
      bc = (/ a(7,1:6), ze /)
!--------------------4th order weights--------------------
      b  = (/5179d0/57600d0, ze, 7571d0/16695d0, &
             393d0/640d0, -92097d0/339200d0, 187d0/2100d0, &
                                1d0/40d0/)
      c(1:star)  = (/ze, 0.2d0, 0.3d0, 0.8d0, 8d0/9d0, one,&
                          one, 8d0/15d0, 1d0/5d0/)
!--------------------Estimator/extrapolator weights--------
      bt = (/ 1d0/20d0, 49d0/180d0, 16d0/45d0, 49d0/180d0,&
```

```
                                                1d0/20d0 /)
      s21 = SQRT(21.0d0)
!-------------------Interior weights for extrapolator-----
      a(11,10) = (7-s21)/14d0
      c(star+1: sgem) = (/ze, a(11,10), 0.5d0,  &
                                 one - a(11,10), one /)
      a(12,10) = -one/10d0
      a(12,11) = 3d0/5d0
      a(13,10) = (9*s21 - 2)/126d0
      a(13,11) = (-13d0)/18d0
      a(13,12) = 26d0/21d0
      a(14,10) = (-749d0)/810d0
      a(14,11) = (315*s21 + 1934)/810d0
      a(14,12) = (-86d0)/27d0
      a(14,13) = (49 - 7*s21)/18d0
      a(10,1:9) = ze
      a(11,1) = (18603*s21 - 110677)/263424d0
      a(11,2) = ze
      a(11,3) = -(9750*s21 - 54250)/381759d0
      a(11,4) = -(4875*s21 - 27125)/131712d0
      a(11,5) = (85293*s21 - 474579)/4653824d0
      a(11,6) = -(429*s21 - 2387)/57624d0
      a(11,7) = (12*s21 - 77)/2744d0
      a(11,8) = 125d0/1176d0
      a(11,9) = -(24*s21 - 56)/1029d0
      a(12,1) = (2592*s21 + 167777)/1505280d0
      a(12,2) = ze
      a(12,3) = (21600*s21 - 114125)/218148d0
      a(12,4) = (21600*s21 - 114125)/150528d0
      a(12,5) = -(1889568*s21 - 9983655)/2.659328d+7
      a(12,6) = (9504*s21 - 50215)/329280d0
      a(12,7) = -(288*s21 - 1973)/31360d0
      a(12,8) = -(2400*s21 - 6125)/18816d0
      a(12,9) = -(96*s21 - 821)/1470d0
      a(13,1) = -(57447*s21 - 118783)/790272d0
      a(13,2) = ze
      a(13,3) = -(35750*s21 - 281750)/381759d0
      a(13,4) = -(17875*s21 - 140875)/131712d0
      a(13,5) = (312741*s21 - 2464749)/4653824d0
      a(13,6) = -(1573*s21 - 12397)/57624d0
      a(13,7) = (55*s21 - 777)/8232d0
      a(13,8) = (1625*s21 + 1125)/10584d0
      a(13,9) = (944*s21 - 15330)/9261d0
      a(14,1) = -(2439*s21 - 1718516)/2540160d0
```

```
      a(14,2) = ze
      a(14,3) = -(27100*s21 + 526000)/490833d0
      a(14,4) = -(6775*s21 + 131500)/84672d0
      a(14,5) = (65853*s21 + 1278180)/1662080d0
      a(14,6) = -(2981*s21 + 57860)/185220d0
      a(14,7) = (271*s21 + 11604)/52920d0
      a(14,8) = (6775*s21 - 330750)/95256d0
      a(14,9) = (2168*s21 + 282462)/59535d0
      ALLOCATE ( yy(neqs), yold(neqs), w(neqs), yt(neqs), &
                                       f(sgem, neqs))
      epw = one/(p + one)
      small4 = two*two*SPACING(one); smallv = 10.0d0*small4
      newstart = .TRUE.
      biggest = smallv; smallest = HUGE(one)
    END SUBROUTINE Init
END MODULE GEM57
!---------------------------------------------------------------
    SUBROUTINE Rkde(neqs, y, yp, t, tout, relerr, abserr, &
                                       flag, mode, ks, yks)
!---------------------------------------------------------------
! Initialises variables, calls the RK stepping subroutine.
! Sets flags for output and modes of operation.
!
!          neqs  is dimension of system
!             y  contains initial value at first call
!                and required solution at every exit.
!            yp  contains derivative when 5th order y
!                and global error of 5th order in other
!                cases.
!             t  is current value of independent variable
!          tout  is output point.
! relerr, abserr are relative and absolute error tolerances
!
!          flag  shows output status
!                2: solution hits tout
!                3: step-size too small
!                4: maximum steps used
!                5: specified solution value
!                6: bad arguments
!                9: normal step solution
!   mode(1 to 7) gives operation mode
!                1: first step
!                2: dense output required at tout
!                3: output at each step
```

```
!                    4: hit output point with discrete step
!                    5: determine specified solution value
!                    6: output global error in mode(3)
!                    7: output extrapolated solution
!           ks    component number of specific solution
!          yks    value of specific solution
!                    (last two arguments are optional)
      USE GEM57
      IMPLICIT NONE
      INTEGER, INTENT(IN) :: neqs
      INTEGER, INTENT(INOUT), OPTIONAL :: ks
      INTEGER, INTENT(INOUT) :: flag
      LOGICAL, INTENT(INOUT) :: mode(7)
      REAL (KIND=2), INTENT(IN) ::  tout, relerr, abserr
      REAL (KIND=2), INTENT(IN), OPTIONAL :: yks
      REAL (KIND=2), INTENT(INOUT) :: t
      REAL (KIND=2), DIMENSION(neqs), INTENT(INOUT) :: y, yp
      LOGICAL :: fatal
      REAL (KIND=2) :: fnorm, xhit
      tol = MAX(relerr, abserr)
      IF(mode(1)) THEN
         IF(.NOT. newstart) CALL Init(neqs)
         yt = y; yy = y
      END IF
      fatal = neqs < 1 .OR. t==tout .OR. relerr < ze &
                      .OR.abserr < ze .OR. tol <= ze
      IF(fatal) THEN
        PRINT*, neqs, t, tout, relerr, abserr, tol, flag
        flag = 9; PRINT*, 'fatal error - bad arguments '
        RETURN
      END IF
      IF (mode(1) .or. (mode(4).AND.flag == 2)) THEN
        start = .TRUE.; first = .TRUE.
        optred = .FALSE.
        natt = 0
        x = t
        f(1, :) = Fcn(x, yy)
        fnorm = MAXVAL(ABS(f(1,: )))
        IF (fnorm > small4) THEN
           h = tol**epw/fnorm
        ELSE
           h = 0.5d0*(tout - t)
        END IF
      END IF
```

```
    mode(1) = .FALSE.
    DO
      IF(x > tout .AND. mode(2)) THEN
! ------------- Computing dense output -------------------
          CALL Dense(neqs, y, yp, tout, ks, yks)
          flag = 5; t = tout
          allstep = allstep + nostep; nostep = 0
          RETURN
      ELSE IF(mode(4)) THEN ! ------ Reduce step to hit ----
          h = MIN(h, tout - x); flag = 4
      ENDIF
! ------------- Computing new Step ----------------------
      CALL Rkstep(abserr, relerr, mode(6).OR.mode(7))
      dout = mode(6).OR.mode(7)
      IF(mode(5)) THEN
        IF (ks > 0) THEN
            yhit = (yks - yold(ks))*(yks - yy(ks)) < ze
        END IF
      END IF
      t = x
      IF(mode(6)) THEN    ! ----------Global error estimation
        y = yy; yp = y - yt        ! yp is error estimate
        flag = 2
      ELSE IF(mode(7)) THEN
        y = yt; yp = yy - yt       ! y is extrapolated
        flag = 3
      ELSE
        y = yy; yp = f(s,:)        ! no extrapolation
        flag = 1
      END IF
      IF(yhit) THEN   ! ------Hit specified value y(ks) = yks
        CALL Dense(neqs, y, yp, xhit, ks, yks)
        t = xhit;   ks = 0;   yhit = .FALSE.
        flag = 6
      ELSE IF (toosmall) THEN
        PRINT*, 'Stepsize too small '
        flag = 8
      ELSE IF(ABS(x - tout) < small4) THEN
        flag = 4
      ELSE IF(nostep > maxstep) THEN
        PRINT*, maxstep, ' Steps computed '
        flag = 7
      ELSE IF(.NOT.mode(3)) THEN
        CYCLE  ! No output this step
```

```fortran
      END IF
      allstep = allstep + nostep; nostep = 0
      RETURN
    END DO
  END SUBROUTINE Rkde
! ------------------------------------------------------------
  SUBROUTINE stats
  USE GEM57
  PRINT 100, allstep, nrej, natt, smallest, biggest
100 FORMAT(1x, 'Total number of steps:', i6/&
            1x, 'Number of rejects     :', i6/&
            1x, 'First attempt rejects:', i6/&
            1x, 'Step-sizes in [',e13.5,',',e13.5,']')
  END SUBROUTINE stats
! ------------------------------------------------------------
  SUBROUTINE Rkstep(abserr, relerr, extrap)
                              ! Compute one acceptable step
! ------------------------------------------------------------
  USE GEM57
  IMPLICIT NONE
  LOGICAL, INTENT(IN) :: extrap
  REAL (KIND=2), INTENT(IN) ::   abserr, relerr
  INTEGER ::  i, k
  LOGICAL ::  done, failed, success
  REAL (KIND=2) :: errmax, alpha, beta
  done = .FALSE. ; success = .FALSE.; failed = .FALSE.
  IF(start) Then
    start = .FALSE.
  ELSE
    f(1,:) = f(s,:)                      ! -------FSAL assumed
  END IF
  DO WHILE(.NOT.done)                    ! ---Main Loop starts
    errmax = ze
    DO i = 2, s
        w = yy + h*MATMUL(a(i, 1:i-1), f(1:i-1, :))
        f(i,:) = Fcn(x + c(i)*h, w)
    END DO ! i
    yold = yy + h*MATMUL(b, f(1:s, :))
    errmax = MAXVAL(ABS(w - yold)/(relerr*w + abserr))
    IF(errmax > one) THEN                ! ---Rejected Step
      nrej = nrej + 1
      IF(first) THEN
          IF (success) THEN
              optred = .TRUE.
```

```
              alpha = sffail*(one/errmax)**epw
              alpha = MAX(alpha, 0.01d0)
         ELSE
              alpha = 0.1d0
         END IF
    ELSE
       IF(failed) THEN
              alpha = 0.5d0
       ELSE
              failed = .TRUE.
              alpha = sffail*(one/errmax)**epw
              alpha = MAX(0.1d0, alpha)
       ENDIF
    ENDIF
    h = h*alpha
ELSE                        ! ---error less than tolerance
    success = .TRUE.
    beta = errmax**epw/sfpred
    IF(first) THEN
       alpha = one/MAX(0.001d0, beta)
    ELSE
       alpha = one/MAX(0.1d0, beta)
       IF (failed) alpha = MIN(one, beta)
    END IF
    hold = h
    h = alpha*h
    IF(first .AND. .NOT.optred .AND. alpha>10.0d0) THEN
       hold = ze
       natt = natt + 1
       failed = .FALSE.
    ELSE                              ! Accepted step
       done = .TRUE. ; first = .FALSE.
       yold = yy
       xold = x
       yy = w
       biggest = MAX(biggest, hold)
       smallest = MIN(smallest, hold)
       IF(extrap) THEN ! - Compute extrapolated solution
         DO i = s + 1, sgem
            IF(i > star) THEN
              w = yt + hold*MATMUL(a(i, 1:i-1), &
                                        f(1:i-1, :))
            ELSE
              w = yold + hold*MATMUL(a(i, 1:i-1), &
```

```
                                                      f(1:i-1, :))
                   ENDIF
                   f(i, :) = Fcn(x + c(i)*hold, w)
                 END DO
                 dout = .TRUE.
                 yt = yt + hold*MATMUL(bt, f(star+1:, :))
               END IF
               x = x + hold
               nostep = nostep + 1
             END IF
           END IF
         END DO !--------------------------------- Main loop ends
         toosmall =  h < x*smallv
       END SUBROUTINE Rkstep
!--------------------------------------------------------------
       SUBROUTINE Dense(neqs, y, yp, xout, ks, yks)
                                 ! Use continuous extension to
                                 ! find solution at xout, or
                                 ! estimate xout such that
                                 ! y(ks)(xout) = yks

!  -------------------------------------------------------------

         USE GEM57
         IMPLICIT NONE
         INTEGER :: neqs, i, j, ks
         REAL (KIND = 2) :: y(neqs), yp(neqs), xout, su, sup, &
                         sigma, gs, gps, delsig, yks
!  ------------ Compute two extra stages if not already done
         IF(.NOT.dout) THEN
           DO i = s+1, star
             y = yold + hold*MATMUL(a(i, 1:i-1),f(1:i-1, :))
             f(i, :) = Fcn(xold + c(i)*hold, y)
           END DO
         END IF
!  ------------- Compute the interpolant and its derivative
!  ------------- (with yhit if specified)
         IF(yhit) THEN
           delsig = 2d0*tol
           sigma = (yks - yold(ks))/(yy(ks) - yold(ks))
           j = 0
!  --------Newton iteration ---------------------------------
           DO WHILE(ABS(delsig) > tol .AND. j < 20)
             gs = yold(ks); gps = ze; j = j + 1
             DO i = 1, star
               CALL Bstar(sigma)
```

```
          gs =  gs + hold*sigma*su*f(i,ks)
          gps = gps + hold*sup*f(i,ks)
       END DO
       delsig = (gs - yks)/gps
       sigma = sigma - delsig
     END DO
   xout = xold + sigma*hold
   IF (j == 20) PRINT*, ' yhit solution not converged '
ELSE
     sigma = (xout - xold)/hold
END IF
y = yold;   yp = ze
DO i = 1, star
   CALL Bstar(sigma)
   y  = y  + su*sigma*hold*f(i, :)
   yp = yp + sup*f(i, :)
END DO
dout = .TRUE.
CONTAINS
  SUBROUTINE Bstar(sigma) ! -- Evaluate b* and derivative
    REAL (KIND=2) :: sigma
    su = ad(deg, i); sup = deg*su
    DO j = deg-1, 1, -1
       su  = su*sigma  +   ad(j, i)
       sup = sup*sigma + j*ad(j, i)
    END DO
  END SUBROUTINE Bstar
END SUBROUTINE Dense !----------------------------------
```

D.3   A driver program for GEM90

The program below will be a basis for using GEM90 to solve various ODE systems.

```
PROGRAM Testgempak ! Driver for GEM90. Uses the same
                   ! MODULE structure as earlier
                   ! programs.
! --------------------------------------- J R Dormand, 1/96
  USE System
  IMPLICIT none
  LOGICAL, DIMENSION(7) :: mode
```

```
LOGICAL :: notdone
INTEGER :: flag, i, neqs, ks
REAL(KIND = 2), Allocatable :: y(:), w(:)
REAL(KIND = 2) :: t, relerr, abserr, tout, tinc, yks, &
                  tend
CHARACTER :: dash(44), ch, modes*6
dash = '-'
mode = .FALSE. ; mode(1) = .TRUE.
PRINT *, '                      MODES'; PRINT*, dash
PRINT '(1x, a, i4, 5i7, 3x, a)', '|', (i,i=2, 7), '|'
PRINT '(1x, a)', &
            '| Dense  Every Hit at  Hit y Global Extrap|'
PRINT '(1x, a)', &
            '|  outpt  step  tout    soln   error  soln  |'
PRINT*, dash
PRINT*, 'Enter required modes'; READ '(a)', modes
DO i = 2, 7
   WRITE(ch, '(i1)') i
   mode(i) = INDEX(modes, ch) > 0
END DO
PRINT*, 'Enter end point '; READ*, tend
IF(mode(5)) THEN
   PRINT*, 'Component number & value? '; READ*, ks, yks
END IF
neqs = neq
ALLOCATE(y(neqs), w(neqs))
CALL Initial(t, y)  ! -------- Initial values from System
tout = t;    tinc = tend - t;    notdone = .true.
IF(mode(2) .OR. mode(4)) THEN
  PRINT*, 'Output Interval? '
  READ*, tinc
END IF
PRINT*, 'Enter relerr and abserr '; read*, relerr, abserr
DO WHILE(tout < tend .AND. notdone)
   tout = tout + tinc
   DO WHILE (t < tout)
     IF(tout > tend) THEN
        tout = tend
        mode(4) = .TRUE.
     END IF
     CALL RkDe(neq, y, w, t, tout, relerr, abserr, flag,&
               mode, ks, yks)
     PRINT 100, t, y; PRINT 101, flag, w
     CALL Output(t, y)  ! ---------- Output unit in System
```

```
      IF(flag > 6) THEN
        notdone = .FALSE.
        EXIT
      END IF
    END DO
END DO
CALL stats
100 FORMAT(1x, f10.4, 5e13.4)
101 FORMAT(1x, i10, 5e13.4)
END PROGRAM Testgempak ! -------------------------------
```

Appendix E

A Runge–Kutta Nyström program

The program given below implements both the special and the general RKN methods. The example system is the restricted three body problem in the rotating coordinates. A similar program in which the special RKN is used is included on the disk accompanying this book.

```
MODULE r3bodr     ! Defines the restricted 3-body
                  ! gravitational problem in the rotating
                  ! coordinate system
!----------------------------------------J R Dormand,   1/1996
   LOGICAL :: printsol, deriv1 = .TRUE.
   INTEGER, PARAMETER :: neq = 2
   INTEGER :: dev, nfunc
   REAL (KIND = 2) :: x0, yd0, mu, mud, period, gerrmax,xend
   REAL (KIND = 2) :: yi(neq), ydi(neq)
   CHARACTER(33) :: fmt, outfil
CONTAINS
   !
   SUBROUTINE Initial(x, y, yd)      ! Sets initial values and
      REAL (KIND = 2) :: x, y(:), yd(:) ! output format
         WRITE (fmt, '(a,2(i1,a))' ) '(1x, f10.4, 2x,', &
                     2*neq,'f10.4,', 2*neq,'e12.4)'
         gerrmax = SPACING(1.0d0)
         PRINT*, 'Print solution(t/f) ' ; READ*, printsol
         IF(printsol) THEN
         PRINT*, 'Output filename ("vdu" for screen) '
         READ '(a)', outfil
```

```
        IF(outfil == 'vdu') THEN
           dev = 6
        ELSE
           dev = 14
           OPEN(dev, FILE = outfil)
        ENDIF
        ENDIF
        x = 0.0d0
        period = 7.4345573d0
        mu = 1.0d0/81.45d0
        x0 = 1.0d0 - mu - 0.120722529000d0
        yd0 = -0.024511616261d0
        mud = 1.0d0 - mu
        xend = period
        y(1) = x0
        y(2) = 0.0d0
        yd(1) = 0.0d0
        yd(2) = yd0
        yi = y; ydi = yd
        nfunc = 0
   END SUBROUTINE Initial
!
   SUBROUTINE Output(x, y, yd)        ! Outputs solution with
      REAL (KIND = 2) :: x, y(:), yd(:) ! error using True
         IF(printsol) THEN
            WRITE(dev, fmt) x, y, yd
         END IF
   END SUBROUTINE Output
!
   FUNCTION Fcn(x, y, yd)                ! Defines derivative
      REAL (KIND = 2) :: x, y(:), yd(:), Fcn(SIZE(y))
      REAL (KIND = 2) :: d1, d2
        nfunc = nfunc + 1
        d1 = SQRT((y(1) + mu)**2 + y(2)**2)**3
        d2 = SQRT((y(1) -mud)**2 + y(2)**2)**3
        Fcn = (/ y(1) + 2.0d0*yd(2) -mud*(y(1)+mu)/d1 -    &
               mu*(y(1)-mud)/d2,  y(2) - 2.0d0*yd(1) -     &
                              mud*y(2)/d1 - mu*y(2)/d2 /)
   END FUNCTION Fcn
!
END MODULE r3bodr
!
PROGRAM RKNG  ! Solves a system of 2nd order ODEs with
              ! variable step RKNG
```

```
                    ! Takes standard RKNG data file
! -------------------------------- J R Dormand,        1/1996
USE r3bodr
  IMPLICIT NONE
  LOGICAL :: done, lgen, fsal
  INTEGER :: s, i, p, q, neqs, nrej, neq2
  REAL (KIND = KIND(1.0D0)), PARAMETER :: z = 0.0D0, &
                          one = 1.0D0, sf = 0.6D0
  REAL (KIND = 2) :: x, h, tol, ttol, delta, alpha,&
                oop, hs, deltay, deltayd
  REAL (KIND = 2), ALLOCATABLE :: y(:), yd(:), w(:), &
                                       wd(:), f(:, :)
  REAL (KIND = 2), ALLOCATABLE :: a(:, :), abar(:,:), &
                          b(:), bl(:),  d(:), dl(:), c(:)
  neqs = neq
  neq2 = 2*neq
  CALL RKNGcoeff
  ALLOCATE( y(neqs), yd(neqs), w(neqs), wd(neqs), &
                                       f(s, neqs))
  PRINT*, 'Enter starting step, tolerance '
  READ*, hs, tol
  CALL Initial(x, y, yd)
  PRINT '(a)', '% logtol  nfunc nrej   max error'
  ttol = tol*sf
  h = hs
  f(1, :) = Fcn(x, y, yd);       nrej = 0
  done = .FALSE.
  DO                    ! Loop over steps
    DO i = 2, s
      w = y + h*(c(i)*yd + h*MATMUL(abar(i, 1:i-1), &
                                      f(1:i-1, :)))
      IF(lgen) wd = yd + h*MATMUL(a(i, 1:i-1),f(1:i-1, :))
      f(i, :) = Fcn(x + c(i)*h, w, wd)
    END DO
    deltay = h*h*MAXVAL(ABS(MATMUL(bl - b, f))) ! Max y err
    deltayd = h*MAXVAL(ABS(MATMUL(dl - d, f))) ! Max y' err
    delta = MAX(deltay, deltayd)
    alpha = (delta/ttol)**oop
    IF(delta < tol) THEN   ! Accepts step
      y = y + h*(yd + h*MATMUL(b, f))
      yd = yd + h*MATMUL(d, f)
      x = x + h
      CALL Output(x, y, yd)
      IF(done) EXIT
```

```fortran
      IF(fsal) THEN
         f(1, :) = f(s, :)
      ELSE
         f(1, :) = Fcn(x, y, yd)
      END IF
      h = h/MAX(alpha, 0.1D0)
      IF(x + h > xend) THEN
         h = xend - x
         done = .TRUE.
      ENDIF
   ELSE
      nrej = nrej + 1
      h = h/MIN(alpha, 10.0D0)
      IF(done) done = .FALSE.
   END IF
END DO
! The example problem has a periodic solution and so the
! end point error is obtained from the initial values
deltay = MAXVAL(ABS(yi - y))
deltayd = MAXVAL(ABS(ydi - yd))
gerrmax = MAX(deltay, deltayd)
PRINT 9999, LOG10(tol), nfunc, nrej, LOG10(gerrmax)
9999 FORMAT(f7.3, 2i8, f10.3)
CONTAINS
SUBROUTINE RKNGcoeff !--Read RKN(G) coeffs from data file
CHARACTER(LEN = 30) :: rkngfile
REAL (KIND = 2) :: cden
REAL (KIND = 2), ALLOCATABLE :: to(:)
PRINT*, 'Enter file containing RK coefficients'
READ '(a)', rkngfile
OPEN(13, FILE = rkngfile)
READ(13, *) s, q, p, lgen
ALLOCATE(a(2: s, 1: s-1), abar(2: s, 1: s-1), b(s), &
                bl(s), d(s), dl(s), c(2:s), to(s))
c = z
IF(lgen) THEN        ! ----- Reads aij for general RKNG
   DO i = 2, s
      READ(13, *) cden, to(1: i-1)
      a(i, 1: i-1) = to(1: i-1)/cden
      c(i) = SUM(a(i, 1:i-1))
   END DO
ELSE IF(deriv1) THEN
   PRINT*, 'System requires general RKN and so formula'
   PRINT*, 'referenced here is invalid'
```

```
    STOP
END IF
DO i = 2, s           ! ------ Reads abar(ij) in RKN or RKNG
    READ(13, *)  cden, to(1: i-1)
    abar(i, 1: i-1) = to(1: i-1)/cden
    IF(.NOT. lgen) c(i) = SQRT(2*SUM(abar(i, 1:i-1)))
END DO
READ(13, *) cden, to
b = to/cden
READ(13, *) cden, to
d = to/cden
READ(13, *) cden, to
bl = to/cden
READ(13, *) cden, to
dl = to/cden
fsal =  SUM(ABS(b(1:s-1) - abar(s, 1:s-1))) < 1.0D-10
PRINT 9998, q, p, s, fsal, lgen
9998 FORMAT(1x, 'RKN',i2,'(',i2,') in',i3,' stages', &
            ' FSAL = ',l1, '  G = ',l1)
oop = one/(1 + p)
CLOSE(13)
END SUBROUTINE RKNGcoeff
END PROGRAM RKNG ! -----------------------------------
```

Bibliography

Alexander, R. (1977): *Diagonally implicit Runge-Kutta methods for stiff ODEs.* SIAM J. Numer. Anal., **14**, 1006–1021.

Brankin, R.W., Dormand, J.R., Gladwell, I., Prince, P.J. and Seward, W.L. (1989): *A Runge-Kutta-Nyström code.* ACM Trans. Math. Softw., **15**, 31–40.

Butcher, J.C. (1963): *Coefficients for the study of Runge-Kutta integration processes.* J. Austral. Math. Soc., **3**, 185–201.

Butcher, J.C. (1987): *The numerical analysis of ordinary differential equations.* John Wiley & Sons Inc., New York.

Calvo, M., Montijano, J.I., and Randez, L. (1990): *A 5th order interpolant for the Dormand & Prince RK method.* J. Comput. Appl. Math., **29**, 91–100.

Dormand, J.R., Duckers, R.R. and Prince, P.J. (1984): *Global error estimation with Runge-Kutta methods.* IMA J. Num. Anal. **4**, 169–184.

Dormand, J.R., El-Mikkawy, M.E.A., and Prince, P.J. (1987a): *Families of Runge-Kutta-Nyström formulae.* IMA J. Num. Anal. **7**, 235–250.

Dormand, J.R., El-Mikkawy, M.E.A., and Prince, P.J. (1987b): *High-order embedded Runge-Kutta-Nyström formulae.* IMA J. Num. Anal. **7**, 423–430.

Dormand, J.R., Gilmore, J.P., and Prince, P.J. (1994): *Globally embedded Runge-Kutta schemes.* Annals of Num. Math. **1**, 97–106.

Dormand, J.R., Lockyer, M.A., McGorrigan, N.E., and Prince, P.J. (1989): *Global error estimation with RK triples.* Computers Math. Applic. **18**, 835–846.

Dormand, J.R. and Prince, P.J. (1978): *New Runge-Kutta-Nyström algorithms for simulation in dynamical astronomy.* Celestial Mechanics **18**, 223–232.

361

Dormand, J.R. and Prince, P.J. (1980): *A family of embedded Runge-Kutta formulae.* J. Comput. Appl. Math. **6**, 19–26.

Dormand, J.R. and Prince, P.J. (1985): *Global error estimation with Runge-Kutta methods II.* IMA J. Num. Anal. **5**, 481–497.

Dormand, J.R. and Prince, P.J. (1986): *Runge-Kutta triples.* Computers Math. Applic. **12**, 1007–1017.

Dormand, J.R. and Prince, P.J. (1987): *Runge-Kutta-Nyström triples.* Computers Math. Applic. **13**, 937–949.

Dormand, J.R. and Prince, P.J. (1989): *Practical Runge-Kutta processes.* SIAM J. Sci. Stat. Comput. **10**, 977–989.

Dormand, J.R. and Prince, P.J. (1992): *Global error estimation using Runge-Kutta pairs.* Computational ordinary differential equations, eds. Cash, J. & Gladwell, I., Oxford University Press, 451–457.

Ellis, T.M.R., Philips, I.R., and Lahey, T.M. (1994): *Fortran 90 programming.* Addison Wesley, Wokingham.

England, R. (1969): *Error estimates for Runge-Kutta type solutions to systems of ordinary differential equations.* The Computer J., **12**, 166–170.

Enright, W.H., Jackson, K.R., Nørsett, S.P. and Thomson, P.G. (1986): *Interpolants for Runge-Kutta formulas.* ACM Trans. Math. Softw., **12**, 193–218.

Fehlberg, E. (1968): *Classical fifth, sixth, seventh, and eighth order Runge-Kutta formulas with step-size control.* NASA Technical Report 287.

Fehlberg, E. (1969): *Low-order classical Runge-Kutta formulas with step-size control and their application to some heat transfer problems.* NASA Technical Report 315.

Fehlberg, E. (1974): *Classical seventh, sixth, and fifth order Runge-Kutta-Nyström formulas with step-size control for general second order differential systems.* NASA Technical Report 432.

Fike, C.T. (1968): *Computer evaluation of mathematical functions.* Prentice–Hall, Englewood Cliffs, N.J.

Fine, J.M. (1985): *Low order Runge-Kutta-Nyström methods with interpolants.* Technical Report 183/85, Department of Computer Science, University of Toronto.

Fine, J.M. (1987): *Interpolants for Runge-Kutta-Nyström methods.* Computing **39**, 27–42.

Hairer, E., Nørsett, S.P. and Wanner, G. (1994): *Solving ordinary differential equations I (2nd ed.).* Springer-Verlag, Berlin.

Henrici, P. (1962): *Discrete variable methods in ordinary differential equations.* John Wiley & Sons Inc., New York.

Herrick, S. (1972): *Astrodynamics (Volume 2).* Van Nostrand Reinhold, London.

Hildebrand, F.B. (1974): *Introduction to numerical analysis.* McGraw-Hill, New York.

Jackson, J. (1924): *Note on the numerical integration of $\frac{d^2x}{dt^2} = f(x,t)$.* Monthly Notices of the R. Astr. Soc., **84**, 602–612.

Krogh, F.T. (1973): *Algorithms for changing the step size.* SIAM J. Numer. Anal. **10**, 949–965.

Lambert, J.D. (1991): *Numerical methods for ordinary differential systems.* John Wiley & Sons Ltd., Chichester.

Merson, R.H. (1957): *An operational method for the study of integration processes.* Proc. Symp. Data Processing, Weapons Research Establishment, Salisbury, Australia, 110-1–110-25.

Merson, R.H. (1974): *Numerical integration of the differential equations of celestial mechanics.* Royal Aircraft Establishment Technical Report 74184.

Mitchell, A.R. and Griffiths, D.F. (1980): *The finite difference method in partial differential equations.* Wiley–Interscience, New York.

Morgan, J.S. and Schonfelder, J.L. (1993): *Programming in Fortran 90.* Alfred Waller Ltd., Henley-on-Thames.

Morton, K.W. and Mayers, D.F. (1994): *Numerical solution of partial differential equations.* Cambridge University Press.

Peterson, P.J. (1986): *Global error estimation using defect correction techniques for explicit Runge-Kutta methods.* Technical Report 192/86, Department of Computer Science, University of Toronto.

Prince, P.J. (1979): *Runge-Kutta processes and global error estimation.* Ph.D. Thesis, C.N.A.A.

Prince, P.J. and Dormand, J.R. (1981): *High order embedded Runge-Kutta formulae.* J. Comput. Appl. Math. **7**, 67–75.

Robertson, H.H. (1967): *The solution of a set of reaction rate equations.* Numerical analysis—an introduction, ed. Walsh, J., Academic Press, London.

Seidelmann, P.K.(ed.) (1992): *Explanatory supplement to the astronomical almanack.* University Science Books, Mill Valley, CA.

Shampine, L.F. (1985): *Interpolation for Runge-Kutta methods.* SIAM J. Numer. Anal. **22**, 1014–1027.

Shampine, L.F. and Gordon, M.K. (1975): *Computer solution of ordinary differential equations.* W.H.Freeman, San Francisco.

Sharp, P.W. and Fine, J.M. (1992): *Some Nyström pairs for the general 2nd order initial value problem.* J. Comput. Appl. Math. **42**, 279–291.

Sharp, P.W. and Fine, J.M. (1992): *A contrast of direct and transformed Nyström pairs.* J. Comput. Appl. Math. **42**, 293–308.

Skeel, R.D. (1986): *Thirteen ways to estimate global error.* Numer. Math., **48**, 1–20.

Smith, G.D. (1985): *Numerical solution of partial differential equations (3rd ed.).* Oxford University Press.

Van Der Houwen, P.J. (1972): *Explicit Runge-Kutta formulas with increased stability boundaries.* Numer. Math. **20**, 149–164.

Van Der Houwen, P.J. and Sommeijer, B.P. (1987): *Explicit Runge-Kutta (-Nyström) methods with reduced phase errors for computing oscillating solutions.* SIAM J. Numer. Anal. **24**, 595–617.

Verner, J.H. (1978): *Explicit Runge-Kutta methods with estimates of the local truncation error.* SIAM J. Numer. Anal. **15**, 772–790.

Zadunaisky, P.E. (1976): *On the estimation of errors propagated in the numerical integration of ordinary differential equations.* Numer. Math., **27**, 21–39.

Index

Printed and bound by CPI Group (UK) Ltd, Croydon, CR0 4YY

22/10/2024

01777605-0015